We dedicate this text to

Kathe Houpt, Hans Klingel and George Waring

who have done more than they will ever realize
to inspire a generation of scientists concerned
about horses.

The Domestic Horse

The Origins, Development and Management of its Behaviour

Humans have had a profound influence on the horse since its domestication in the late Neolithic period. Used for transport, labour, food and recreation, it has become important in many facets of our societies.

Daniel Mills and Sue McDonnell have produced an exceptional account of our current knowledge of the development and management of the behaviour of the domestic horse from its wild roots.

The Domestic Horse brings together, for the first time, an unrivalled collection of international scientific authors to write on the latest work concerning the behaviour and welfare of the domestic horse.

Beautifully illustrated throughout, this book will appeal to animal scientists, those working with horses in a professional capacity and the owner enthusiast. It also provides sound complementary reading for animal/equine science courses and veterinary students.

DANIEL S. MILLS is Professor of Veterinary Behavioural Medicine and Director of the Animal Behaviour Referral Clinic at the University of Lincoln. He is the first specialist in veterinary behavioural medicine to be formally recognized by the Royal College of Veterinary Surgeons and has published widely on behaviour problems in the domestic horse, and is co-author of *Equine Behaviour, Principles and Practice*.

SUE M. MCDONNELL is an associate professor and founding head of the Havemeyer Equine Behavior Programme at the University of Pennsylvania School of Veterinary Medicine, New Bolton Center. She has published widely on stallion sexual behaviour and dysfunction, and is author of *The Equid Ethogram, A Practical Field Guide to Horse Behavior*.

The Domestic Horse:

The Origins, Development and Management of its Behaviour

Edited by

D. S. Mills
University of Lincoln, UK

and

S. M. McDonnell
*University of Pennsylvania
School of Veterinary Medicine, USA*

CAMBRIDGE UNIVERSITY PRESS
Cambridge, New York, Melbourne, Madrid, Cape Town, Singapore, São Paulo

Cambridge University Press
The Edinburgh Building, Cambridge CB2 2RU, UK

www.cambridge.org
Information on this title: www.cambridge.org/9780521814146

First published 2005

Printed in the United Kingdom at the University Press, Cambridge

A catalogue record for this book is available from the British Library

Library of Congress Cataloguing in Publication data
The domestic horse : the origins, development, and management of its
behaviour / D. S. Mills, S. M. McDonnell (editors).
   p.   cm.
Includes bibliographical references and index.
ISBN 0 521 81414 6 (alk. paper) – ISBN 0 521 89113 2 (alk. paper)
1. Horses–Behavior.   2. Horses.   I. Mills, D. S.   II. McDonnell, Sue M.
SF281.D66   2005
636.1 – dc22   2004048883

ISBN-13 978-0-521-81414-6 hardback
ISBN-10 0-521-81414-6 hardback
ISBN-13 978-0-521-89113-2 paperback
ISBN-10 0-521-89113-2 paperback

Contents

III The impact of the domestic environment on the horse

List of contributors

Melissa Albentosa, *Animal Behaviour, Cognition & Welfare Group, University of Lincoln, Lincoln LN2 2LG, UK*

Lee Boyd, *Department of Biology, Washburn University, Topeka, KS 66621, USA*

Rachel A. Casey, *Anthrozoology Institute, Department of Clinical Veterinary Science, University of Bristol, Langford, Bristol BS40 5DU, UK*

Janne W. Christensen, *Animal Health and Welfare, Danish Institute of Agricultural Sciences, 8830 Tjele, Denmark*

Jonathan J. Cooper, *Animal Behaviour, Cognition & Welfare Group, University of Lincoln, Lincoln LN2 2LG, UK*

Sharon L. Crowell-Davis, *College of Veterinary Medicine, University of Georgia Athens, GA 30602, USA*

Claudia Feh, *Association pour le cheval de Przewalski, Station Biologique de la Tour du Valat, 13200 Arles, France*

Deborah Goodwin, *University of Southampton, School of Psychology, Animal Behaviour Programme, New College Campus, Southampton SO17 1BG, UK*

Stephen J. G. Hall, *Department of Biological Sciences, University of Lincoln, Riseholme Hall, Lincoln LN2 2LG, UK*

M. Hausberger, *Ethologie-Evolution-Ecologie, UMR CNRS 6552, Université de Rennes 1, France*

Katherine A. Houpt, *College of Veterinary Medicine, Cornell University, Ithaca, New York 14853-6401, USA*

Carys F. Hughes, *University of Southampton, School of Psychology, Animal Behaviour Programme, New College Campus, Southampton SO17 1BG, UK*

Ronald Keiper, *Department of Science, Valencia Community College, Orlando, FL 32802, USA*

Jan Ladewig, *Ethology, The Royal Veterinary and Agricultural University, Grønnegårdsvej 8, 1870 Frederiksberg C, Denmark*

Marsha A. Levine, *McDonald Institute for Archaeological Research, University of Cambridge, Cambridge CB2 3ER, UK*

Sue M. McDonnell, *Equine Behavior Laboratory, New Bolton Center, University of Pennsylvania School of Veterinary Medicine, Kennett Square, PA 19348, USA*

Paul D. McGreevy, *Faculty of Veterinary Science, University of Sydney, NSW 2006, Australia*

Andrew N. McLean, *Australian Equine Behaviour Centre, Clonbinane, Victoria 3658, Australia*

June McNicholas, *Department of Psychology, University of Warwick CV4 7AL, UK*

Daniel S. Mills, *Animal Behaviour, Cognition and Welfare Group, University of Lincoln, Lincoln LN2 2LG, UK*

Christine J. Nicol, *School of Veterinary Science, University of Bristol, Langford BS40 5DU, UK*

Marie-Annick Richard-Yris, *Ethologie-Evolution-Ecologie, UMR CNRS 6552, Université de Rennes 1, France*

Eva Søndergaard, *Animal Health and Welfare, Danish Institute of Agricultural Sciences, 8830 Tjele, Denmark*

Natalie Waran, *School of Natural Sciences, Unitec New Zealand, Mount Albert, Auckland New Zealand*

Jennifer Weeks, *College of Veterinary Medicine, University of Georgia Athens, GA 30602, USA*

Preface

Holar in Northern Iceland in 2002 saw probably the greatest gathering in recent times of equine behaviour and welfare scientists, with expertise spanning five decades of research. The workshop was sponsored by the Dorothy Russell Havemeyer Foundation and co-organized by the editors of this text.

It was a unique opportunity for scientists to come together from across the globe and present their work and review the current state of knowledge on all aspects of the science underpinning our understanding of the impact of management practices on the horse. The meeting began with a dozen scientific poster presentations on the latest research ranging from the effect of different handling and training techniques to data on conflicting physiological measures that are often used to assess stress in the horse. The first invitation session was chaired by Professor George Waring and focused on the foundations of equine behaviour, with presentations critically examining current theories on its domestication, behavioural genetics and development. State of the art reviews on equine com-

munication and the behavioural ecology of the horse highlighted the horse's normal adaptive range and the extent to which this is challenged in domestic situations. In the second session, chaired by Professor Hans Klingel, attention focused further on specific behaviour patterns in the horse and the relevance of this information to improving our management of horses in captivity. Special emphasis was placed on the importance of diet and feeding practices, play and sexual behaviours. The third session, chaired by Professor Frank Ödberg, examined management practices and the problems that commonly arise as a result. Invited presentations focused on problems of the ridden horse, training and behavioural rehabilitation in the horse and recent advances in treatment of equine stereotypies (stable vices). The final session was chaired by Professor Katherine Houpt and focused specifically on equine welfare. Information on the scientific techniques and approaches used to assess welfare was followed by presentations on practical welfare issues faced in a range of countries from Australia to mainland Europe and Iceland to Brazil and North America. The issues facing various cultures with different socio-economic pressures and where horses were used in very different contexts were also highlighted. Practical welfare issues ranging from ethologically sound housing to training techniques and the assessment of pain were also discussed in greater depth. Throughout the meeting concern was expressed through the discussions at the harm that could be done from well-meaning intention, which unfortunately does not always equate with good welfare for the horse. Closer examination of many popular practices revealed that there was still insufficient evidence in many cases for the beneficial claims made, and in some cases real harm might occur. It was recognized that education of the public has a key role to play in bringing about a global improvement in equine welfare, but this must be based on sound science more than popular opinion. Regrettably, at present, funding opportunities for research into these matters are severely limited.

The meeting finished with delegates agreeing on a consensus statement on the importance of collective responsibility for equine welfare and the relationship between scientists and the media for improved welfare:

The group proposes that the welfare of the domestic horses is both an individual and societal responsibility. Research is required into both the fundamental and applied aspects of equine behaviour to facilitate the development and dissemination of soundly based scientific knowledge to help provide optimal welfare in practice. In order to achieve this, it is important that good research is widely communicated and not misrepresented in public interpretations of the work.

To this end there have been two initiatives from this meeting to increase the public understanding of equine science. First, the abstracts from the conference are available online at: http://www2.vet.upenn.edu/labs/equinebehavior/hvnwkshp/hv02/hvwk6-02.htm, and second, and perhaps most importantly, this text has been produced, based largely but not solely around the talks given and updated to 2004. This publication was always seen as an essential product of this meeting, bridging the gap not only between the public and scientists but also between the scientific groups working on equine behaviour. Naturally these groups have developed along their own paths and publish in the journals most appropriate to their discipline. This can mean that there is not always as great an awareness of the current state of the art as there should be, even amongst experts in the field. The aim of this text is to bring this information to an even wider audience, to educate and, we hope, to inspire.

Acknowledgements

We are indebted to the Dorothy Russell Havemeyer Foundation and especially Gene Pranzo for facilitating the original workshop which led to this text. Our hosts in Iceland Vikingur Gunnarsson of Hólar Agricultural College and Hrefna Sigurjónsdóttir of Iceland University of Education, as well as the delegates who all contributed to the discussions and so ultimately our better understanding of horse behaviour. We are also indebted to the numerous referees who worked anonymously and have given us so much of their time freely to review and enhance the work of their peers, which is presented here. A work such as this involves commitment from many, not least our families who have supported us throughout. Finally we would like to thank the staff at CUP who have worked on its production, especially Tracy Sanderson, Sarah Jeffery and Martin Griffiths.

Introduction

Perhaps no animal more than the horse has played and continues to play such an important role in shaping society. Horses were instrumental to economic development as well as the success of large-scale human invasions and the cultural changes these brought around the world. They remained a vital instrument of industry and war across the globe until well into the twentieth century, when they were largely replaced by machinery in the industrialized world. None the less, the vast utility of this animal has ensured its survival, and their greater use in leisure has meant that they continue to make a significant contribution to those economies that have otherwise sought to replace them as a beast of burden. There are no longer any wild ancestors of the domestic horse; the closest we have to these are the feral progeny of domesticated animals and the re-introduced Przewalski's horse. Domestication and captive breeding processes bring about certain changes dependent on the selective forces involved, none the less these horses appear to adapt easily to their natural environment, suggesting that they have retained much of their natural biology. Given the importance of the horse's behaviour and management to its utility it may seem surprising that research into these aspects of its biology remains largely piecemeal, with strong but relatively small research groups scattered around the world. For the first time, this book brings together much of this knowledge in a single text, in three parts.

The first, 'Origins and selection of horse behaviour', not only examines how the horse has been integrated into human society, but also emphasizes how every horse has an individual identity and role. Levine highlights how difficult it is to unravel the historical evidence which led to domestication and the basis for the ubiquitous presence of the horse today. This utility of the horse is examined further in the following chapter by Hall on the role of the horse in society across the world. Whilst it may seem obvious that horses must vary in their behaviour to fulfil their diverse roles, it is perhaps surprising how little work has been done on the heritability of horse behaviour and the genetic basis of individual variability. Hausberger and

The Domestic Horse: The Origins, Development, and Management of its Behaviour, ed. D. S. Mills & S. M. McDonnell.
Cambridge University Press. © Cambridge University Press 2005.

Richard-Yris provide a comprehensive review of these topics in the final chapter of this section.

Part two emphasises what we can learn from the wild and so starts with a chapter on how the free-roaming behaviour of the horse has evolved to be adapted to these environments (Boyd and Keiper) and another on the natural communication patterns used by the horse (Feh). These are followed by a series of chapters which contrast the behaviour of horses in natural social and environmental conditions with their behaviour in domestic environments. Emphasis is given to maintenance activities (Houpt), sexual behaviour (McDonnell), mare–foal interaction (Crowell-Davis and Weeks), the process of behavioural development (Ladewig, Søndergaard & Christensen) and play (Goodwin and Hughes).

Part three considers the demands and effect of the domestic environment on horse behaviour and welfare. Mills and McNicholas start with a review of the current literature and present data from recent studies that have sought to elucidate what attracts humans to the horse. The utility of the horse is related to its ability to be conditioned for a variety of functions and so Nicol provides a critical review of what we actually know about the learning ability of the horse. This is followed by a chapter which considers how this knowledge is applied in general horse training and the increasingly popularized techniques of 'natural horsemanship' (Waran and Casey). McGreevy and McLean then discuss how problems arise during the training of the ridden horse and how these can be resolved through 'the installation of clear operant basics'. Another conspicuous behaviour problem arising from a general lack of appreciation of the natural behavioural biology of the domestic horse is the occurrence of a variety of repetitive behaviours. Mills evaluates the current body of research into these and their management. Finally, Cooper and Albentosa examine the methods used to evaluate equine welfare and emphasize the need for a multidisciplinary approach if we are to understand how domestic practices impact on their well-being.

I

Origins and selection
of horse behaviour

1

Domestication and early history of the horse

Marsha A. Levine

Introduction

Before the development of firearms, the horse was crucial to warfare and before the invention of the steam engine, it was the fastest and most reliable form of land transport. Today its importance has scarcely diminished in parts of South America, Asia, Africa and Eastern Europe, and even elsewhere it is of great economic importance to sport and leisure industries. Nevertheless, in spite of intensive investigations over many years, researchers know very little about the origins of horse domestication and the evolution of horse husbandry.

The origins of horse domestication

Throughout the course of the twentieth century a variety of theories have been developed purporting to explain where, when and for what purposes the horse was first domesticated. The basic positions can be summarized as, that it was first domesticated:

- during the Neolithic, Eneolithic or Early Bronze Age (Table 1.1);
- for meat, riding or traction;
- in Ukraine, Kazakhstan, Eastern Europe or Western Europe;
- possibly in response to contacts with the Near East;
- at a single locus or at a number of different loci, more or less simultaneously.

In some situations it is, of course, easy to show how horses had been used in ancient times. For example, the horses found in some of the Altai Early Iron Age kurgan burials (Figure 1.1) – such as Pazyryk, Bashadar and Ak-Alakha – as a result of their burial in permafrost, were accompanied by well-preserved equipment such as bridles, saddles and harnessing (Rudenko, 1970; Polos'mak, 1994). Because of Rudenko's publication *The Frozen Tombs of Siberia*, Pazyryk is especially well known, but many other Early Iron Age sites from the Ukraine and south Siberia are equally spectacular. Many of the richer graves contained objects made of gold and silver – for example, jewellery, vessels, harness ornaments, weapons and belt buckles. Carpets, wall hangings,

The Domestic Horse: The Origins, Development, and Management of its Behaviour, ed. D. S. Mills & S. M. McDonnell. Cambridge University Press. © Cambridge University Press 2005.

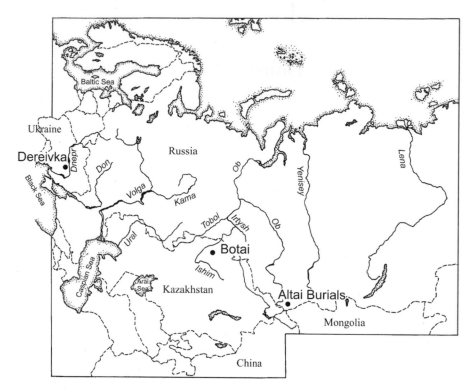

Figure 1.1. Map showing locations of Dereivka, Botai and Early Iron Age Altai sites: Ak-Alakha, Bashadar and Pazyryk.

Table 1.1. *Chronology of the west Eurasian steppe*

Approximate Dates (BC)	Period
900–300	Early Iron Age
1000–900	Transition to Early Iron Age
1800/1700–1200/1000	Late Bronze Age
2000/1900?–1800/1700	Middle Bronze Age
3000/2900 (2750?)–2300/1900?	Early Bronze Age (EBA)
3500/3400–3000/2900 (2750?)	Final Eneolithic – beginning of EBA
4800/4700?–3500/3400	Eneolithic
6000–4800/4000	Neolithic

clothing made of felt, furs and even silk from China have also been recovered (Cahen-Delhaye, 1991). It is intriguing to note that the jointed snaffle bit was in wide use in central Eurasia during the Early Iron Age (first millennium BC).

At most sites, however, especially those dating from the period when horses were first domesticated for riding and traction, determining whether an animal was domesticated is much more difficult, sometimes impossible. Organic materials such as leather and wood are only very rarely recoverable from the archaeological record. In unfavourable soil conditions even bone is eventually destroyed. Moreover, not only is it possible to ride a horse without the use of a saddle or bridle, but also, during the early stages of horse domestication, it is likely that they were usually ridden that way.

In recent years two sites have come to the fore in debates concerning the origins of horse domestication: Dereivka and Botai. Dereivka, an Eneolithic Ukrainian settlement site, has been central to the problem of the origins of horse domestication, because for the past three decades it has been regarded as the site with the earliest evidence of horse husbandry (e.g. Bökönyi, 1978; Bibikova, 1986b; Telegin, 1986; Mallory, 1989; Anthony & Brown, 1991; Gimbutas, 1991). More recently Botai, an Eneolithic settlement site from Kazakhstan, has also been associated with this question (Brown & Anthony, 1998) (Figure 1.1). Because of the enormous numbers of horse bones found at Botai, it was inevitable that this site would be considered in such discussions.

However, upon further examination such as that below, it should be clear that the evidence backing the claims for both sites is deeply flawed.

Types of evidence for the origins of horse domestication

Two types of evidence are referred to in discussions of the origins of horse domestication: direct and indirect. *Direct evidence* relates to artistic, textual and funerary evidence, where there is virtually no doubt both that the horses were caballine and that they were ridden or used for traction.

Indirect evidence is inferred from characteristics of bones and artefacts. It includes evidence derived from analytical methods such as population structure, biogeographical distribution and artefact studies. It is invariably the case that any one pattern manifested by these types of data could have more than one explanation. On its own, no one type of data can provide satisfactory evidence of horse domestication. Indirect evidence must have corroboration from as many directions as possible.

Some types of indirect evidence are frequently confused with direct evidence. That is, data – whose association with horse husbandry is only inferred – are treated as if they could only be explained by horse domestication. Eneolithic bit wear and cheekpieces are examples of *false direct evidence*.

Direct evidence

The earliest unambiguous direct evidence – that is, dateable textual and artistic evidence – for horse domestication probably only dates back to the end of the third millennium BC (Postgate, 1986; Zarins, 1986; Piggott, 1992; Kuz'mina, 1994a,b, 1996; Littauer & Crouwel, 1996). Evidence of horses in graves, accompanied by artefacts unambiguously associated with riding or traction is even more recent, dating to probably no later than the beginning of the second millennium BC (Postgate, 1986; Piggott, 1992; Kuz'mina, 1994a,b, 1996; Anthony, 1995; Littauer & Crouwel, 1996). By the middle of the second millennium BC horses were widely used to pull chariots – for example, in the Near East, Greece, and on the Eurasian steppe (Piggott, 1992; Littauer & Crouwel, 1996; Renfrew, 1998). There is apparently no reliable textual or artistic evidence for horse riding earlier than the end of the second millennium BC (Renfrew, 1987, 1998; Piggott, 1992). There are

earlier representations of people riding equids in the Near East. However, because of the extreme difficulty of distinguishing artistic representations of horses from those of asses, it is impossible to identify the earliest evidence for horse riding itself (Piggott, 1992). It is highly improbable, however, that traction horses could have been herded either on foot or from a vehicle. Therefore it seems almost certain, as far as the horse is concerned, that riding would have preceded traction.

One interpretation of this evidence is that the horse was first domesticated for traction around the end of the third millennium BC and for riding a little earlier (Khazanov, 1984; Renfrew, 1987; Kuz'mina, 1994a,b). However, it is almost certain that horse husbandry must have developed well before its earliest unambiguous manifestations in art and burial ritual. As Piggott pointed out, these representations are not merely portrayals of everyday life, they are closely connected with the delineation of power and prestige (Piggott, 1992, p. 69). If horse riding, at its inception and during its early development, did not have high status, it would have been unlikely to have been represented in art or burials. It might not, in fact, have left any direct evidence at all. This evidential 'invisibility' seems to suggest that, whatever its practical value, the horse was of little or no political or social significance until the end of the third millennium BC. This point of view is, of course, in direct conflict with the picture of horse-powered migration and warfare during the Eneolithic and Early Bronze Age proposed by Gimbutas (e.g. 1970; 1991), and supported by many others (Telegin, 1986; Mallory, 1989; Anthony, 1991).

False direct evidence

Four types of evidence, conventionally accepted as proof of horse domestication, fall into the category of false direct evidence:

- Horse-head sceptres.
- Horse burials not associated with tack.
- Cheekpieces.
- Bit wear.

Horse-head 'sceptres' or 'maces' (Figure 1.2)

Horse-head 'sceptres' (Telegin, 1986; Gimbutas, 1991) or 'maces' are found in Eneolithic burials from the Volga to the Lower Danube. Only a few have

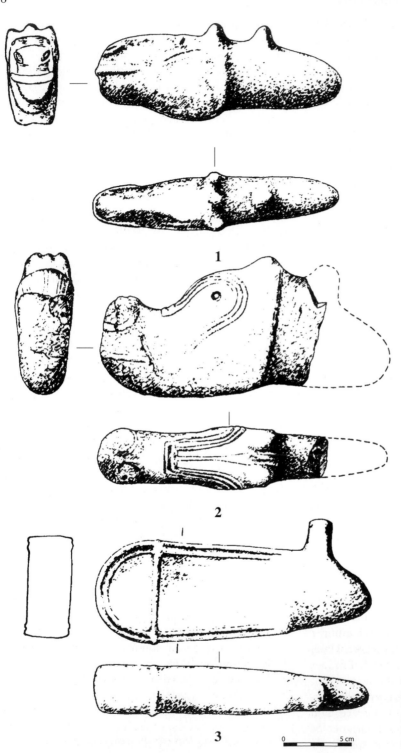

Figure 1.2. Horse-head 'sceptres':
(1) Suvorovo II, Kurgan 1, Burial 7;
(2) Kasimcea; (3) Khutor
Shlyakhovsky, Kurgan 3 Burial 3
(from Rassamakin, 1999, figure 3.14;
© McDonald Institute).

more than a passing resemblance to horses' heads and those are found west of the Dneiper, mainly in the Balkan and Lower Danube regions (Telegin, 1986; Häusler, 1994a). These sculptures are conventionally regarded as symbols of the power wielded by the male occupants of the graves in which they were found (Anthony, 1991, p. 267). It was, however, the archaeologists not the Eneolithic people who described them as 'maces' or 'sceptres'. Their association with power is largely based upon the fact that they are made of exotic stone such as porphyry. The markings carved on some of the sculptures, which have been described as depictions of harnessing, are too schematic to be used as evidence of such (Anthony, 1991). The sculptures are not, in fact, found with any direct or even indirect evidence of horse husbandry.

Horse bones in human burials

Mallory (1981) and Anthony and Brown (2000), describe cemeteries from the Pontic-Caspian region in which horse bones are associated with human burials from Eneolithic (Khvalynsk) and Early Bronze Age (Yamnaya and Catacomb) cultures (Chernykh, 1992; Mallory, 1981, 1989; Y. Y. Rassamakin, pers. comm.). Domestic animal bones are relatively rare in graves from these periods and both ovicaprids and cattle are much more frequently found than horses (Mallory, 1981). Cattle skulls from these cultures are found in human burials with wagons, but horses are not. Neither are they associated with riding tack or harnessing (Mallory, 1981; Piggott, 1992; Renfrew, 1998). Complete horse skeletons are very rarely found in these burials. Often only a few or even only one horse bone will be included (Mallory, 1981; Anthony & Brown, 2000). The skull, teeth or jaw are the most frequently represented anatomical elements, followed by foot bones. However, skull and foot bones are rarely found together. According to Mallory (1981), there is no correlation between horse bones and other symbols of wealth or ranking in these graves.

That Eneolithic and Early Bronze Age peoples went to the trouble of burying horses attests to their symbolic significance, but it cannot be taken as evidence of domestication. Hares were also found in Yamnaya and Catacomb culture graves, but no one claims, on that basis, that they were domesticated (Mallory, 1981).

Cheekpieces (Figure 1.3)

Perforated antler tines, found at Dereivka and a few other Eneolithic or Early Bronze Age sites, have been widely interpreted as cheekpieces and thus taken as evidence for Eneolithic horse riding (e.g. Mallory, 1981, 1989; Anthony, 1986, 1991; Telegin, 1986). This identification has been questioned in recent years on a number of grounds (Levine, 1990; Uerpmann, 1990; Dietz, 1992; Häusler, 1994b; Rassamakin, 1999). That is, these objects have never been found in place on a horse's skull; rarely are they even found in association with horses. There is no contextual support for the notion that they were bridle cheekpieces. Those at Dereivka were found in association with other bone tools, pottery, and flaked and ground stone tools (Telegin, 1986). Moreover, their form is so general that they could have served a variety of purposes.

Bit wear

Although bit wear had been described earlier by Bökönyi (1968) and Clutton-Brock (1974), its use to investigate the origins of horse riding, has been pioneered by Anthony and Brown (1991; Brown & Anthony, 1998). They define bit wear as: 'the damage that occurs on the occlusal . . . surfaces of the second premolar teeth . . . , particularly the lower second premolars . . . , when a horse chews the bit' (Brown & Anthony, 1998, p. 331). They state that the pattern of wear that they define as bit wear is direct evidence for horse riding or traction.

Useful though it certainly is, their approach has a number of problems and limitations which they have glossed over:

• Brown and Anthony (1998) describe two types of bit-wear evidence: microscopic and macroscopic. The microscopic evidence is problematic in the archaeological context, since it is likely to be lost under most burial conditions. Even at Botai, an Eneolithic settlement site, where tooth preservation was very good, Brown and Anthony were not able to find any microscopic wear evidence.
• Bit wear would seem, by definition, to provide direct evidence for the use of the horse for transport. However, Anthony and Brown have not demonstrated that the wear pattern that they describe as bit wear could not have had other causes (Anthony & Brown, 1991; Brown & Anthony, 1998). Indeed, there is

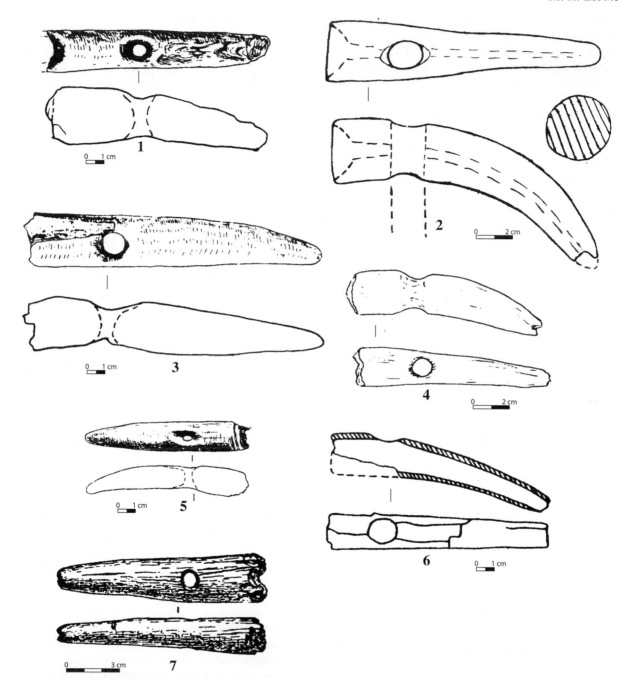

Figure 1.3. 'Cheekpieces': (1)–(5) Dereivka; (6) Mayaki;
(7) Vulkaneshty (from Rassamakin, 1999, Figure 3.55;
© McDonald Institute).

evidence that such wear can result from abnormal occlusion (Levine *et al.*, 2002; A. von den Driesch, pers. comm.). Moreover, the median bevel of the domestic population is only 0.5 mm greater than the maximum bevel of their comparative feral sample.
- Brown and Anthony (1998) have recently described a series of experiments they carried out to prove: (1) that a metal-bitted horse, which is ridden regularly, will have bit wear; and (2) that soft bits – leather, rope and bone – will also cause bit wear. However, the results of both tests were, in fact, inconclusive. Most significantly, after 150 hours of riding, none of the soft-bitted horses showed a significant bevel.

Two conclusions can be drawn from this: (1) that it has not been demonstrated that soft bits can cause archaeologically visible bit wear; and (2) that it has not been proved that a bone bit could result in 'significant' bit wear. Bit-wear studies are useful, but they do not on their own provide conclusive or direct evidence of the use of the horse for transport.

Indirect evidence

None of the genuinely direct types of evidence reach back far enough to be informative about the nature of human–horse relationships during the Eneolithic and Early Bronze Age, the period upon which arguments about the origins of horse domestication focus. It is therefore necessary to use indirect evidence to reconstruct relationships that took place probably at least 1000 years before the earliest direct evidence for horse husbandry.

Inappropriate use of indirect evidence

Much of the indirect evidence, that is, evidence based upon inference and interpretation, has been unsatisfactory. Some important problems include:

- confusion of intensification with domestication;
- use of a single type of evidence as proof of domestication;
- inadequate testing of theories.

As most of these points have been discussed in detail elsewhere (Levine, 1990, 1993, 1999a), some of the central issues will only be briefly reviewed here.

Criteria used as evidence (e.g. Bökönyi, 1984; Bibikova, 1986a; Telegin, 1986) that horses from Dereivka were domesticated include:

- absence of old horses;
- presence of a high proportion of male horse skulls;
- presence of objects identified as bridle cheekpieces;
- results of a morphological analysis comparing the Dereivka horses with other equid material;
- their association with other domesticates;
- relatively high percentage of horse bones and teeth in the deposit.

These are not good criteria for horse domestication. In fact, on the basis of archaeological, ethnographic and ethological comparisons, the absence of old individuals is much more likely to indicate hunting than herding. Males would outnumber females either if bachelor groups or stallions protecting their harems were targeted in the hunt. Morphological studies have involved very small and disparate samples and produced contradictory results. The association of horses with other assumed domesticates is not evidence of horse domestication. In any case, they were also associated with wild animals. In fact, the most important criterion used was the relatively high proportion of horse bones and teeth present at the site (e.g. Bökönyi, 1984; Petrenko, 1984; Bibikova, 1986b; Matyushin, 1986; Telegin, 1986; Gimbutas, 1988; Dergachev, 1989; Makarova & Nurumov, 1989; Mallory, 1989; Anthony, 1991; Anthony & Brown, 1991). However, throughout the Palaeolithic the archaeological record shows that horse meat was almost always an important component of the human diet.

Uerpmann claims that: 'Reduction in size on the one hand and increase in variability on the other are classic indicators of domestication' (Uerpmann, 1990, p. 127). However, horse populations have exhibited this kind of variability throughout the Pleistocene and into the post-Pleistocene. Environmental change, geographical isolation and genetic drift are all connected with size change.

Additionally any morphological changes brought about by domestication would almost certainly have appeared too late to signal its earliest stages. In any case, there are no indisputable osteological differences between wild and domesticated horses.

Biogeographical range is also problematic as evidence of early horse domestication. During the Upper Pleistocene (*c.* 130 000–10 000 BP; Otte, 1996) wild caballine horses (the ancestors of both the domestic horse and the Przewalski's horse) were found throughout most of northern Eurasia, ranging from 75° N to

35° N and from 130° E to 10° W (Eisenmann, 1996). During the following Holocene period the archaeological evidence for horses, particularly in Western and central Europe, is much sparser – fewer remains were found at fewer sites (Clutton-Brock, 1992). This has been interpreted as meaning that the horse had become extinct throughout large parts of its original range. The natural Holocene range of the horse was thus taken to comprise Eastern Europe and central Asia.

However, horse numbers in western and central Europe during this period are greatly underestimated for a variety of reasons. For one thing, relatively few faunal assemblages, dating to this period, have been submitted to detailed analysis. Additionally, when horse remains are identified outside their expected geographical range, it is frequently assumed that they must have been either domesticated or intrusive from later levels (e.g. Grigson, 1993; Curci & Tagliacozzo, 1995). Such material is frequently excluded from faunal reports (K. Boyle, pers. comm.).

Recent research suggests that the natural distribution of the Holocene horse was much wider than had formerly been believed. Neolithic remains from putatively wild horses have, for example, been found in Sweden, Denmark, the Netherlands, France, Spain, Italy, Germany, Switzerland, Hungary and Serbia – in addition to Ukraine, Russia and Kazakhstan (Azzaroli, 1985; Groves, 1986; Zarins, 1986; Cabard, 1987; Clason, 1988; Schibler & Steppan, 1999). Because the origins of the earliest domestic horses are not known, it is not certain that all these horses were, in fact, wild (Uerpmann, 1990; Benecke, 1999). Moreover, it should not be assumed that the absence of horses from archaeological assemblages is evidence that they were not present. They might well have been available, but not hunted for either logistic or cultural reasons.

The belief that the horse became extinct in western Europe and relatively rare in central Europe underlies the assumption that its earliest domestication must have taken place in eastern Europe or central Asia. However, the fact that wild horses were more common in those regions does not prove that they were first domesticated there.

Population structure analysis

The study of population structure, that is, the age and sex structure of an archaeozoological assemblage, can offer useful insights into the nature of ancient relationships between people and animals. Each pattern of behaviour or method of exploitation is characterized by its own typical, though not necessarily unique, population structure. These structures can be used as models to which the archaeological data can be compared. The raw material for this analytical method is the aged teeth from archaeological deposits. Determination of an individual's age at death is based upon measurements of crown height and assessments of eruption and wear (Levine, 1982, 1999b).

Population structure of wild horses The anchor for this method is the population structure of the wild horse. The natural reproductive unit of the horse is the family group, composed of a stallion, his mares and their young up to the age of about two to four years (Klingel, 1969, 1974; Berger, 1986; Boyd and Keiper Chapter 4). It may comprise from 2 to 17 individuals. The average number of mares is around two to four. The stallion normally starts his own family group at the age of five or six years, although he might not be successful at holding one against attacks from other males until he is older. The second natural horse social unit is the bachelor group, composed entirely of males, too young or too old to belong to a family group. Its average size is typically two to four horses.

The Attritional Assemblage Model (Figure 1.4a) The mortality distributions for natural attrition, scavenging, coursing on foot and livestock husbandry, where meat production is of secondary importance, are all similar to the Attritional Model. Mortality is low for adults during their reproductive years, and high for juveniles and senescent individuals (Caughley, 1966).

The Carnivorous Husbandry Model (Figure 1.4b) A mortality curve resembling the Carnivorous Husbandry Model might be generated if the slaughter of individuals at around the age of two to four years were superimposed upon the pastoral nomadic attritional pattern (Levine, 1999a). A very similar age distribution was produced from data provided by Yuri Shavardak, a semi-traditional horse herder from northern Kazakhstan.

The Life Assemblage or Catastrophe Model (Figure 1.4c) The Life Assemblage Model is representative

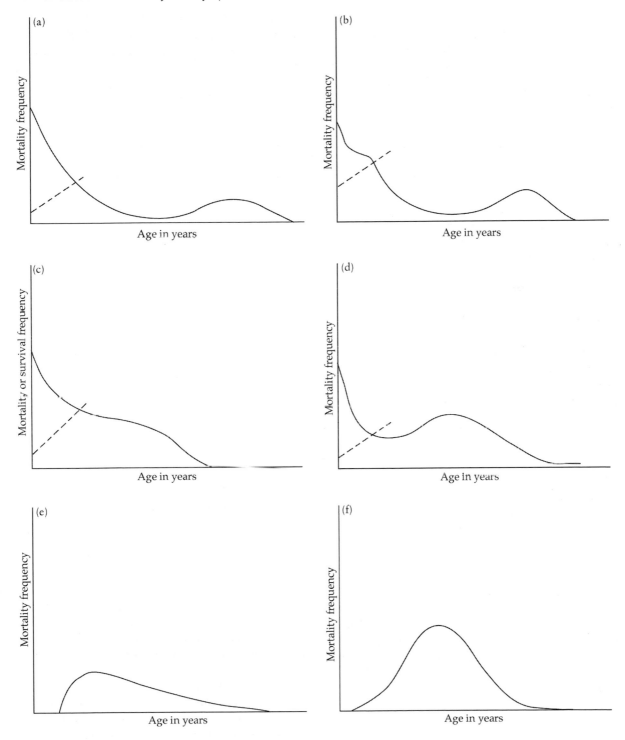

Figure 1.4. Age structure models. (a) The Attritional
Assemblage Model. (b) The Carnivorous Husbandry Model.
(c) The Life Assemblage or Catastrophe Model. (d) Family
Group Model. (e) Bachelor Group. (f) The Stalking Model.

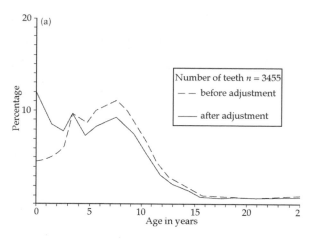

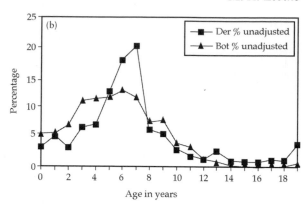

Figure 1.5. Archaeological age structures. (a) Pooled Western European, Upper Pleistocene archaeological assemblages. (b) Age structure at Botai (Bot) and Dereivka (Der).

either of a living population, a catastrophe assemblage, or an assemblage in which all age classes are represented as they would be in the living population because of completely random sampling (Levine, 1983). Herd driving, or any other non-selective hunting technique, should produce this mortality pattern or that of the Family Group Variant.

Social group models (variants of the Life Assemblage Model) The main difference between the Life Assemblage Model and the *Family Group Model* (Figure 1.4d) is the relatively low proportion in the latter of individuals three to six years of age, marking the absence of bachelor males (Levine, 1983). This is the kind of pattern produced by the Western European, Upper Pleistocene material previously studied – particularly when adjusted to compensate for the probable under-representation of immature animals[1] Figure 1.5a.

Figure 1.4e illustrates a hypothetical *Bachelor Group* age distribution. Its most archaeologically visible characteristic would be the absence of any individuals younger than about two years of age. Bachelor-group hunting might, in the archaeological context, be indistinguishable from the stalking of prime adults (Levine, 1983).

The Stalking Model (Figure 1.4f) Stalking is a selective hunting technique in which the prey is approached by stealth and killed. Hunting mainly

prime adults should also produce this distribution (Levine, 1983).

Botai and Dereivka (Figure 1.5b) The age distributions for both Botai and Dereivka fit hunting models, but the differences between them strongly suggest that different hunting techniques were used. For example, although the mortality rates for both Botai and Dereivka are very similar from the age of eight years and onwards, the rates for younger horses are distinctly divergent. At Dereivka mortality is concentrated between the ages of five to eight years, while at Botai it extends at least back to the age of three years. The difference is even greater when the distribution is adjusted. While the horses from Dereivka were probably stalked, it seems that those from Botai were killed in herd drives. This conclusion is supported by the very different sex ratios at the two sites. At Dereivka the ratio of males to females is 9:1, which is compatible with stalking; while at Botai the ratio is almost 1:1, which is best explained by a non-selective technique, such as herd driving (Levine, 1999b).

Palaeopathology

Palaeopathological analysis provides one of the most promising approaches for the study of the evolution of horse husbandry. The results of a recent palaeopathology project carried out by the author with Leo Jeffcott and Katherine Whitwell indicate that the types and incidences of certain abnormalities of the caudal

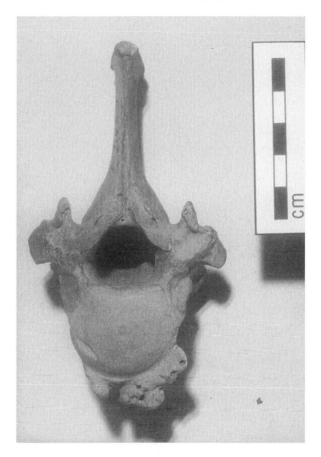

Figure 1.6. Deposition of spurs of new bone on the ventral and lateral surfaces of the vertebral bodies adjacent to the intervertebral space. Ak-Alakha 5, kurgan 3, horse 4; thoracic vertebra 14.

thoracic vertebrae could be connected with riding (Levine, 1999b; Levine *et al.*, 2000). Four Early Iron Age, Scytho-Siberian skeletons from Ak-Alakha 5 (Altai), dated fifth to third century BC, were found buried with bits between their teeth. Although their bones were well preserved, burial conditions were not good enough for the survival of saddles. However, the context of the burials suggests that they were riding horses. Most of the anatomical elements from all four Early Iron Age horses are normal. However, all of them have similar abnormalities of the caudal thoracic vertebrae:

(1) Deposition of spondylotic spurs of new bone on the ventral and lateral surfaces of the vertebral bodies adjacent to the intervertebral space (Figure 1.6).

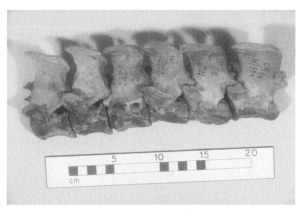

Figure 1.7. Overriding or impinging dorsal spinous processes. Ak-Alakha 5, kurgan 3, horse 1; thoracic vertebrae 14–18, lumbar vertebra 1.

(2) Overriding or impinging dorsal spinous processes (Figure 1.7).
(3) Horizontal fissures through the epiphysis (Figure 1.8).
(4) Periarticular osteophytes: the deposition of new bone on and above adjacent articular processes between vertebrae (Figure 1.9).

This work is still in progress, but the initial results of comparisons of the Early Iron Age horses, which wore pad saddles, with free-living modern Exmoor ponies, which were never saddled, and with Medieval Turkic horses (Ak-Alakha 1, Altai), which wore frame saddles, strongly suggest that these abnormalities, as a complex, are associated with the use of pad saddles and, most probably, with riding bareback (Levine *et al.*, 2000; Levine *et al.* in press.) (Tables 1.2a–c).

These types of abnormalities are entirely absent from the sample of vertebrae studied from the Eneolithic site, Botai (Figure 1.10), supporting the results of the population structure analysis, which concluded that the horses from Botai were wild. Unfortunately the vertebrae from Dereivka had all been discarded before they could be examined.

Taming and domesticating horses

Despite our ignorance of the origins of horse domestication, it is nevertheless possible – using archaeological, ethnographic and ethological data – to consider how horse husbandry might have originated.

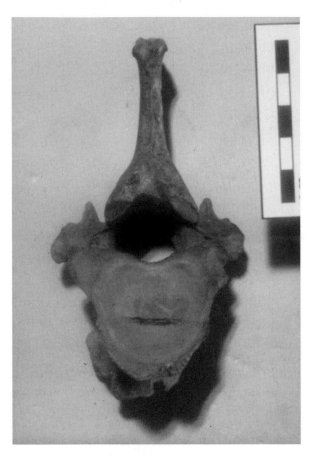

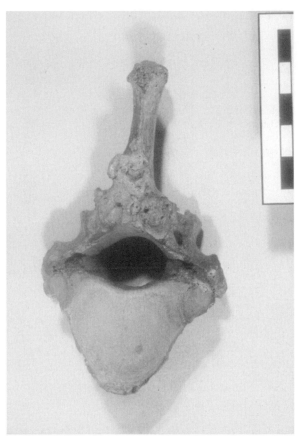

Figure 1.8. Horizontal fissure through the epiphysis.
Ak-Alakha 5, kurgan 3, horse 4, thoracic vertebra 14.

Figure 1.9. Periarticular osteophytes: the deposition of new
bone on and above adjacent articular processes between
vertebrae. Ak-Alakha 5, kurgan 3, horse 4; thoracic
vertebra 16.

According to Juliet Clutton-Brock, 'A tame animal
differs from a wild one in that it is dependent on man
and will stay close to him of its own free will' (Clutton-
Brock, 1987, p. 12)). Aboriginal hunter–gatherers and
horticulturists throughout the world are known to
tame all kinds of wild animals to keep as pets. There
is no reason to think that this would not have been the
case at least from the time of the earliest anatomically
modern *Homo sapiens* and, when the need arose, taming would probably have been the first step towards
domestication (Galton, 1883; Clutton-Brock, 1987;
Serpell, 1989). Wild horses, particularly as foals, can
be captured and tamed and, as such, ridden or har-
nessed and, at the end of their lives, if necessary,
slaughtered and eaten. During historical times both
the North American Plains tribes and the Mongols
used the arkan, lasso or herd drive to capture wild or
feral horses to eat or to tame them (Levine, 1999a).
Horse taming was regarded as a skill most successfully carried out by specialists, whose most important
tool was their intimate knowledge of horse behaviour.
On this basis, a possible scenario for the development
of horse domestication may be proposed.

As a working hypothesis, the author would like to
suggest that horse taming probably first arose as a by-
product of horse hunting for meat. Orphaned foals,
captured between the ages of perhaps two months
and one year, or possibly somewhat older – when
they were no longer dependent upon their mother's
milk – would sometimes have been adopted and raised
as pets. Eventually, and perhaps repeatedly, the dis-
covery was made that these pets could be put to
work. This knowledge could have been acquired and
lost many times from the Pleistocene onwards. But it

Table 1.2a. *Description of thoracic vertebrae 11–19 abnormalities: Early Iron Age horses*

	Ak-Alakha 5 (Altai)			
Horse number	1	2	3	4
Age (years)	16+	10–15	$10^1/_2$	7–10
Sex	Male	Male	Male	Male (possible gelding)
Number of thoracic vertebrae	18	19	18	18
(1) Osteophytes/spondylosis on the ventral and lateral surfaces of the vertebral bodies adjacent to the intervertebral space	T11 to 18	Increasing from T11 to T14 (11 & 12 caudal; 13 & 14 caudal + cranial)	T14, 17, 18 strongly developed; T15 weakly developed	T13 to 15 most strongly developed, but extends to T17
(2) Impinging or overriding spinous processes	T16–18 (possibly T15 also)	T14–15 probably; T15–19 possibly	Unclear because of poor preservation	T10–12 probably
(3) Horizontal fissure through epiphysis	T13 and 15, caudal	T13 and 14, (most developed on T14), caudal	T14, caudal; T18, cranial	T13 and 14, (most developed on T14), caudal
(4) Periarticular changes	T16–17 small exostoses	T15–18 small exostoses	T17, 18 small exostoses	Exostoses, increasingly from T14 to T17, then decreasing[a]

[a]At T16–T17 and, to a lesser extent at T15–T16, these changes were pronounced and extended dorsally to involve the adjacent vertebral arches and lower regions of the spinous processes. This had not, however, resulted in the fusion of the vertebrae.

Table 1.2b. *Description of thoracic 11 to 19 abnormalities: modern Exmoor ponies*

	Exmoor	
Horse number	97/2	97/7
Age (years)	12	27
Sex	Female	Female
Number of thoracic vertebrae	18	18
(1) Osteophytes/spondylosis on the ventral and lateral surfaces of the vertebral bodies adjacent to the intervertebral space	Not present	Sightly developed, T12–14
(2) Impinging or overriding spinous processes	Not present	Not present
(3) Horizontal fissure through epiphysis	Not present	Not present
(4) Periarticular changes	Not present	T11–18 small exostoses

Furthermore, considering the problems encountered by modern collectors trying to breed Przewalski's horses, it seems likely that horse-keeping would have had to have been relatively advanced before controlled breeding over successive generations, and thus domestication, would have been possible. As Boyd and Houpt point out: 'Failure to consider the typical social organization of the species can result in problems such as pacing, excessive rates of aggression, impotence and infanticide' (Boyd & Houpt, 1994, p. 222). Thus, in order to breed wild horses successfully in captivity, their environmental, nutritional and social requirements must be met. Again quoting Boyd and Houpt (1994, p. 226):

In zoos, juvenile male Przewalski's horses should be left in their natal bands for at least a year so that they can observe mating behaviour. They should be placed in bachelor herds when removed from the natural band, and not given harems until they are at least four or five years of age. The first mares placed with the stallion should be younger than he and the harem size should be kept small until the stallion gains age and experience.

The cognitive and logistical difficulties involved in creating such an environment at the time of the earliest horse domestication should not be underestimated. Although it is not possible to know for sure that the ancestor of the domestic horse would have been more amenable to captive breeding than the Przewalski's

was, apparently, only during the Holocene – possibly between the Neolithic and the Early Bronze Age – that it began to influence human social developments.

Initially the difficulties involved in keeping captured wild horses alive would have set limits to their impact – as work animals – on human society.

Table 1.2c. *Description of thoracic 11 to 19 abnormalities: Medieval Turkic horses*

	Ak-Alakha 1 (Altai)	
Horse number	1	2
Age (years)	11 years	10.5 years
Sex	male	male
Number of thoracic vertebrae	18	18
(1) Osteophytes/spondylosis on the ventral and lateral surfaces of the vertebral bodies adjacent to the intervertebral space	T14, small ventral spondylotic spur	Not present
(2) Impinging or overriding spinous processes	Spines crowded together and T13–16 have been touching, no significant overriding	Not present
(3) Horizontal fissure through epiphysis	Not present	Not present
(4) Periarticular changes	Slight erosion and pitting of articular facets between T14 & 15, T15 & 16, T17 & 18	Some fairly insignificant bony thickening on the processes between T13 & T14, T14 & T15

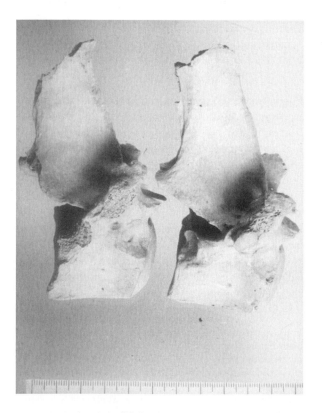

Figure 1.10. Botai caudal thoracic vertebrae.

horse, it seems unlikely. That capturing wild horses and stealing tamed or domesticated ones was regarded by the Plains tribes as preferable to breeding them supports the scenario proposed here. If it is correct, it seems likely that there would have been a relatively long period of time when new horses would have been recruited from wild populations. This could have been carried out by trapping, driving and chasing, as documented for the Mongols and North American Plains tribes (Levine, 1999a).

Consideration of the ethological and ethnographic evidence further suggests a scenario in which the initial taming and perhaps the earliest breeding of the domestic horse progenitors could have been confined to the 'horse whisperers' of the ancient world (Levine, 1999a). That the breeding of domestic horses today is so straightforward seems likely to be the consequence of the selective breeding of particularly amenable beasts some thousands of years ago. As Hemmer (1990, pp. 187–8) points out, behavioural change is an inherent part of the process of domestication:

The environmental appreciation of the domestic animal is innately reduced when compared to that of wild animals. This is expressed in a lower intensity or even disappearance of particular patterns of behaviour . . . A singular intensification of sexual activity contrasts with the general attenuation of other behaviour.

Such behavioural changes, in fact, probably do have a genetic basis (Hemmer, 1990; Hausberger and Richard-Yris Chapter 3). Given the complexity of both horse and human behaviour, as described above, it seems rather unlikely that these kinds of changes would have taken place many times in many places. However, once controlled breeding was established and the horse domesticated, it is likely that its value as a work animal would have been appreciated and its population could have increased quite rapidly (Berger, 1986).

This leads to a hypothesis that horse domestication would have taken a relatively long time to develop and might well have depended upon chance genetic changes that would have predisposed some horses to breed in captivity. Another possibility is that the human understanding of horse behaviour might have developed to such a degree that horses finally would have been able to reproduce in captivity. Perhaps the most likely scenario is that the human and equine parts of the equation would have evolved together.

Mitochondrial DNA and the origins of horse domestication

At first glance the above scenario might appear to conflict with the most recent evidence from mitochondrial DNA (mtDNA) research. All three recent investigations of horse mtDNA have revealed that, in contrast to cattle and goats, the domestic horse is characterized by high genetic diversity (Jansen *et al.*, 2002; Hall Chapter 2). Lister *et al.* (1998) explains this with a model that envisages: 'domestic horses having arisen from wild stock distributed over a moderately extensive geographical region, large enough to have contained within it considerable pre-existing haplotype diversity' (Lister *et al.*, 1998, p. 276). Vilà *et al.* (2001) propose two hypotheses: (1) the 'selective breeding of a limited wild stock from a few foci of domestication'; and (2) 'a large number of founders recruited over an extended time period from throughout the extensive Eurasian range of the horse' (Vilà *et al.*, 2001, p. 474). They finally reject the first explanation and choose a modified version of the second: 'initially, wild horses were captured over a large geographic area and used for nutrition and transport' (Vilà *et al.*, 2001, p. 477).

The study by Jansen *et al.* (2002), with its much larger sample of 652 horses, expands upon these results. The analysis revealed that at least 77 successfully breeding mares were recruited from the wild. Moreover, several of the phylogenetic clusters correspond to breeds or geographic areas. This study confirms that the Przewalski's horse mtDNA types are closely related and are not found in any other breed and that this animal is very unlikely to be ancestral to the domestic horse. Jansen *et al.* (2002) also examine both the 'restricted origin' and 'multiple

founder' hypotheses. They conclude that the present mtDNA diversity could not have originated from a single wild population. However, they point out that this finding does not necessarily support the 'multiple founder' hypothesis. Taking archaeological, ethological, ethnographic and genetic evidence into account, at present the most likely scenario for the origin of horse domestication is one in which the knowledge of horse-breeding and the concomitant spread of the horse diffused from an origin, localized both spatially and temporally. The overall data seem to suggest that, as the knowledge of horse-breeding spread, additional horses from wild populations were incorporated into the domestic herds, thus forming the regional mtDNA clusters. It must be said, however, that this field of study is only at a very early stage. Further research – archaeological, palaeopathological and genetic – hold great promise for our understanding of the origins of horse domestication and the evolution of horse husbandry.

Acknowledgements

I would like to thank Yuriy Rassamakin and Yakov Gershkovich for help with the archaeological chronology and Katie Boyle and Peter Forster for comments about the text. I am grateful to Dora Kemp who helped to prepare the illustrations for this chapter. I would like to thank the McDonald Institute for Archaeological Research for allowing me to use figures 3.14 and 3.55 from Rassamakin (1999). I would like to express my gratitude to all my CIS collaborators who made this work possible. I acknowledge the following organizations for funding the research discussed here: the McDonald Institute for Archaeological Research, the Natural Environmental Research Council, the Wenner-Gren Foundation, the British Academy and the Leakey Foundation. I am grateful to the University of Cambridge and the British Academy for providing travel expenses to enable me to attend the Havemeyer Workshop.

Endnote

1 A hypothetical adjustment factor has been applied to compensate for the probable under-representation of immature animals (for details see Levine, 1983). To obtain the average adjusted frequency of teeth for each age class from birth to the age of five years, the frequency of teeth in each age class (from 0 to 5 years) is multiplied by $1/0.23 + 0.17$(age). 'Age' refers to average age; for example, 0.5 is used for 0–1 years.

References

Anthony, D. W. (1986). The 'Kurgan Culture', Indo-European origins and the domestication of the horse: a reconsideration. *Current Anthropology*, **27**(4): 291–313.

(1991). The domestication of the horse. In *Equids in the Ancient World*, ed. R. H. Meadow and H.-P. Uerpmann. Wiesbaden: Dr. Ludwig Reichert Verlag, pp. 250–77.

(1995). Horse, wagon and chariot: Indo-European languages and archaeology. *Antiquity*, **69**(264): 554–565.

Anthony, D. W. & Brown, D. R. (1991). The origins of horseback riding. *Antiquity*, **65**: 22–38.

(2000). Eneolithic horse exploitation in the Eurasian steppes: diet, ritual and riding. *Antiquity*, **74**: 75–86.

Azzaroli, A. (1985). *An Early History of Horsemanship*. Leiden: E. J. Brill.

Benecke, N. (1999). The domestication of the horse. In *Domestication of Animals: Interactions between Veterinary and Medical Sciences*, ed. J. Schäffer. Giessen: Deutschen Veterinärmedizinischen Gesellschaft, pp. 9–22.

Berger, J. (1986). *Wild Horses of the Great Basin, Social Competition and Population Size*. Chicago: University of Chicago Press.

Bibikova, V. I. (1986a). On the history of horse domestication in south-east Europe. In *Dereivka a Settlement and Cemetery of Copper Age Horse Keepers on the Middle Dnieper*, ed. D. Y. Telegin BAR International Series, **287**: 163–82.

(1986b). A study of the earliest domestic horses of Eastern Europe. In *Dereivka Settlement and Cemetery of Copper Age Horse keepers on the Middle Dnieper*, ed. D. Y. Telegin. BAR International Series, **287**: 135–62.

Bökönyi, S. (1968). Mecklenburg Collection. Part I: Data on Iron Age horses of Central and Eastern Europe, *American School of Prehistoric Research, Peabody Museum, Harvard University, Bulletin*, **25**: 3–71.

(1978). The earliest waves of domestic horses in East Europe. *Journal of Indo-European Studies*, **1–2**: 17–73.

(1984). Horse. In *Evolution of Domesticated Animals*, ed. I. L. Mason. London: Longman, pp. 162–73.

Boyd, L. & Houpt, K. A. (1994). Activity patterns. In *Przewalski's Horse, the History and Biology of an Endangered Species*, ed. L. Boyd & K. A. Houpt. Albany: State University of New York Press, pp. 195–228.

Brown, D. & Anthony, D. (1998). Bit wear horseback riding and the Botai site in Kazakstan. *Journal of Archaeological Science*, **25**: 331–347.

Cabard, P. (1987). La faune Néolithique de la région d'Averdon Vallée de la Cisse (Loir-et-Cher). Paper read at 14e Colloque Interrégional sur le Néolithique, at Blois, France.

Cahen-Delhaye, A. (Ed.) (1991). *L'Or des Scythes, Trésors de l'Ermitage, Leningrad*. Brussels: Musées royaux d'Art et d'Histoire.

Caughley, G. (1966). Mortality patterns in mammals. *Ecology*, **47**: 906–18.

Chernykh, E. N. (1992). *Ancient Metallurgy in the USSR: the Early Metal Age*. Cambridge: Cambridge University Press.

Clason, A. T. (1988). The equids of Gomolava. In *Gomolava, Cronologie und Stratigraphie der Vorgeschichtlichen und Antiken, Kulturen der Donauniderung und Südosteruopas*, ed. N. Tasic & J. Petrovic. Belgrade: Novi Sad.

Clutton-Brock, J. (1974). The Buhen horse. *Journal of Archaeological Science*, **1**: 89–100.

(1987). *A Natural History of Domesticated Mammals*. London: British Museum of Natural History.

(1992). *Horse Power: a History of the Horse and the Donkey in Human Societies*. Cambridge, MA: Harvard University Press.

Curci, A. & Tagliacozzo, A. (1995). Il pozzetto rituale con scheletro di cavallo dell'abitato, Eneolitico di le Cerquete-Fianello, Maccaresse -RM. *Origini*, **18**: 297–350.

Dergachev, V. (1989). Neolithic and Bronze Age cultural communities of the steppe zone of the USSR. *Antiquity*, **241**: 793–802.

Dietz, U. L. (1992). Zur Frage vorbronzezeitlicher Trensenbelege in Europa. *Germania*, **70**(1): 17–36.

Eisenmann, V. (1996). Quaternary horses: possible candidates to domestication. In *Proceedings of the XIII Congress Forlí, Italy 8–14 September 1996*, vol. 6, no. 1. *Workshop 3 The Horse: Its Domestication, Diffussion and Role in Past Communities*. Forlí: International Union of Prehistoric and Protohistoric Sciences, pp. 27–36.

Galton, F. (1883). *Inquiries into Human Faculty and its Development*. New York: Macmillan.

Gimbutas, M. (1970). Proto-Indo-European culture: the Kurgan Culture during the fifth, fourth and third millennia B.C. In *Indo-European and Indo-Europeans*, ed. B. Cordona, H. M. Hoenigswald and A. Senn: University of Pennsylvania Press, pp. 155–97.

(1988). Review of archaeology and language by C. Renfrew. *Current Anthropology*, **3**: 453–6.

(1991). *The Civilization of the Goddess*. San Francisco: Harper.

Grigson, C. (1993). The earliest domestic horses in the Levant? – New finds from the fourth millennium of the Negev. *Journal of Archaeological Science*, **20**: 645–55.

Groves, C. P. (1986). The taxonomy, distribution and adaptations of recent equids. In *Equids in the Ancient World*, ed. R. H. Meadows and H.-P. Uerpmann. Wiesbaden: Dr. Ludwig Reichert Verlag, pp. 11–65.

Häusler, A. (1994a). Archäologische Zeugnisse für Pferd und Wagen in Ost- und Mitteleuropa. In *Die Indogermanen und das Pferd*, ed. B. Hänsel and S. Zimmer. Budapest: Archaeolingua, pp. 217–57.

(1994b). The North-Pontic region and the beginning of the Eneolithic in south-east and central Europe. In *The Archaeology of the steppes. Methods and strategies*, ed. B. Genito. Naples: Istituto Universitario Orientale, pp. 123–147.

Hemmer, H. (1990). *Domestication: The Decline of Environmental Appreciation.* 2nd edn. Cambridge: Cambridge University Press.

Jansen, T., Forster, P., Levine, M., Oelke, H., Hurles, M., Renfrew, C., Weber, J. & Olek, K. (2002). Mitochondrial DNA and the origins of the domestic horse. *Proceedings of the National Academy of Sciences* 99 (16): 10905–10.

Khazanov, A. M. (1984). *Nomads and the Outside World.* Cambridge: Cambridge University Press.

Klingel, H. (1969). The social organisation and population ecology of the Plains Zebra, *Equus quagga. Zoologica Africana,* 4(2): 249–63.

(1974). A comparison of the social behaviour of the Equidae. In *The Behaviour of Ungulates and its Relation to Management,* ed. V. Geist & F. Walther. Morges, Switzerland: IUCN Publications, pp. 124–32.

Kuz'mina, E. E. (1994a). *From where did the Indo-Aryans come [Otkuda prishli indoarii].* Moscow: Russian Academy of Sciences.

(1994b). Stages of development of stock-breeding husbandry and ecology of the steppes in the light of the archaeological and palaeoecological data, 4th millennium BC–8th century BC. In *The Archaeology of the Steppes. Methods and Strategies,* ed. B. Genito. Naples: Istituto Universitario Orientale, pp. 31–71.

(1996). The ecology of the Eurasian steppe and the origins of nomadism [Ekologiia stepei Evrazii i problema proiskhozhdeniia nomadizma]. *Bulletin of Ancient History [Vestnik drevnei istorii],* pp. 73–85.

Levine, M. A. (1982). The use of crown height measurements and eruption-wear sequences to age horse teeth. In *Ageing and Sexing Animal Bones from Archaeological Sites,* ed. B.Wilson, C. Grigson and S. Payne. Oxford: BAR British Series, pp. 223–250.

(1983). Mortality models and the interpretation of horse population structure. In *Hunter-Gatherer Economy in Prehistory,* ed. G. N. Bailey. Cambridge: Cambridge University Press, pp. 23–46.

(1990). Dereivka and the problem of horse domestication. *Antiquity,* 64: 727–40.

(1993). Social evolution and horse domestication. In *Trade and Exchange in Prehistoric Europe,* ed. C. Scarre and F. Healy. Oxford: Oxbow, pp. 135–141.

(1999a). Botai and the origins of horse domestication. *Journal of Anthropological Archaeology,* 18: 29–78.

(1999b). The origins of horse husbandry on the Eurasian steppe. In *Late Prehistoric Exploitation of the Eurasian Steppe,* ed. M. A. Levine, Y. Y. Rassamakin, A. M. Kislenko & N. S. Tatarintseva. Cambridge: McDonald Institute, pp. 5–58.

Levine, M. A., Bailey, G. N., Whitwell, K. E. & Jeffcott, L. B. (2000). Palaeopathology and horse domestication. In *Human Ecodynamics and Environmental Archaeology,* ed. G. Bailey, R. Charles & N. Winder. Oxford: Oxbow, pp. 123–133.

Levine, M. A., Whitwell, K. E. & Jeffcott, L. B. (2002). A Romano-British horse burial from Icklingham, Suffolk. *Archaeofauna,* 11: 63–102.

(in press). Abnormal thoracic vertebrae and the evolution of horse husbandry. *Archaeofauna.*

Lister, A. M., Kadwell, N., Kaagan, L. M., Jordan, W. C., Richards, M. B. & Stanley, H. F. (1998). Ancient and modern DNA in a study of horse domestication. *Ancient Biomolecules,* 2: 267–80.

Littauer, M. A. & Crouwel, J. H. (1996). The origin of the true chariot. *Antiquity,* 70: 934–9.

Makarova, L. A. & Nurumov, T. N. (1989). The problem of horse breeding in Neolithic–Eneolithic Kazakhstan [*K probleme konevodstva v Neolit – Eneolite Kazakhstana*]. In *Interactions between Nomadic Cultures And Ancient Civilisations [Vzaimodeistvie Kochevikh Kul'tur I Drevnikh Tsivilizatsii].* Alma-Ata: 'Nauka', pp. 122–131.

Mallory, J. P. (1981). The ritual treatment of the horse in the early Kurgan tradition. *The Journal of Indo-European Studies,* 9(3–4): 205–226.

(1989). *In Search of the Indo-Europeans.* London: Thames and Hudson.

Matyushin, G. (1986). The Mesolithic and Neolithic in the southern Urals and Central Asia. In *Hunters in Transition,* ed. M. Zwelebil. Cambridge: Cambridge University Press, pp. 133–50.

Otte, M. (1996). *Le paléolithique inférieur et moyen en Europe.* Paris: Masson & Armand Colin.

Petrenko, A. G. (1984). *Ancient and Medieval Animal Husbandry in the Volga and Ural regions [Drevnee i Spednevekoboe Zhivotnobodstvo Srednevo Povolzhia i Preduralia].* Moscow: Academy of Sciences.

Piggott, S. (1992). *Wagon, Chariot and Carriage.* London: Thames and Hudson.

Polos'mak, N. V. (1994). *Griffins watching over gold, the Ak-Alakha kurgans [Steregyshchie zoloto grify, ak-alakhinskie kurgany].* Novosibirsk: Hauka.

Postgate, J. N. (1986). The equids of Sumer, again. In *Equids in the Ancient World,* ed. R. H. Meadow and H.-P. Uerpmann. Wiesbaden: Dr. Ludwig Reichert Verlag, pp. 194–206.

Rassamakin, Y. Y. (1999). The Eneolithic of the Black Sea steppe: dynamics of cultural and economic development 4500–2300 BC. In *Late Prehistoric Exploitation of the Eurasian Steppe,* ed. M. A. Levine, Y. Y. Rassamakin, A. M. Kislenko and N. S. Tatarintseva. Cambridge: McDonald Institute, pp. 59–182.

Renfrew, C. (1987). *Archaeology and Language: the Puzzle of Indo-European Origins.* London: Jonathan Cape.

(1998). All the king's horses: assessing cognitive maps in later European prehistory. In *Creativity in Human Evolution and Prehistory,* ed. S. Mithen. London: Routledge, pp. 260–284.

Rudenko, S. I. (1970). *The Frozen Tombs of Siberia.* London: J. M. Dent and Sons.

Schibler, J. & Steppan, K. (1999). Human impact on the habitat of large herbivores in eastern Switzerland and southwest Germany in the Neolithic. *Archaeofauna*, 8: 87–99.

Serpell, J. (1989). Pet-keeping and animal domestication: a reappraisal. In *The Walking Larder: Patterns of Domestication, Pastorialism and Predation*, edited by J. Clutton-Brock. London: Unwin Hyman, pp. 10–21.

Telegin, D. Y. (1986). *Dereivka, a settlement and cemetery of Copper Age horse keepers on the Middle Dnieper. BAR International Series*, 287.

Uerpmann, H.-P. (1990). Die Domestikation des Pferdes im Chalkolithikum West-und Mitteleuropas. *Madrider Mitteilungen*, 31: 109–153.

Vilà, E., Leonard, J. A., Götherström, A., Marklund, S., Sandberg, K., Lidén, K., Wayne, R. K. & Ellegren, H. (2001). Widespread origins of domestic horse lineages. *Science*, 291: 474–77.

Zarins, J. (1986). Equids associated with human burials in third millennium BC Mesopotamia: two complementary facets. In *Equids in the Ancient World*, ed. R. H. Meadow and H.-P. Uerpmann. Wiesbaden: Dr. Ludwig Reichert Verlag, pp. 164–193.

2

The horse in human society

Stephen J. G. Hall

Introduction

For most of history the horse has been the key to power, conquest and the wealth of nations. Within human societies, those who have possessed superior horses have acquired social prestige as a result (Clutton-Brock, 1992; Diamond, 1997).

When agricultural production has been high enough to sustain production of fodder, horses and other draught animals have been made available for work as motive power for a wide range of machinery and vehicles and have thus contributed to industrialization.

In developed countries horses are no longer necessary for work but they clearly provide other services, in the broadest sense, for which people are willing to pay. In developing countries they continue to provide power, along with other work animals.

While the military value of the horse has largely disappeared, except for ceremonial purposes, many new functions have emerged. As a first example, in the developing world, horses are more and more widely used for transport, both rural and urban, and the contribution of these animals to assuring national food production and local food distribution is as deserving of attention as that of the horses that work in the field. Some other new roles are numerically far less important, but are still of economic significance and are also of great interest in demonstrating the range and depth of the horse's role in society. For example, in rich countries, horses are increasingly seen as companion animals. And, again in rich countries, many valued landscapes need to be grazed to maintain their biodiversity and visual appeal and there is growing interest in the use of horses for this purpose as well as in the conservation of populations of feral horses as emblems of wildness.

The pivotal role of the horse in the development of culture and civilization has been eloquently described by Clutton-Brock (1992). The steppe nomads of Eurasia, the classical civilizations of Greece and Rome, and the medieval world have all left clear records, whether in writing, art or archaeological remains, of the significance of the horse in their

The Domestic Horse: The Origins, Development, and Management of its Behaviour, ed. D. S. Mills & S. M. McDonnell. Cambridge University Press. © Cambridge University Press 2005.

societies. The horse was the key to the military conquest of the Americas and to their agricultural colonization (Diamond, 1997; Smil, 2000).

Hundreds of books on the farm horse, fox hunting, other sports and horse racing attest to the importance of these animals in the production, ostentatious display and redistribution of wealth. Far less well documented is the story of how the availability of ample horse power contributed to industrialization. Even in societies that we might regard as quite highly mechanised, such as the UK and the USA in the late 1800s and early 1900s, horses were of massive importance.

Horses in the distant past
Domestication

The archaeological record suggests horses were domesticated at least 5000 years ago (Levine, Chapter 1) and the study of mitochondrial DNA (mtDNA) of present-day horses and of fossils enables competing hypotheses about this process to be tested. As mtDNA is maternally inherited, there is no recombination. Also, selection on this DNA is thought to be weak. Thus, any difference in DNA sequences between the mtDNA of groups of animals is due to the accumulation of mutations, which would be expected to proceed independently (and at a predictable rate) during the history of the groups as separate entities ('lineages'). The extent of the sequence difference indicates how long it is since the groups diverged from their common ancestor. If there had been a single domestication of the horse, with one population alone contributing founder animals, then today's mtDNA lineages should reveal that the common ancestor lived around 5000 years ago, however, Vilà *et al.* (2001) found that the common ancestor of modern horse lineages must have lived within the period 630 000–320 000 years ago. This means that today's domesticated horses are not all descended from the same population: many different wild populations must have been involved which had been separated long enough to have evolved distinctive mtDNA. The observed diversity of mtDNA suggests that domestication took place in many districts, perhaps as a gradual assimilation of wild horses into human society rather than as the abrupt process usually hypothesized for other livestock species, and with contributions from many wild populations and subsequent exchange of breeding animals (Vilà *et al.*, 2001; Levine, Chapter 1).

Jansen *et al.* (2002) found that most of today's horse mtDNA lineages group into one of 17 distinct clusters, suggesting that this may be the approximate number of populations of the wild ancestor that contributed to the domesticated horse. The northern European ponies form a cluster, as do the Iberian and north African horses. These groups could each be descended from a distinct domestication event. However, it is still possible that there was a single such event – as Jansen *et al.* (2002) point out, domesticated stallions could have been mated with wild mares (termed 'secondary domestication' by Hemmer, 1990).

Horses in recent times
Power, wealth and prestige

A small number of important inventions – the stirrup during the first centuries AD, the horse collar and the nailed horseshoe (in universal use by the eleventh century) – increased the potency of the horse in war. The importance of the horse in the armies of the great civilizations and empires of ancient Egypt, Greece and Rome, the Orient, western Asia and the steppes, is clear (Clutton-Brock, 1992). The agricultural use of equines is much less well documented, but general patterns are known, for example in Rome, oxen, mules and donkeys were the main beasts of burden and plough animals.

The break-up of the Roman Empire was followed in Europe by a phase of relatively weak, decentralized and often short-lived kingdoms which lived under the shadow of attack by their neighbours and also by warriors such as the Muslims from the south and the Gothic and other cavalry-based peoples from the east. With the development of the feudal system, under which the lord was granted, by the king, absolute power over his tenantry and serfs, it became possible to maintain or be able to mobilize at short notice, armies based on cavalry.

By 1200 AD the supremacy of horses in military power was well understood in Europe and Asia. Each ruler would aspire to have cavalry that was at least as effective as that of his enemies.

Innovation was not only technological. Horses were bred for new conformations and body sizes, the first instance being the larger horses necessary for cavalry. The medieval war horse was, in fact, quite small up to the fourteenth century. It was rather like a modern Cob (and incidentally nothing like today's Shire horse). In England, to assure the supply of large war horses, in 1535 King Henry VIII required all big landowners to keep at least two mares of 13 hands or so, to be covered by stallions of at least 14 hands, while in 1540 he forbade the keeping of stallions of less than 15 hands on most common grazings (Russell, 1986).

Around 1200, in England, horses provided only 5% of the draught animals but by 1400, 20% of the draught animals on the estates and 50% of those on peasant farms were horses. In Norfolk, a mainly arable county, by 1400 there were three times as many horses as oxen while in England as a whole there were four oxen for each horse (Campbell, 1988). The change from ox to horse tillage must have resulted from the interplay of many factors – 'several reasons can be advanced . . . but none can be substantiated by written evidence. A heavier and stronger type of horse may well have become available as a result of Norman importations from the Continent; some advance in [plough] design may have made it possible to use a faster draught animal; and an increase in the demand for beef to feed the growing number of non-producers in army, garrison, court, town and monastery may have made [cattle] more valuable for meat than for work' (Trow-Smith, 1959). While oxen with their steady, slow tractive effort were better at pulling a wooden plough through heavy land, every technical development in plough design culminating in the steel ploughshare, in the mid-nineteenth century, favoured the horse (Smil, 2000).

Horse-drawn societies

In the nineteenth century, the cities of the richest countries were heavily dependent on horses. Equine census data tend, however, to be incomplete. Certain classes of horse were apparently almost invisible to census enumerators – there were probably 65 000 ponies working in coal mines in the UK in 1924 which were not counted in the survey of Thompson (1976) while the numbers of horses kept to drive machinery in the UK do not seem to have been estimated. In the USA, in 1900 there were 24 million horses of which three million were in cities. In 1920, urban horses were in the process of being replaced by motor vehicles, their numbers having dropped to 1.7 million (8% of the total) while there were still 21.4 million in the rural areas (McShane & Tarr, 1997). This reflects the speed of urban motorization in the USA. In the UK in 1920 there were 2.2 million horses (774 934 draught

farm horses, 411 307 horses in 'trade', 19 743 Thoroughbreds, 620 766 light horses, 254 707 draught horses under three years old and 110 708 army horses: Chivers, 1988). Thus, about 19% of the horses in the UK were used in transport, a much higher proportion than in the USA.

The incentive to replace human power by horse power has always been strong. A horse eating 4 kg of oats a day can do as much work as 10 strong men who together would consume nearly 7 kg of grain daily (Smil, 2000). Horses were widely used to drive machinery but this has been less thoroughly studied than their uses in tillage and transport. McShane and Tarr (1997) describe how many ingenious machines were built, generally using horses working on a treadmill or on a circular track around a central point, with belts, drive shafts and gearing to transmit the power to pumps, sawmills, lathes, grinding machines, lifting equipment, etc. Their use appears better documented in America than in the UK; whether this is because this function was less important in the latter does not seem to have been studied.

The horse today
How many horses are there today?

By the end of 2004, each individual equine in the UK is projected to have a 'passport', in order to comply with an EU Directive and also to give the basis of a health database (www.horse-passports.co.uk/news/defra.html). With the force of law behind it, this differs from previous modern enumerations of the horse population. It will make other attempts at achieving a census of academic interest only, except in that some of the problems encountered during censuses, and the incidental information gathered, do shed light on aspects of the horse in UK society. For example Chivers (1988) reported how the 1975 British Horse Society pilot survey in Wiltshire and Avon was extremely difficult to organize, with census enumerators often getting a 'less than welcoming' reception. Maybe a proportion of the UK horse population is kept by relatively marginalized elements of society. Other attempts have been made, including estimates based on the production figures for shoeing steel, but Chivers (1988) concluded that detailed census information for horses can 'never be obtained by a voluntary body'. A current estimate (National Equestrian Survey, 1999) is of 900 000 horses and 2.4 million riders in the UK in 1998/99, representing an increase since 1995. France has about 500 000 horses, or about eight horses per thousand people (Berger, 2001) while the UK has 15 horses per thousand people. Presumably ecological, social and historic factors contribute to such differences.

In an extension to a national ecological survey based on representative 1×1 km sites during summer 1984, Barr *et al.* (1986) deduced a UK population of 558 400 horses. No distinction was made as to type and only horses that were outside at the time of the survey were censused. Horses were recorded on 31% of the sample sites and were thought to be using up to one million hectares of land (15% of the lowland grass of the UK). The densest population was in East Anglia (7.7 horses per km^2); here the highest numbers of horses were usually in the urban fringes and also in equine establishments. The lowest density was in north-west Scotland (0.1 horses per km^2).

Equine breed biodiversity

There is not a single definition for the term 'breed'. In the developed world the term is associated with the existence of a breed society, usually with a published breed standard and a pedigree system. In the developing world it may be applied to the livestock of a particular area, provided they have some uniformity and cultural identity. According to Lauvergne (1993) the term 'breed', or its counterpart in translation, first began to appear in print in the seventeenth century (in relation to cattle, in France in 1600; and to horses, in Germany in 1672). A world definition of the term became particularly desirable when the 1993 Rio Convention on Biological Diversity gave countries sovereignty over their genetic resources and obliged them to accept responsibility for their stewardship. Breeds are recognized as components of world biodiversity (Heywood & Watson, 1995). The international inventory of breeds is in the custodianship of the Food and Agricultural Organization of the United Nations (FAO) and to qualify for inclusion, all that a government has to do is claim that a particular population of domesticated animals is unique.

Worldwide to date, 682 breeds of horse are recognized, in addition to which there are 107 breeds that are known to have gone extinct: Table 2.1.

The number of animals of each livestock species can be divided by the number of breeds to give an index of the extent to which each species has been differentiated into breeds. This has been done for

Table 2.1. *Numbers of horse breeds in each region. Deduced from Porter (2002), a compendium of the breed names recognized in the literature*

	Total breeds	Of which rare	Of which extinct
Africa	51	10	4
Asia	166	32	5
Europe	357	144	72
North and Central America	84	29	6
South America	34	5	0
Oceania	6	2	0
Former USSR	91	31	20
Totals	789	253	107

the main mammalian livestock species (cattle, donkey, goat, horse, pig, sheep and water buffalo) by Hall and Ruane (1993), who found that with 62 million horses worldwide and 682 breeds, this index has the value of 91 000. Of course this is not an average value for the size of each breed because the world total includes large numbers of crossbreds and of horses of undefined breed. But it can lead to comparisons between species and continents; Hall and Ruane (1993) showed that worldwide, the horse is the livestock species with the lowest value of this index, showing that this is the species where development of specialized breeds has been most widely practised. Donkeys have a value of 545 000 and water buffalo (the least differentiated species), 1 902 000 (Hall & Ruane, 1993).

A breed that exists today can be seen as an expression of a history of genetic selection and genetic drift. Its genotype will include genes and gene combinations coding for particular characteristics and it will have lost from its genotype, due to the random genetic processes which accompany restricted mating, many of the genes which are present in other breeds.

Breeds exist and maintain their distinctiveness because they have distinct functions and identity. Considering all livestock species worldwide, it is clear that countries with large human populations have large numbers of breeds (Hall & Ruane, 1993), implying that breeds are numerous when there are many different functions that livestock are expected to perform. This is a reflection of cultural and social complexity; thus the term 'function' when applied to breeds has both cultural and economic connotations.

Table 2.2. *Some recent studies of breed origin and affinities, using molecular genetics (breed names as in Porter, 2002)*

Brazil	Pantaneiro horse shown to be related to Iberian breeds; though well adapted to the Pantanal wetland there has been some crossing with other breeds	Cothran *et al.* (1998)
Croatia	Indigenous Posavina breed has closest resemblance to Polish draught horse, and high genetic variation consistent with long history of influence from foreign breeds	Cothran & Kovac (1997)
South Africa	Basuto pony, Boer and Nooitgedacht pony all closely related. Influence of imports from USA in 1782 and 1808 is still detectable; affinity with Thoroughbred confirms historical accounts	Cothran & van Dyk (1998)
Czech Republic	Kladruby has grey and black subpopulations which are genetically distinct, though clearly related	Horín *et al.* (1998)
Spain	Studbooks for seven 'Celtic' breeds from north coast were only established recently, there is evidence of crossbreeding, patterns of variation suggest that it is reproductive isolation that has contributed most to breed distinctiveness	Cañon *et al.* (2000)
Slovenia	Several maternal lineages in Lipitsa, consistent with known history of contributions from many other breeds	Kavar *et al.* (1999)
World	Historical accounts of origin of Thoroughbred confirmed and relative importance of the different founders in today's population deduced	Cunningham *et al.* (2001)
Germany	Senne horse has distinct mtDNA, which is unusual elsewhere, consistent with documented descent from one mare (born 1725) and suggesting distinct history. Dülmen pony has mtDNA suggesting ancient origin, and they come from an area where in 1316 'wild' horses were recorded	Jansen *et al.* (2002)
Portugal	Sorraia horse has vastly different mtDNA from other Iberian breeds and history of descent from 'wild' horses of the area	Jansen *et al.* (2002)

Breeds become extinct because their function disappears or because some other breed performs better; extinction is usually by a process of repeated crossbreeding whereby one breed becomes incorporated into another. It is relatively unusual for a breed to die out for any other reason – in livestock, extinction usually means loss of identity rather than extermination.

A breed will continue to have an independent existence as long as people place a value on its distinctiveness. Borrowing from ecological terminology, it must have an economic or cultural function or specialization or else it will be replaced by another (Hall & Ruane, 1993). The role of the horse can differ radically even between geographically close countries, with broadly similar economic systems and national wealth, such as the UK and France. Horses are not eaten in the UK (32 700 tonnes consumed in France in 2000: Berger, 2001); in Britain large numbers (63 000: National Equestrian Survey, 1999) are kept primarily for hunting foxes while in France far fewer, about 10 000, are kept for hunting, some for fox but mainly for hare, deer and wild boar (Pinet, 1987).

Testing cherished traditions

Other genetic studies on particular breed groups have also thrown light on to historical events and it is now possible to test, objectively, many treasured histories of breed origin. Microsatellites came into widespread use in livestock genetics shortly after mtDNA was adopted and the former can be used to investigate hypotheses about breed origin and evolution. They are apparently neutral genes inherited in a mendelian manner (there is no sex linkage and recombination is possible, unlike the situation with mtDNA). Microsatellites show a high allelic diversity and can be used to calculate genetic distances between breeds. Vilà *et al.* (2001) found that using microsatellites, 95% of horses could be correctly assigned to breed.

The horse in the twenty-first century
Food security

Today's human population is about 6 billion; it is expected to level off at 9 billion by 2050. While

800 million are undernourished, this number is expected also to decline, perhaps to 600 million (mainly in sub-Saharan Africa) by 2025 (Trewavas, 2002). The major cause of undernourishment is poverty, and the modern view of world food policy regards access to food (food security) as necessitating reduction in poverty. Food security is becoming better understood and is now emphasized more than food production in social policy. It is defined as having access at all times to enough food for an active, healthy life. Smith *et al.* (2000) summarize the factors which interact to assure food security. For the rural poor, who make up over 80% of the poorest 20% of the people of the developing world (Conway & Toenniessen, 1999), food security requires local production of food because of their limited purchasing power and distance from markets. Draught animals clearly have an especially important part to play here. Today, about 400 million animals are used for draught (Sansoucy, 1995). Most of these are cattle and water buffalo, the numbers of which are increasing while those of working equines has not changed greatly, worldwide. There are about 43 million donkeys, 13.5 million mules and 58 million horses worldwide (FAO statistics cited by CAB International, 2002).

Traction animals can help to make these interactions efficient. National food production must be made available to the community, which at a more local level, household income must be sufficient to give access to food.

In developing countries, employment opportunities are usually greatest in the cities with many country households depending on remittances from family members working there, and there can be labour shortages back in the village. One argument for encouraging draught animal power has been provision of manure, but it is now clear that inorganic fertilizer is necessary for the large-scale crop production that is essential if cities are to be provisioned (Carr, 1989; Jones, 1991; Wiggins, 1999).

Draught animals also provide a means to utilize crop residues. For each kg of grain produced there is at least 1 kg of straw and Smil (1999) calculated that the world's annual harvest of 2750 million tonnes of dry matter was accompanied by production of 3750 million tonnes of crop residues. Equines, with their ability to digest roughages of high bulk, are particularly well adapted to this diet but will also need feed supplements.

Much research has been done on the biology of the draught animal especially as used for tillage, and on the constraints on its effectiveness (e.g. Pearson, 1998b). This research will help to improve food production but will not automatically lead to better access to food. Only a few developing-world farmers are rich enough to own draught animals and the extent to which the poor in society benefit from them, may be debated. However, the growth of demand for transport is obvious and there is a strong case for more work on the haulage technologies. Considering the horse from this angle, its role in tillage may be becoming less important than its role in transport. If so, there may be a shift in the type of conformation that is preferred, which could affect the fortunes of particular breeds.

The present status of draught animals and their future prospects vary greatly between countries and ultimately they appear largely dictated by the price and availability of motor power and the availability of fodder and grazing land. As more land is brought into cultivation, and cropping regimes are intensified, less grass and conserved forage will be available, but there will be an increase in crop residues and by-products of food processing. The existence of a range of species suitable for draught gives flexibility. For example, in parts of west Africa where soils are light and ploughing is not common, horses have replaced oxen because the former can work more quickly (Pearson, 1998a). Some quite closely integrated systems have been described, for example the horses of Senegal, west Africa, which are used for tillage, transport and racing (Dehoux *et al.*, 1996). Generally, in areas where equine diseases are not a serious problem, such as temperate, subtropical and high altitude areas, horses are often the work animal of choice (Starkey, 1997).

Whether changes in patterns of use of horses will lead to changes in the ways they are bred and traded, is not known. Increased use may lead to more and more horses working in a different district, region or even country from where they were born. This could result in breeds losing distinctiveness, mainly because of the movement of breeding animals to new areas. Perhaps also, breeders will make mating decisions with sale to the trade in mind, which would reduce the value placed on pure breeding.

Although there are several charities addressing the welfare of working horses in the developing world, horses do not attract so much interest from aid agencies, who have traditionally been more involved in research and knowledge transfer about oxen, water buffalo and, more recently, working cows. The appeal of large ruminants to the livestock specialists who advised these agencies was probably the apparent ease of the integration of these animals into farming systems (Sansoucy, 1995) and their contribution to self-sufficiency. From the perspective of the rich world, animal welfare in developing countries is often poor and there is a strong motivation to help. However, projects should be sensitively designed. If drugs and nutritional treatments are provided for draught animals which are not so freely available for sick or malnourished children, most people's sense of fairness would be offended. So it may be best in these countries to concentrate on interventions that do not have a human analogue, such as farriery, the treatment of injuries and the design of harnessing (Hall, 2001). See Cooper and Albentosa, Chapter 16, for a discussion of the approaches to horse welfare used in the developed world.

As information accumulates about the growing importance of donkeys (see, e.g. Pearson, 1998a; Starkey, 2000) some of the attention previously given to large ruminants may well shift to this species. Patterns of use of donkeys worldwide would make this species a stronger candidate than the horse for a role in achieving poverty reduction and gender equity. For example, in Ethiopia there are 5 million donkeys that are 'extremely important' as pack animals in both town and country (Starkey, 2000), while mules and horses are primarily used for riding (mainly by men) in the country and for pulling passenger carts in the towns. Freight is seldom hauled by horses and mules. With this background, one might expect development agencies to see horses and mules as being a matter for the private sector while work with donkeys would bring general social benefits.

Ecosystem engineering

In the western world, globalization of food production and changing views on rural development are leading former agricultural land to be reallocated to nature conservation. In many such areas, horses are an ideal 'ecosystem engineer', i.e. they are able to maintain the environment in a state that is of nature conservation value. They do this by grazing and are especially suited for this because of the large bulk of food they eat daily (Menard *et al.*, 2002). The horses may be of especial interest in their own right, for example feral herds of horses in the USA may be seen as emblems of wildness, and society is evidently willing to meet the opportunity costs (estimated as \$1900 per year per horse, and incurred because grazing for cattle and space for game animals is reduced) of allowing these horses to remain (Bastian *et al.*, 1999).

Practical experience is accumulating on the management of equines to maintain areas of biodiversity interest. In many ways horses are better than cattle in this respect; horses tend to create vegetation mosaics with 'lawns' and areas of tall, undisturbed vegetation. In French wetlands, horses were found to prefer swards less than 5 cm in height, and to concentrate on grass species, while cattle prefer swards 9–16 cm in height and will eat more dicotyledonous plants (Menard *et al.*, 2002).

Controlled experiments are lacking, but evidence is growing that suggests, breed differences. The Konik, a Polish pony of 12–13 hands which may reach 500 kg under UK conditions (Tolhurst & Oates, 2001), is used in a number of wetland systems where this pony now has a reputation for being free of laminitis. A herd of 600 of these ponies grazes the 5600 ha of the Oostvaardersplassen in the Netherlands, with 800 red deer and 600 cattle (Beije & Dekker, 2002); this is probably the largest conservation grazing herd of equines in the world. Experience in Britain with a total of about 60 much smaller operations using a diversity of horse and pony breeds is summarized by Tolhurst and Oates (2001).

Companionship, therapy and leisure

Greater personal wealth means that horses can be kept as companions after their economic life is over, raising ethical and welfare issues such as are already of interest in relation to dogs and cats (McDonnell, 2002). In addition to having general health benefits (Mills and McNicholas, Chapter 11), this companionship can be of specific therapeutic value to people. Interaction with horses can also play a part in raising the quality of life of disadvantaged people (see, e.g. Garrigue, 2001). Much of the evidence in this area is anecdotal – for example, when the 'world's largest horse', a Shire gelding named Goliath who stood 196 cm at the shoulder, died near Lincoln, UK, in 2001, the press report (Rooth, 2001) emphasized the comfort wheelchair-using and arthritic people had derived from his companionship.

Less anecdotal is the growing body of evidence that children with cerebral palsy and adults with spasticity resulting from injury to the spinal cord can benefit from 'recreational horseback riding therapy' – also known as hippotherapy (Sterba *et al.*, 2002; Lechner *et al.*, 2003). In the USA this is seen as a medical therapy provided under a physician's prescription (Borzo, 2002). It is distinct from supervised recreational riding for people with disabilities, such as has been practised for many years in Europe – in the UK the charity Riding for the Disabled was registered in 1965.

The importance of the horse for leisure and recreation, at least in the wealthier nations, is clear (summarized for the UK by the National Equestrian Survey, 1999).

Population sizes of horse breeds

In spite of the wealth of information on their economic significance, there is often a remarkable lack of detailed information on the composition of this important resource. For example, in the UK survey of animal genetic resources (DEFRA, 2002), population estimates were obtained for all 72 native and imported cattle breeds, for 66 of 85 sheep breeds, but for only 13 of 44 equine breeds. Surprisingly, there seem to be no population estimates for two of the UK's best known breeds, the Welsh pony and the Shetland pony.

For many, probably most breeds of horse, authoritative estimates of population size are lacking. Worldwide the Thoroughbred is probably the most numerous breed, with a total of 122 045 in its five main European populations (UK, Ireland, France, Italy and Germany: http://www.eftba.com, November 2003 data). Of these, 19 450 are foals. The foal crop in North America in 1993 was almost twice that figure (36 700: http://www.ansi.okstate.edu/breeds/horses) so the total population of Thoroughbreds there is probably around a quarter of a million. These numbers are known because of the requirement to register animals of this breed.

Some famous breeds are in fact surprisingly rare. In the UK there are only 1800 Shire horses, 500 Clydesdales and 69 Suffolks (DEFRA, 2002). This emphasizes the importance of conservation efforts like that, in the UK, of the Rare Breeds Survival Trust (http://www.rbst.org.uk) and of international breed inventories, most notably that of the Food and Agri-

cultural Organizations of the United Nations (FAO) (http://www.fao.org/dad-is). Ideally, new uses would be found for these breeds to safeguard their numbers; failing that, they should be conserved either at public expense or (a more likely outcome) by private individuals. Ideas on how breeds should be prioritized for conservation are currently under development (Ruane, 2000). This is rather difficult for equines, because it is harder to demonstrate a potential benefit to the public of breed conservation than it is with livestock that are mainly bred for food.

New research needs

These changing roles necessitate new research, especially in welfare (Cooper and Albentosa, Chapter 16). Development of the transport role means that horses will be kept by people who do not have several generations of horse-keeping experience behind them, but who, being urbanized and in the public eye, may well be receptive to education on welfare-oriented husbandry.

Greater understanding is needed of the patterns of trade in horses in the developing world. Probably most welfare interventions take place in the cities and tourist resorts, and are directed at the end users of the animals, but attention should also be paid to welfare issues arising elsewhere in the horse trade. Appreciation of the behaviour of free-ranging horses is also needed, whether these horses are feral or are kept under extensified conditions, and further study is needed of adaptation to these conditions.

Continual attention needs to be paid to the conservation of rare equine breeds. While the general principle, that livestock biodiversity is worthy of conservation in its own right, is becoming more and more widely accepted, the practical issues of deciding where limited conservation resources should be applied are more likely to be resolved to general satisfaction if objective scientific criteria for conservation funding are developed.

Reference List

Barr, C. J., Benefield, C., Bunce, B., Ridsdale, H. & Whittaker, M. (1986). *Landscape Changes in Britain*. Grange-over-Sands, Cumbria: Institute for Terrestrial Ecology.

Bastian, C. T., Van Tassell, L. W., Cotton, A. C. & Smith, M. A. (1999). Opportunity costs related to feral horses: a Wyoming case study. *Journal of Range Management*, **52**, 104–12.

Beije, H. & Dekker, H. (2002). *Grazing and Grazing Animals.* Special issue, *Vakblad Natuurbeheer.* The Hague: Ministry of Flemish Community.

Berger, Y. (2001). The role of the horse in rural policies from the French viewpoint. Paper presented at *EU Equus 2001 Conference, Skara, Sweden, 12–13 June 2001* and published online http://www.equus-2001. (Downloaded 3 November 2003.)

Borzo, G. (2002). *Horse Power: When Riding Turns into Treatment.* Amednews.com, 17 June 2002. Published online at http://www.ama-assn.org/amednews/2002/06/17/hlsa0617.htm. (Downloaded 3 November 2003.)

CAB International (2002). Animal health and production compendeum, Wallingford: CAB International.

Campbell, B. M. S. (1988). Towards an agricultural geography of medieval England. *Agricultural History Review,* **36,** 87–98.

Cañon, J., Checa, M. L., Carleos, C., Vega-Pla, J. L., Vallejo, M. & Dunner, S. (2000). The genetic structure of Spanish Celtic horse breeds inferred from microsatellite data. *Animal Genetics,* **31,** 39–48.

Carr, S. J. (1989). *Technology for Small-Scale Farmers in sub-Saharan Africa. Experience with Food Crop Production in Five Major Ecological Zones.* World Bank Technical Paper 109. Washington DC: World Bank.

Chivers, K. (1988). *History with a Future. Harnessing the Heavy Horse for the 21st Century.* Peterborough: Shire Horse Society/Royal Agricultural Society of England.

Clutton-Brock, J. (1992). *Horse Power. A History of the Horse and the Donkey in Human Societies.* London: Natural History Museum Publications.

Conway, G. & Toenniessen, G. (1999). Feeding the world in the twenty-first century. *Nature,* **402** (Suppl.), C55–8.

Cothran, E. G. & Kovac, M. (1997). Genetic analysis of the Croatian Trakehner and Posavina horse breeds. *Zivocisna Vyroba,* **42,** 207–12.

Cothran, E. G. & van Dyk, E. (1998). Genetic analysis of three South African horse breeds. *Journal of the South African Veterinary Association,* **69,** 120–5.

Cothran, E. G., Santos, S. A., Mazza, M. C. M., Lear, T. L. & Sereno, J. R. B. (1998). Genetics of the Pantaneiro horse of the Pantanal region of Brazil. *Genetics and Molecular Biology,* **21,** 343–9.

Cunningham, E. P., Dooley, J. J., Splan, R. K. & Bradley, D. G. (2001). Microsatellite diversity, pedigree relatedness and the contributions of founder lineages to thoroughbred horses. *Animal Genetics,* **32,** 360–4.

DEFRA (2002). *UK Country Report on Farm Animal Genetic Resources.* London: Department for Environment, Food and Rural Affairs.

Dehoux, J.-P., Dieng, A. & Buldgen, A. (1996). The Mbayar horse in the central part of the groundnut zone of Senegal. *Animal Genetic Resources Information,* **20,** 35–54.

Diamond, J. (1997). *Guns, Germs and Steel. The Fates of Human Societies.* London: Jonathan Cape.

Garrigue, R. (2001). Equestrian activities for handicapped and disadvantaged people. Paper presented at *EU Equus 2001 Conference, Skara, Sweden, 12–13 June 2001* and published online http://www.equus-2001. (Downloaded 3 November 2003.)

Hall, S. J. G. (2001). Trends in third-world agriculture with relevance to traction animals. In *Traction Animal Health and Technology. Proceedings of Two Seminars Organised by World Association for Transport Animal Welfare and Studies (TAWS) and British Veterinary Association (Overseas Association),* ed. by R. J. Connan. Wheathampstead, Hertfordshire: Universities Federation for Animal Welfare, pp. 52–61.

Hall, S. J. G. & Ruane, J. (1993). Livestock breeds and their conservation: a global overview. *Conservation Biology,* **7,** 815–25.

Hemmer, H. (1990). *Domestication: The Decline of Environmental Appreciation.* Cambridge: Cambridge University Press.

Heywood, V. H. & Watson, R. T. (1995). *Global Biodiversity Assessment.* Cambridge: United Nations Environment Programme/Cambridge University Press.

Horín, P., Cothran, E. G., Trtková, K., Marti, E., Glasnák, V., Henney, P., Vyskocil, M. & Lazary, S. (1998). Polymorphism of Old Kladruber horses, a surviving but endangered baroque breed. *European Journal of Immunogenetics,* **25,** 357–63.

Jansen, T., Forster, P., Levine, M. A., Oelke, H., Hurles, M., Renfrew, C., Weber, J. & Olek, K. (2002). Mitochondrial DNA and the origins of the domestic horse. *Proceedings of the National Academy of Sciences of the USA,* **99,** 10905–10.

Jones, D. W. (1991). How urbanization affects energy use in developing countries. *Energy Policy,* **19,** 621–30.

Kavar, T., Habe, F., Brem, G. & Dovc, P. (1999). Mitochondrial D-loop sequence variation among the 16 maternal lines of the Lipizzan horse breed. *Animal Genetics,* **30,** 423–30.

Lauvergne, J. J. (1993). Breed development and breed differentiation. In *Proceedings of CEC Workshop on Data Collection, Conservation and Use of Farm Animal Genetic Resources, 8–9 December 1992,* ed. D. Simon & D. Buchenauer. Tierärztliche Hochshule. Hannover, Germany, pp. 53–64.

Lechner, H. E., Feldhaus, S., Gudmundsen, L., Hegemann, D., Michel, D., Zach, G. A. & Knecht, H. (2003). The short-term effect of hippotherapy on spasticity in patients with spinal cord injury. *Spinal Cord,* **41,** 502–5.

McDonnell, S. M. (2002). Behaviour of horses. In *The Ethology of Domestic Animals. An Introductory Text,* ed. P. Jensen. Wallingford: CABI Publishing.

McShane, C. & Tarr, J. A. (1997). The centrality of the horse in the nineteenth-century American city. In *The Making of Urban America,* 2nd edn, ed. R. A. Mohl. Wilmington, Delaware: SR Books, pp. 105–130.

Menard, C., Duncan, P., Fleurance, G., Georges, J.-Y. & Lila, M. (2002). Comparative foraging and nutrition of horses and cattle in European wetlands. *Journal of Applied Ecology,* **39,** 120–33.

National Equestrian Survey (1999). *The National Equestrian Survey 1999.* Summary published online http://www.beta-uk.org/EqSurvey/EqSurvey.asp. (Downloaded 3 November 2003.)

Pearson, R. A. (1998a). The future of working equids – prospects and problems. In *3er Coloquio Internacional sobre Equidos de Trabajo, Mexico 1998*. Mexico City: Universidad Nacional Autónoma de México, pp. 1–20.

  (1998b). Draught animals and their management: the future in rain-fed agriculture. *Annals of Arid Zone*, **37**, 233–51.

Pinet, J.-M. (1987). *The Economy of Hunting*. Paris: Laboratoire d'Ecologie et de la Faune Sauvage de l'Institut National Agronomique.

Porter, V. (2002). *Mason's World Dictionary of Livestock Breeds, Types and Varieties*, 5th edn. Wallingford: CABI Publishing.

Rooth, B. (2001). World's biggest horse dies. *Lincolnshire Echo*, 23 July 2001.

Ruane, J. (2000). A framework for prioritizing domestic animal breeds for conservation purposes at the national level: a Norwegian case study. *Conservation Biology*, **14**, 1385–93.

Russell, N. (1986). *Like Engend'ring Like. Heredity and Animal Breeding in Early Modern England*. Cambridge: Cambridge University Press.

Sansoucy, R. (1995). Livestock – a driving force for food security and sustainable development. *World Review of Animal Production*, **84**, 5–17.

Smil, V. (1999). Crop residues: agriculture's largest harvest. *Bioscience*, **49**, 299–308.

  (2000). Horse power. *Nature*, **405**, 125.

Smith, L. C., El Obeid, A. E. & Jensen, H. H. (2000). The geography and causes of food insecurity in developing countries. *Agricultural Economics*, **22**, 199–215.

Starkey, P. (1997). Transport animals: world-wide trends and issues. In *Traction Animal Health and Technology. Proceedings of a symposium organised by World Association for Transport Animal Welfare and Studies (TAWS) and British Veterinary Association (Overseas Association). Royal Veterinary College, 11 April 1996*, ed. S. J. G. Hall. Wheathampstead, Hertfordshire: Universities Federation for Animal Welfare, pp. 5–16.

  (2000). *Local Transport Solutions: People, Paradoxes and Progress. Lessons Arising from the Spread of Intermediate Means of Transport*. Working paper no. 56. Rural Travel and Transport Program. Sub-Saharan Africa Transport Policy Program. Washington DC: World Bank. Published online: www.worldbank.org/afr/ssatp/Working%20Papers/SSATPWP56.pdf.

Sterba, J. A., Rogers, B. T., France, A. P. & Vokes, D. A. (2002). Horseback riding in children with cerebral palsy: effect on gross motor function. *Developmental Medicine and Child Neurology*, **44**, 301–8.

Thompson, F. M. L. (1976). Nineteenth-century horse sense. *Economic History Review*, **5**, 64–7.

Tolhurst, S. & Oates, M. (2001). *The Breed Profiles Handbook. A Guide to the Selection of Livestock Breeds for Grazing Wildlife Sites*. Norwich: Grazing Animals Project/English Nature.

Trewavas, A. (2002). Malthus foiled again and again. *Nature*, **418**, 668–70.

Trow-Smith, R. (1959). *British Livestock Husbandry to 1700*. London: Routledge and Kegan Paul.

Vilà, C., Leonard, J. A., Gotherstrom, A., Marklund, S., Sandberg, K., Liden, K., Wayne, R. K. & Ellegren, H. (2001). Widespread origins of domestic horse lineages. *Science*, **291**, 474–7.

Wiggins, S. (1999). Setting the scene: recent change in west African farming and natural resource management. In *Natural Resource Management in Ghana and its Socio-Economic Context*, ed. R. M. Blench. London: Overseas Development Institute, pp. 1–20.

3

Individual differences in the domestic horse, origins, development and stability

M. Hausberger and M. A. Richard-Yris

Introduction

Horse behaviour and temperament are major aspects of concern to horse use. All cultures tend to insist on the importance of having a good relationship with the horse, (e.g. Franchini, 2001; Mills and McNicholas, Chapter 11). Despite this, little mention seems to be made of behaviour or 'character' in the choice of breeding animals. Native North Americans, for example, insisted on size, colour and resistance (Franchini, 2001), whereas Arab horses were selected for speed and morphology, with some characteristics like chest width or eye size supposed to indicate intelligence or courage (Dossenbach & Dossenbach, 1983). It has been suggested that environmental conditions in early ages in different latitudes and geographic regions resulted in inevitable changes which are reflected in the diversity of present breeds. These selective forces would have acted both on morphology and behaviour to create a northern 'line' of horses leading to quieter, larger barreled and shorter legged horses (and present ponies) living in forests or mountains and a southern

line closer to the Arabian type living in open habitats where speed, flightiness and a lean morphology would have been necessary for survival (Dossenbach & Dossenbach, 1983). To some extent, the local people would have had to deal with these local 'horse types' and one may speculate that they adapted their skills and equipment to these features.

Assessing behavioural 'character' poses particular problems and this has been recognized since early times. Xénophon in antiquity (1995) in his recommendations to a future horse buyer, says that one can only choose the future morphology of a horse when it is young as one can not assess its personality until it is ridden. Although he underlined the importance of behaviour in the adult horse, since 'In dangers, the rider entrusts his own body to his horse', he makes no mention of the possible selection of breeding stock for such criteria. However it is likely that undesirable extremes would have been eliminated (Levine, Chapter 1).

Recent breeding programmes have been successful in estimating and selecting horses on the basis

The Domestic Horse: The Origins, Development, and Management of its Behaviour, ed. D. S. Mills & S. M. McDonnell. Cambridge University Press. © Cambridge University Press 2005.

Table 3.1. *Estimates of heritability (h_2) of variables of German Warmblood stallions performance test at station*

			h_2
Character	TR	———	0.21
Temperament	TR	———	0.26
Performance	TR	———	0.29
Constitution	TR	———	0.35
Ease to ride	TR	———	0.42
Jumping ability	TR	———	0.62
Trot	TR	———	0.54
Canter	TR	———	0.42
Walk	TR	———	0.33
Ease to ride (unfamiliar rider)	TE	———	0.44
Free jumping	TE	———	0.47
Parcours jumping	TE	———	0.38
Trot	TE	———	0.45
Canter	TE	———	0.36
Walk	TE	———	0.34
Eventing	TE	———	0.20

Modified from Brockmann and Bruns (2000).
($n = 2815$), TR: training, TE: test.

of performance (Best Linear Unbiased Prediction), or morphology (e.g. Langlois, 1984; Zechner *et al.*, 2001), but measuring 'psychological characteristics' is considered difficult and largely based on subjective assessments which show little heritability (e.g. Brockman & Bruns, 2000; Zeiler & Distl, 2000) (Table 3.1). Terms like temperament, winning character and trainability have been used in descriptions without clear definitions and their importance for a possible selection has only recently been recognized contrary to jumping ability, speed or gait traditionally included as selection criteria.

Scientific research on behaviour and especially temperament in domestic horses has grown slowly, compared to other species. Real difficulties are linked to the 'special' status and use of horses that make it almost impossible to compare animals in controlled conditions for a certain duration and to have large samples of progeny raised in specific conditions (Heird & Grandin, 1998). Divergent selection experiments for specific behavioural traits such as those which have been undertaken in mink, foxes and quail, is also impractical in horses. For all these reasons, the 'genetics of behaviour in horses is in its infancy' (Houpt & Kusunose, 2000). We have to rely on comparisons between individuals, families or breeds to generate indirect evidence for genetic transmission. Data are still scarce and we review here the current evidence from both observational studies in the field and temperament assessment through a variety of methods. It should be emphasized that these only provide indirect evidence as, for example, breed differences may result from other correlated factors such as prenatal and postnatal maternal environment and management conditions including type of work (McGreevy *et al.*, 1995a,b). The complexity of the relationship between genes and behaviour is frequently not appreciated and so we begin with a brief overview of this subject area before considering the work on genetic influences over specific behaviours. This is followed by a consideration of the broader subject of temperament and specifically its assessment in the horse.

Genes and behaviour

At a time when we are closer than ever to a precise knowledge of genes and approach an old dream of controlling nature via genetics, it is a paradox that we now see how far we are from understanding how genes code for characters like behaviour. Much work is needed to fill the gap between molecular biology and organismic approaches to the study of behaviour. Behavioural traits are polygenic and a number of different genes are able to produce similar phenotypes but also a given set of genes can produce different phenotypes in different environments (Clément *et al.*, 2001).

Whereas a few decades ago, the difficulty was in identifying genes possibly involved in behaviour, nowadays we are confronted with the difficulty that a large number of genes are involved in even basic elements of a behavioural sequence (Roubertoux *et al.*, 1998). Different behaviours appear to share the same physiological processes and unexpected correlations have been found: e.g., weight gain, upright posture and eyelid opening share the same chromosomal area in mice (Le Roy *et al.*, 1999). Such phenomena may explain surprising relationships between behavioural and morphological traits: coat colour and aggressiveness (Pontier *et al.*, 1995) or sociability (Robinson, 1977 cited in Mendl & Harcourt, 2000) in cats, hair-whorl position and temperament in cattle (Lanier *et al.*, 2001) and supposedly in horses (Tellington-Jones, 1996).

Indirect selection of behaviour may also have occurred as a result of selection for other traits (weight gain, morphology . . .) in domestic animals. According to Grandin and Deesing (1998), reactive temperaments would be associated with a lean, fine-boned physique, and a calm temperament would be more common in animals selected for heavy bones. This may explain some of the observed differences in the behaviour of, for example, breeds selected primarily for weight gain or milk productivity (Le Neindre, 1989).

Conversely, selection on the basis of behaviour may have surprising side effects. Selections made on behavioural traits like fear or sociability (quails: Faure & Mills, 1998) or tameness (silver foxes: Belyaev & Trut, 1975, mink: Hansen, 1996) have been successful and strongly suggest genetic effects on behaviour. However, experiments like that of Belyaev (1979) draw our attention to the side effects of such selection: silver foxes selected for tameness gave, after only a few generations, a progeny that was totally friendly with humans but had piebald fur, a rolled tail and several behavioural problems among which were abnormal maternal behaviour.

McCune (1995) has shown that genetic influences could be modulated by environmental factors such as handling, since she found that friendly-fathered unhandled kittens resemble unfriendly-fathered handled animals, while the friendly-fathered handled animals had the highest scores of friendliness.

Interactions between genes and the cell environment affect the type of phenotype that will appear (Roubertoux & Carlier, 2002). But of course, genotypes are always expressed in an environment and, especially where behaviour is concerned, mainly early experiences (including prenatal interferences), but also those throughout life (Magnusson & Cairns, 1995), will add to or interact with genetic expressions (see also Crabbe *et al.*, 1999).

The genetics of individual differences in behaviour
Normal behaviour

The occurrence of individual differences in behaviour is a general phenomenon (Slater, 1981). In horses, this appears in social interactions (Feh, Chapter 5), food selectivity (Marinier & Alexander, 1992), maternal behaviour (Crowell-Davis, 1986, Chapter 8) and time budget distribution (Boyd, 1988). A variety of factors determine these individual differences like sex,

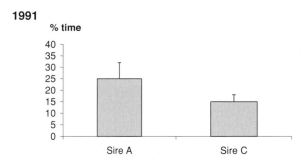

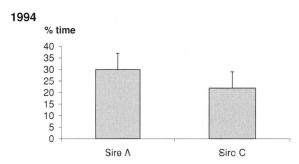

Figure 3.1. Percentage of time spent at 1–5 m from the dam by three-month-old foals issued from two different sires and over two generations.

age and the mother (Houpt & Wolski, 1980; Waring, 1983), but there have been few reports considering the influence of breed or sire on field behaviour. It has been suggested that there is probably a genetic basis to some of the differences shown by foals at three months in the time they spend suckling, playing and the distance kept from the dam (Wolff & Hausberger, 1994). At this age, the foals appear to be responsible for their proximity to the dam, which is in accordance with the findings of Crowell-Davis (1986).

Comparisons of 32 French Saddlebred foals, all born and raised in the same stud, and belonging to two generations, revealed that some of the differences seemed to be related to the sire and were consistent over generations (M. Hausberger and co-workers. unpub. data). The main differences associated with the sire were in the time spent suckling, playing and in dam–foal distance (Figure 3.1). Since offspring of a given sire had different mothers, this would seem to exclude nutritional need as an explanation for these behaviours and it was suggested that enhanced suckling observed in some foals may reflect a greater need for comfort or reassurance (see also

Martin & Bateson, 1988). Distance from the dam may also reflect a reluctance for exploration, social contact with other foals and/or a need for reassurance (Bowlby, 1978; Fairbanks, 1996). As sires were not present with the foals, there seems indeed to be a genetic effect on some essential aspects of foal behaviour. It remains uncertain whether this genetic effect relates to specific behavioural traits or a more general state of reactivity to the environment.

Problem behaviour

Foal rejection Two studies, one based on surveys in horse magazines (135 cases), the other based on questionnaires to horse owners through breed associations (about 1300 replies), have revealed that a higher proportion of foal rejection occurred in Arabian as compared to Quarter Horses and Thoroughbreds (Houpt & Lieb, 1994) or Paint horses (Juarbe-Diaz et al., 1998). In the latter study, 5% of the Arabian foals had been rejected and only 2% of Paint foals. Moreover, some sires (father and son) were found frequently in the origins of the Arabian rejecting mares (Juarbe-Diaz et al., 1998).

Stereotypic behaviour A number of studies have suggested a genetic influence on the occurrence of stereotypic behaviour. In a pioneering study, Hosoda (1950) showed differences in the frequencies of windsucking between families of Thoroughbreds and Japanese native horses, with some families having up to 7% of their members exhibiting this behaviour despite a variety of life environments. Later, Vecchiotti and Galanti (1986) interviewed trainers of more than 1000 Thoroughbreds aged 3–8 years. Whereas the incidence of cribbing, weaving and stall-walking was about 2.5% in the whole population, the frequencies was as much as 30% in particular families. According to Luescher et al. (1998), if one of the parents exhibits a stereotypic behaviour, its offspring has a 60% chance of also developing such behaviour and 89% if both parents do. Interestingly, the offspring does not necessarily develop the same type of stereotypic behaviour as the parents, which suggests transmission of a 'susceptibility' rather than of a particular behavioural trait (Luescher et al., 1998). There seems also to be family or breed influences on the occurrence of self-mutilation (Dodman et al., 1994), which is considered by some to be a stereotypic behaviour.

In a study made on 120 gelding French Saddlebreds at the French national riding school, Hausberger et al. (1996, unpub. data) observed that Angloarab horses showed more stereotypies and spent less time sleeping than French Saddlebreds. As all these horses were kept in the same conditions, with the same feeding and bedding routines, and similar types of work, breed differences per se seem indeed to exist.

All these studies point to the conclusion that there is a genetic basis to the transmission of at least a susceptibility to develop stereotypies, which suggests that care should be given to this in terms of selection.

Stereotypies are discussed in more detail by Mills in Chapter 15 but it should be emphasized that both genetic and environmental factors are involved and neither should be considered as the sole cause of the problem. Horses with a genetic susceptibility may not express these tendencies in favourable environments but may express them at the slightest disturbance. This is illustrated well by the case described by Houpt and Kusunose (2000) of a young Przewalski stallion out of a strongly stereotypic dam, that never expressed stereotypy until he was removed from his family group.

Temperament and personality
Definition

It is surprising that despite the importance of temperament for the daily use of horses, literature has been scarce on this topic until very recently. General interest in temperament has increased in a wide variety of animals (e.g. Gosling & Bonnenburg, 1998) as individual differences have again become a centre of interest rather than some undesirable factor for animal scientists to cope with (Slater, 1981).

Definitions vary but Hall (1941) defined the temperament of animals as 'the raw material of animals'. For this author, when environment and culture are added, it becomes personality ('behavioural style': Feaver et al., 1986; Lyons, 1989) (Figure 3.2). Bates (1989) defines temperament as 'an ensemble of individual biological differences in behavioural tendencies that appear early in life and remain stable across time and situations'.

The various lines of research in earlier human studies have differed, particularly with regard to the number and definition of temperament/personality traits. Buss (1989), for example, distinguished three traits: emotionality–activity–sociability (E–A–S). Most

Environmental factors

-pre/postnatal maternal influence
-housing
-type of work

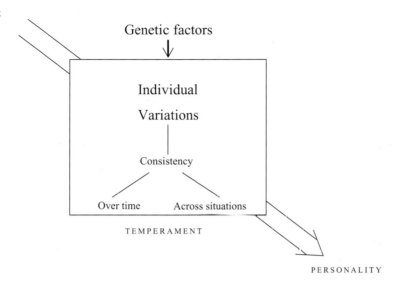

Figure 3.2. An illustration of definitions of temperament by Hall (1941) and Bates (1989) (see text).

animal studies have concentrated on emotionality, which can be divided further into 'fear' as a short-term avoidance reaction to aversive stimulation (Jones, 1987) and 'anger', expressed as aggressiveness towards others (Buss, 1989). Fearfulness as a personality trait would be 'the general susceptibility of an individual to react to a variety of potentially threatening situations' (Boissy, 1995) whereas fear responses would be the immediate behavioural expression of the individual's present internal state. Thus, individual differences in behavioural responses can be called 'personality traits' if they show some consistency across time and different situations (Francis, 1990; Jensen, 1995).

Temperament and personality are multi-trait concepts: individuals can be ordered in 'categories' or 'types' according to how they rank along each trait (Feaver *et al.*, 1986; Buss, 1989; Erhard *et al.*, 1999).

Clearly, selection may operate on these temperamental traits but, in our daily relationship with animals, we have to deal with their personality, the result of this 'raw stuff' and all its experiences. Disentangling the two is not an easy task and

certainly explains the debates about definitions and measures (Goldsmith *et al.*, 1987). Thus, terms like temperament or emotionality are often used interchangeably as unitary concepts (see Boissy, 1995) while learning abilities are rarely included as part of temperament.

Examining the genetic basis of temperament is a difficult task given the diversity of life experiences of individuals within a species. Gerlai and Csañyi (1990) have shown that differences between genotypes of macropodes are best revealed in unusual situations, as they are much more difficult to assess in the home environment. Donkeys do not express the same responses when they change location (French, 1993). Experimental tests where animals are presented with new challenges may increase the chances of identifying these 'basic behavioural tendencies' in individuals and indeed it is the preferred technique in studies of temperament. Knowledge of the daily behaviour of animals via observations (scoring) and questionnaires however may be complementary and enable us to study the personality as a result of the interaction between temperament and environment.

Methods of assessment

Although most riders have an opinion about the personality of their horses, and this is of major importance for the horses' use, the development of scientific methods to assess individual differences has only recently taken off. A variety of approaches have been devised with a clear emphasis on behavioural tests, as observations alone in the home situation may not be sufficient to assess personality/temperament traits (Manteca & Deag, 1993). Houpt and Rudman (2002) suggest an 'amalgamation of tests' in order to evaluate the suitability of horses for their use.

Observer ratings Two types of studies in horses have both used assessments by evaluators. The first group used methods designed to see whether results in a behavioural test were correlated with observations in the working situation. Correlations were found between general learning abilities in the working situation and performance in a learning test (Fiske & Potter, 1979; Le Scolan *et al.*, 1997); diverse emotional scales (fear, flightiness) and reactions to behavioural tests of emotionality (Le Scolan *et al.*, 1997; Visser *et al.*, 2002; Momozawa *et al.*, 2003). The second type of study focused more on evaluations of personality, which were inspired by psychometrics and research on human personalities (Mills, 1998; Morris *et al.*, 2002). Both studies evaluated inter-observer correlations on a subjective rating of horse personalities. Inter-assessor correlations were rather low in Mills' study; this may be related to the use of current terms that were empirical and not reliably defined (e.g. affectionate, bold, confident). Morris *et al.*'s study followed the same line but used questionnaires adapted from the NEO-PI-FFI of Costa and McCrae (1992) originally constructed for humans. In this study, high inter-observer correlations were found in ranking the horses on the Big Five dimensions (neuroticism, extraversion, openness to experience, agreeableness, conscientiousness). The contradictory results observed in terms of inter-observer reliability are probably due to reason of methodological differences (see also Anderson *et al.*, 1999; Seaman *et al.*, 2002). Mills (1998) underlines the importance of clear and objectively described terms and the risks associated with using commonly used terms that may be interpreted differently by different people.

Behavioural tests A variety of tests has been used in order to evaluate emotionality in horses. As

mentioned before, behavioural tests are probably more prone to revealing temperamental traits, as they are usually designed to put the animals in challenging situations where general underlying tendencies may be better revealed (Gerlai & Csányi, 1990). Learning abilities may be another relevant trait (Visser, 2002) and different types of tests have also been developed to test for these. Both aspects have been evaluated in comparisons of breeds or bloodlines. Other aspects, such as dominance or aggressiveness, have been tested less (e.g. Haag *et al.*, 1980).

Different tests have been designed in order to measure emotionality, the horses are: placed in a chute (e.g. McCann *et al.*, 1988); confronted by a novel object which is either static (Wolff *et al.*, 1997) or moving (e.g. Anderson *et al.*, 1999; Visser *et al.*, 2001) while they are released or being led (Mackenzie & Thiboutot, 1997). Reactions to humans can also be measured: when a human stands still (Seaman *et al.*, 2002; Visser *et al.*, 2001); approaches suddenly (Hausberger & Muller, 2002) or slowly (McCann *et al.*, 1988); when a veterinarian inspection occurs (Houpt & Kusunose, 2000); or a spray is given (Seaman *et al.*, 2002).

For example, three tests (Figure 3.3 A, B, C) have been developed by Wolff and Hausberger (1992) and used and adapted by many workers:

- the 'arena test': the horses are released alone in a familiar arena and their behaviour is recorded every 10 seconds for 10 minutes or less (Anderson *et al.*, 1999; Seaman *et al.*, 2002). This is different from a classical open field test as it is a familiar place and it is suggested that it is more a social separation test (Le Scolan *et al.*, 1997).
- the 'novel object test': an object is placed in the arena and the horse is released for 5 minutes. Behaviour, locomotion, gazing and approaches are noted. The object can be static (inflated ball, grid with red ribbons) or moving (umbrella: Lankin & Bouissou, 1998; Visser *et al.*, 2002).
- the 'bridge test': the horse is led over an unknown obstacle constructed with planks (Wolff *et al.*, 1997), a mossy mattress (Le Scolan *et al.*, 1997) or concrete blocks (Visser *et al.*, 2001).

Measurements include latency to approach (object or person) (Visser *et al.*, 2001), scores of reactivity with a predefined scale (McCann *et al.*, 1988; Houpt & Kusunose, 2000; Hausberger & Muller, 2002) or based on the scale of reactivity in natural

A: Arena test

B: Novel object test

C: Bridge test

D: Chest opening test

Figure 3.3. Examples of experimental tests performed to test for emotionality (A, B, C) or for learning abilities (D).

situations and frequency of behaviour patterns (Wolff *et al.*, 1997). Principal components analysis or factor analysis are frequently used in order to define the components of reactivity (Figure 3.4).

Despite some discrepancy in either the tests used or the evaluation of behavioural reactions, these experiments reveal some consistency. Individual differences can clearly be measured and the reactions described, in terms of emotionality, are similar: an increase in locomotion, postural displays (tail raised, vigilance) as opposed to exploration, walking, close looking at objects (e.g. Figure 3.4), which reflects observations of natural behaviour (Waring, 1983).

Tests designed to assess individual tendencies in learning ability have varied greatly and are discussed separately below in relation to their specific findings (see also McCall *et al.*, 1981; Marinier & Alexander, 1994).

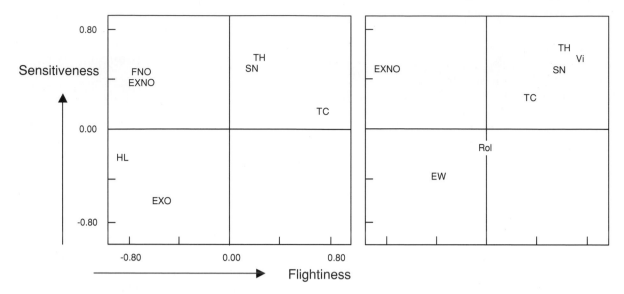

Figure 3.4. Examples of two principal components analysis performed on 42 Dutch Warmblood horses 22 months old (left) and 120 French Saddlebred adult horses (right) in two different studies. The frequency of behavioural patterns has been noted in a novel object test. Note similarity between the graphs despite some differences both in the details of the protocol and type of novel object (umbrella versus grid with ribbons). TH, tail raised; SN, snorting; Vi, vigilance; TC, trot-canter; Rol, rolling; EW, exploratory walk; EXO, exploration; HL, head low; FNO EXNO, exploration and close gazes at novel object. One axis is more related to object, the other to emotional reactions. Modified from Visser *et al.* (2001) and Hausberger *et al.* (unpublished data) respectively.

Physiological measures Heart rate (HR) and other related variables have been measured in several studies. It has been shown that HR and HR variability (HRV) are useful to differentiate individual rats (Sgoifo *et al.*, 1997). Heart rate represents the net effect of parasympathic nerves that slow it down and the sympathetic nerves that accelerate it. In resting conditions, both parts of the autonomic nervous system are thought to be tonically active. Several stressful conditions can result in an increase in HRV and hence low levels of parasympathetic nerve activity (Sgoifo *et al.*, 1997). Porges (1991) showed that parasympathetic activity is suppressed during autonomic and behavioural responses to stress and these responses to challenging situations by the autonomic nervous system can differentiate individuals.

In horses, some studies revealed individual variations partly due to sex and age (Clément & Barrey, 1995) and an influence of early handling experience (McCann *et al.*, 1988, Jezierski *et al.*, 1999).

Several studies (Visser *et al.*, 2002, 2003a) have investigated whether the responses of HR and HRV (standard deviation of beat-to-beat intervals and the root mean square of successive beat-to-beat differences) to challenging situations could differentiate between the temperaments of horses. Heart-rate variables showed consistency between years in young horses (Visser *et al.*, 2002). It was also shown that HR variables measured in a novel object test and in a handling test showed significant correlations with temperamental traits assessed by riders (Visser *et al.*, 2003a).

In other studies, it has been shown that crib-biting seems to be associated with a decreased heart rate compared with other activities (Lebelt *et al.*, 1998; Minero *et al.*, 1999). Minero *et al.* (1999) did find a difference in the baseline HR and HR during treatment between the crib-biting and the control horse but Lebelt *et al.* (1998) and Bachmann *et al.* (2003) did not. Similarly, measures of heart rate in crib-biters, weavers and normal horses suggest that weavers react differently from crib-biters depending on the kind of stimulus used (Meers *et al.*, 2002). Nevertheless, power spectral analysis of HRV revealed significant differences between crib-biters and controls at rest: crib-biters had a lower vagal tone (high frequency component, HF) and a higher sympathetic tone (low frequency component, LF) than controls (Bachmann *et al.*, 2003). During presentation of a food stimulus (i.e. waiting for the food), control horses reacted with a significant increase of the LF and a significant decrease of the HF component, resulting in a significant LF/HF ratio whereas in crib-biting horses, this reaction was not significant (Bachmann *et al.*, 2003). Due to the lower basal parasympathetic and higher sympathetic activity, the reactivity range of the autonomic nervous system of a crib-biting horse may not be capable of reacting as competently to an external stimulus as in a normal horse.

In conclusion, heart-rate measurements are an interesting complementary tool to behavioural measures in order to evaluate individual differences. However, up to now few studies have investigated possible genetic differences (Bagshaw *et al.*, 1994; Clément & Barrey 1995) and have not led to conclusive results.

The other candidates for the physiological assessment of stress reactions are the stress hormones like cortisol. Results of different studies confirm the relationship between stressors and cortisol production but also give rise to contradictory results (Alexander *et al.*, 1988; Martinez *et al.*, 1988; Mal *et al.*, 1991; Mills *et al.*, 1997). Whether or not individual differences in the mean baseline are present in horses remains to be further investigated. One way of broaching this question has consisted of comparing stereotypic and normal horses. According to the study, stereotypic and normal horses did (McGreevy & Nicol, 1995) or did not (Pell & McGreevy, 1999) differ in their mean plasma cortisol level. In the latter study, arousal levels of stereotypic and normal horses appeared to be similar. The two studies differed in their sample sizes and temporal aspects, which may explain the divergence. Obviously, care has to be taken in the interpretation of results of such physiological studies.

Consistency of temperament and personality

The definition of temperament given above implies that it should be evident early in an animal's development, consistent over time and across situations. Work that has examined aspects of individuality, which fulfill each of these criteria, is discussed below.

Early differences among animals raised in homogeneous conditions Two groups of studies have looked at the possible differences in temperamental traits by using young horses from the same breed, living in the same conditions.

Wolff *et al.* (1997) compared the reactions of 42 French Saddlebreds one- to three-years-old to the arena, novel object and bridge tests. Clear interindividual variations appeared in the behaviour patterns expressed (Figure 3.5). Using an index of emotivity, no difference was found according to sex or age in the general reaction to the situation, nor to the bridge test. Half-siblings (same sire), however, tended to show more similarities than unrelated animals (Figure 3.5) and a lower variability still was observed in full sisters compared to half sisters. The offspring of one particular stallion showed fewer reactions in the arena test but more in the novel object and bridge tests than the offspring of other stallions.

The same young horses were also tested in instrumental (chest opening) and spatial (detour) tasks (Wolff & Hausberger, 1996). Individual differences again occurred in both cases: the females tended to be more successful in the spatial task. Some differences according to the sire could again be observed. Thus the offspring of one particular stallion (A) required more time in order to learn to open the chest than unrelated animals. In the detour test, the offspring of another stallion (B) was significantly more successful than the offspring of other sires.

Visser *et al.* (2001) similarly found individual differences in the reactions of young Dutch Warmbloods, raised since an early age in similar conditions on an experimental farm. During tests (at 9, 10, 21 and 22 months), although the response for some variables varied with age, the coefficient of variation increased for the variables where the mean decreased over ages and vice versa.

Consistency over time So far only a few studies have investigated consistency over time for variables measured in behavioural tests. Visser *et al.* (2001, 2003b) tested 41 Dutch Warmbloods (see above) that were presented to a novel object and bridge test at the ages of 9, 10, 21 and 22 months. Some habituation to the novel object seemed to occur: less trotting and cantering and more exploration and contact with the object was observed over time, but by contrast, the horses required more time in order to cross the bridge when 20 months old as compared to 10 months. Most behavioural variables were correlated within one year

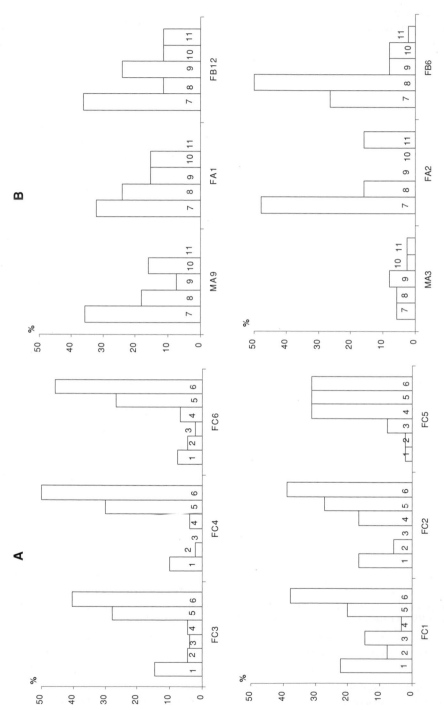

Figure 3.5. The proportion of the different behaviour patterns shown by six one-year-old females in the arena test (A) and six horses in the bridge test (B). Clear individual differences were observed between unrelated animals, whereas half-siblings tended to behave in the same way. M, males; F, females; FC3, FC4, FC6 (A), MA9, FA1 and FB12 (B), offspring of the same father, whereas the other horses are unrelated. 1, standing; 2, hesitant walk; 3, sustained walk; 4, trot; 5, vigilance; 6, others (passage, gallop, tail raised, whinnys, snorts); 7, licking-sniffing; 8, swerving, 9, one or two feet on the bridge; 10, walking backwards; 11, others (snorts, rising, jumping). From: Wolff *et al.* (1997).

(9/10 and 21/22 months), and reluctance, tendency to explore objects other than the novel object, tendency to hold head low, latency to touch the novel object and time spent trotting and cantering seemed to be correlated over the two years.

'Flightiness' (high locomotory responses) in the novel object test and patience in the bridge test were positively correlated over the two years. These results were supported by some of the data on HR variables measured, which revealed some consistency in both the bridge test and novel object test.

The same horses tested in two different learning tasks (avoidance and reward learning) showed consistencies of performance in the short-term, and some consistency between years in the avoidance task. Younger animals (one-year-old) appeared to have more difficulty 'understanding' the task, including taking the food reward. Wolff and Hausberger (1996) found that yearling animals were unable to learn a spatial task. Animals did however perform better in a 'repeat' test one month later suggesting a memory of the task. However, there was no consistency over time in these individual learning abilities for this task. By contrast, correlations have been found between learning and memory in such tasks with older animals (M. Hausberger & C. Muller, unpub. data). This corresponds with results from tests designed to assess emotivity in which no correlation was found between arena and novel object tests in one- to three-year-old animals, whereas individual responses were highly correlated in older animals (Wolff et al., 1997; M. Hausberger & C. Muller, unpub. data). Maturational processes may explain these differences.

An alternative to repeat test measurements is to look for consistency in more general underlying variables that may be expressed differently in young or adult animals. Foals clearly differ from each other at an early age and some of these behavioural differences seem to be related to general aspects of fearfulness, social dependency or exploratory tendencies that may remain throughout life. In order to test this hypothesis, M. Hausberger and co-workers (unpub. data) looked for possible correlations between the reactions of three-year-old horses in tests and their previously observed behaviour in paddocks as three-month-old foals with their dam. The distance the foal kept from the dam was found to be dependent on the paternal origin (sire) of the foal. This was also predictive of some apparent emotional reactions at

three years: foals that tended to keep a close contact with the dam (< 5 m) showed stronger reactions in the arena test, and a correlation was found between the time spent at 5–10 m from the dam and reactions to the novel object and bridge test when adult. Positive correlations were also found between the frequency of attempted suckling and reactions to the arena test, the frequency of observation and the reactions to the novel object test and frequency of locomotion and learning abilities when adult.

Similarly, Visser (2002) found that the behaviour of young horses in the tests above was, to a large extent, a good predictor of jumping performance at the age of three years. This illustrates an important practical implication of being able to measure consistent individual differences at an early age. Performance was predicted better if only the animals that succeeded in the avoidance learning test were taken into account, which raises intriguing questions: what relationship may there be between some learning abilities and later jumping abilities?

Both studies show that looking for early predictors of adult behaviour is a promising line of research and may provide a reliable basis for possible selection. They also raise the question of what measurements are appropriate for revealing consistency. Obviously, comparisons of young and older animals have to use different types of measurements, as the same level of arousal may be expressed differently (see also Mendl & Harcourt, 2000). There seems therefore to be early predictors of adult behaviour (e.g. foals that tended to stay close to their dam and made frequent attempts at suckling (seeking comfort) were, as adults, more reactive in the arena test, i.e. more sensitive to social separation).

Consistency across situations The previous examples show that some form of consistency in behavioural style appears not only over time but also across different situations. However, the question of consistency is complex: although some studies have shown consistency between different tasks (e.g. avoidance test/maze learning: Haag et al., 1980; different maze tests: McCall et al., 1981; Heird et al., 1986a) most studies show that there is no correlation between the results of different tests whether emotional traits (Wolff et al., 1997; Seaman et al., 2002; Visser, 2002) or learning abilities (Wolff & Hausberger, 1996; Nicol, 2002, Visser, 2002) were assessed. Similarly,

Table 3.2. *Correlations found between indices of emotivity or time in behavioural tests and scoring by familiar riding teachers*

	Arena test (indices)	Novel object test (indices)	Bridge test (time, seconds)	Instrumental task: memorisation part (time, seconds)
Gregariousness				
Familiar surroundings	$n = 62$, $r_s = 0.44$, $P < 0.001$			
Unfamiliar surroundings	$n = 53$, $r_s = 0.41$, $P < 0.001$			
Nervousness when ridden		$n = 70$, $r_s = 0.24$, $P < 0.05$		
Fearfulness				
Handled			$n = 72$, $r_s = 0.38$, $P = 0.013$	
Ridden			$r_s = 0.49$, $P = 0.013$	
Learning abilities				$n = 51$, $r_s = -0.42$, $P < 0.001$

From: Le Scolan *et al.* (1997).

n, number of individuals rated by the two methods. There are different *n* because riding teachers could not rate all their horses in all situations.

r_s, value of the coefficient of correlation (Spearman test).

Gregariousnes: difficulty to separate from others.

Nervousness: e.g. stamping.

Anderson *et al.* (1999) did not find any correlation between cortisol levels and different measurements of emotionality but sociable horses had lower levels. In some cases, the reactions to two tests are not correlated in young animals but are correlated in adults (Wolff *et al.*, 1997; Le Scolan *et al.*, 1997). Seaman *et al.* (2002) conclude that behavioural tests are inaccurate for measuring temperament in horses. However, another option is to consider that these different tests just measure different aspects of temperament, which would explain these differences and apparent contradictions (see also Boivin *et al.*, 1992, Hausberger *et al.*, 2004). This view does not however improve the ease with which such a question can be broached.

Studies have compared the reactions to behavioural tests and ratings by 'horse users' in working situations. Le Scolan *et al.* (1997) compared the reactions of 72 adult horses in different tests to scoring by riding teachers. Despite the horses belonging to three different riding schools and therefore different teachers involved, strong correlations appeared between the arena test and gregariousness ('social dependency'), novel object test and nervousness (stamping . . .),

bridge test and fearfulness, instrumental learning and learning abilities (Table 3.2). No correlation was found between a detour task and any teacher's evaluation. These results tend to confirm the idea that different tests measure different temperament traits. They also suggest that some form of consistency does exist, although it is not necessarily expressed the same way in different situations. Similar observations were made by Visser *et al.* (2003a). Eighteen adult Swedish Warmbloods were tested in the novel object and bridge tests while an evaluation of temperament traits was scored by 16 unfamiliar riders after a standardized riding session. A high correlation was found in the ratings of the different evaluators. Correlations were found between HR variables in the novel object test and between behavioural variables and ratings in the bridge test. For example, standing in front of the bridge was positively correlated with 'spooky' and negatively correlated with 'brave'.

In an early study, Fiske and Potter (1979) had also found a correlation between learning performance in a discrimination task and evaluations of learning abilities by riders.

These results are of course indirect evidence for consistency across situations but suggest that some general tendencies are consistent and may be expressed in different situations. We would probably need to have tests assessing reactions to different novel objects, confrontation to different types of bridges, etc., in order both to control for test-specificity and to control similar types of measures. A whole line of tests has to be developed in horses as in other species.

In conclusion, evaluating personality/temperament is not an easy task and obviously it is a line of research that deserves further study. Below are several points for consideration:

- Little correlation has been found between tests (Wolff *et al.*, 1997; Le Scolan *et al.*, 1997; Anderson *et al.*, 1999; Seaman *et al.*, 2002; Visser, 2002) which may suggest that either tests are not valid or rather that they test different facets of temperament/personality.
- Tests used in young animals may not be suitable for older animals and so different but related measurements need to be identified.
- Observer ratings may bring new information, in particular on traits little tested like prosociality and aggressiveness, but should be considered as complementary to behavioural tests (Manteca & Deag, 1993). Inter-observer reliability does not guarantee either that reactions of horses have been seen in a variety of situations or that it reflects real personality traits as, for cultural reasons in particular, evaluators may agree on a trait but be mistaken (Peng *et al.*, 1993).
- Subjective assessments, using questionnaires (Stevenson-Hinde *et al.*, 1980), and behavioural tests should both be used. If selection has to occur, then standardized testing should be designed so that studies can be compared. Surprisingly, despite the variety of tests performed, general agreement can be detected on many points.

Genetics of temperament and personality

Whereas breed differences in behaviour are common sense in the horse 'world', experimental evidence has been rather scarce. In a large survey where 50 trainers and veterinarians gave comments about 10 breeds (Hayes, 1998), Arabian horses were classified as over-reactive, playful, energetic and fast learners; Thoroughbreds as fearful, emotional, hard workers but with difficulty in concentrating on a learning task; whereas Quarter Horses were considered as unfearful, obedient, patient, easy and fast learners and Appaloosas as unfearful, easy going but with more difficulties to learn and dependent on routine (Houpt & Kusunose, 2000).

Similarly, within breeds, some bloodlines are considered as more difficult to handle and more fearful, but clear measurements have been missing until recently. This area of research certainly deserves further consideration.

Emotional reactions Wolff *et al.* (1997), working on French Saddlebreds, and Lankin and Bouissou (1998), working on Welsh ponies, found differences in emotional reactions according to the sire in a limited number of progeny. Kusunose (cited in Houpt & Kusunose, 2000) investigated this sire effect in a large sample of young Thoroughbreds. The reactions to a veterinarian examination of more than 9000 young horses were scored. These young horses were from 62 stallions and each stallion had progeny in two Japanese training centres. Scores (A to C) were given by the veterinarian according to the reactions of the horse to three types of procedure: checking eyes, auscultation and blood sampling. A strong correlation was observed with the sire. This is possibly the best example of a sire effect in horses up to now.

There are few studies where breed differences are specifically addressed. Budzynski *et al.* (1992) submitted more than 300 horses from three different Polish breeds to a series of tests. The 'timidity' test where horses were led between two moving checked wooden squares appeared to be the most discriminating: more Wielkopolski than Silesian females appeared to be quiet in the test.

Heart rate and other physiological variables have been measured in some studies to investigate breed effects. Clément and Barrey (1995) did not observe significant differences between Angloarabs and French Saddlebreds, whereas Bagshaw *et al.* (1994) found higher heart rates in response to isolation in Arabs than in Standardbred mares.

Using the three experimental tests (arena, novel object and bridge) on 120 Angloarab and French Saddlebred adult geldings, all living in the same conditions, M. Hausberger and C. Muller (unpub. data) found that breeds differed in the bridge test. Angloarabs required more time and refused more

X = 1 (39.69 %)   Y = 2 (29.23 %)

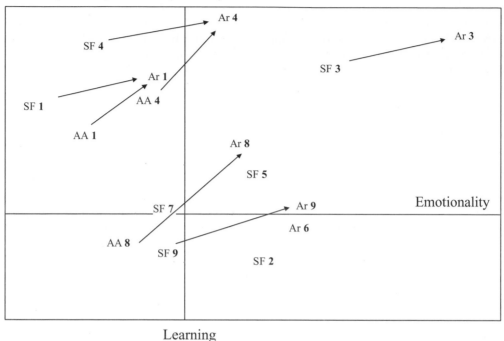

Learning

Figure 3.6. Results of a principal components analysis performed on 98 breeding stallions living in nine different national studs. In order to facilitate reading, only three breeds (AA, Angloarab; Ar, Purebred Arab; SF, French Saddlebred) are represented and one point (e.g. Ar 9 corresponds to the mean point of all Purebred Arabs of stud 9). Arrows, to accentuate the direction of the level of emotional responses between breeds on each site. Modified from Hausberger *et al.* (2004).

often to cross the bridge. Interestingly, no difference according to breed was found for the other two tests. The same experiments done on a large sample of horses living in a variety of environments revealed strong breed effects in all three tests (see Figure 3.7). In this test Angloarabs appear closer to French Saddlebreds compared to other breeds, which may be a reflection of their genetic closeness (frequent cross-breeding) (Moureaux *et al.*, 1995).

In the study of Hausberger and Muller (2002) on more than 200 horses, French Saddlebreds had more often a friendly attitude (the horse looked at the experimenter with upright ears and approached). Finally, Thoroughbreds showed more indifference (no change in behaviour, no gaze) to sudden human approach than Angloarabs and French Saddlebreds. Again, all these horses lived in similar conditions and the breeds were distributed between caretakers, so the actual environmental conditions are unlikely to explain the breed differences.

Learning ability In an early study, Mader and Price (1980) submitted 11 Thoroughbreds and 10 Quarter Horses to a discrimination test. The horses had to identify a door, behind which food had been hidden, among other doors covered with different drawings (checkerboard, heart and horse silhouettes). The Quarter Horses tended to be more successful in this task but the authors suggest that factors other than learning ability per se may have brought about the difference (e.g. emotionality, food motivation). Budzynski *et al.* (1992), in their study on Polish breeds, found that in a maze task Malopolski stallions performed better than Wielkopolski or Silesian horses.

Hausberger and colleagues used an instrumental task (chest opening) in a series of studies, which revealed differences according to sire within young French Saddlebreds (Wolff & Hausberger, 1996), differences between Saddlebreds and Angloarabs in the same conditions (Hausberger *et al.*, 1996) and differences according to the breed in a large sample of animals living in a variety of environments (Hausberger *et al.*, 1998, 2004). In the second of the studies, Angloarabs required more time and fewer of them learned the task. In the third of the studies, breeds like Haflingers or Appaloosas tended to perform better than Thoroughbreds, Purebred Arabs, Angloarabs or French Saddlebreds (Figure 3.6).

The interplay of genetic and environmental factors

It is obvious that numerous environmental factors can affect the personality of horses: early handling may affect learning abilities or emotional reactions when adult (Heird *et al.*, 1986b). Housing and type of work may affect the general emotional state of the animal. These environmental factors may have different impacts on different 'types' of animals. Disentangling the weight of these different factors on the behaviour of an animal at a given stage is challenging and has been of great interest for a long time. Gardner (1937) tested Belgian and Percheron horses in a situation where they had to open a feeder and concluded that age (older and younger horses were slower), breed (Belgian horses showed fewer emotional reactions) and type of work (farm horses were faster than military horses) affected the results. Although there was no statistical testing in this early work, these observations suggest how different factors interact. Additive effects have been found in a study by Hausberger *et al.* (2004) on 98 breeding stallions of different breeds living in nine different national studs. Housing and diet were similar and the main activity for all the stallions was breeding. The same breed differences in the reactions to tests were found at all sites: in particular, purebred Arabs were more reactive than French Saddlebreds. However, when sites were compared, it appeared that a French Saddlebred in one site was as reactive as a Purebred Arab in another (Figure 3.6). The stallions of one stud appeared more reactive than all the other stallions, whatever the breed. Differences between sites remain to be explained but might be due to differences in the human–animal relationship (Hausberger & Muller, 2002). Here, environmental conditions add to breed differences in the expression of emotional reactions.

In a further, larger study on more than 700 horses, Hausberger *et al.* (2004) investigated the extent to which the variance could be explained by breed, sire and/or environmental factors. The horses were 2 to 20 years old, belonged to different breeds and were housed in a large variety of sites. Nine factors were investigated: sire, breed, site, type of work, housing (box/paddock), food, number of riders, sex and age. Multivariate analyses were used in order to test for significance. All horses were submitted to the arena, novel object and bridge tests as well as to the chest opening test (Figure 3.7). Only two factors appeared as non-significant (sex and age) and among the seven other factors, three (sire, breed and site) had a strong weight in explaining the variance. The relative weight of the different factors depended on the test: the sire appeared as the most important factor in the novel object test, breed in the bridge test whereas type of work and site (clearly environmental factors), were predominant in the arena and chest opening tests. The type of work was found to have a significant effect on all four tests. These results suggest that the influence of genetic factors depends on the behavioural trait considered, and differs according to the test. This is not surprising but it also suggests that if selection for temperament is to occur: (1) it is more likely to occur for some traits rather than others; and (2) that one has to be quite clear on the selection looked for: Does it aim to reduce fear, nervousness, gregariousness? Obviously trying to reduce 'emotionality' is too vague a concept. At the same time, these results emphasize the importance of environmental factors, which are sometimes underestimated, for example the type of work, as here the horses were tested outside their working conditions.

Conclusion

Although it cannot be stated that there is genetic determinism of behavioural traits in horses, this review of the evidence of influences of breed and sire on individual variations suggests that selection on the basis of behaviour should be possible in horses as in other species, but some difficulties have to be overcome:

• The aims must be clearly defined: it is probable that the requirements for given behavioural traits may greatly vary not only between the types of work the horse is required to do, but also according to the rider's own personality. To what extent would a horse selected for quietness be satisfactory to any two different riders (see also Mills and McNicholas, Chapter 11)?
• The methods need to be refined: behavioural tests, questionnaires and physiological assessments are all

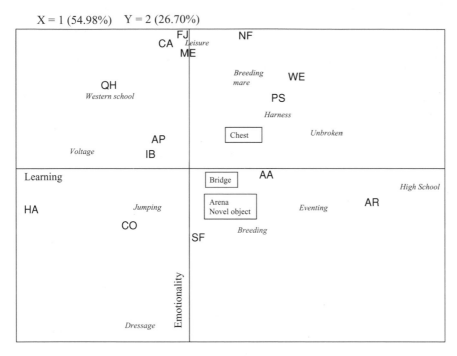

X = 1 (54.98%)   Y = 2 (26.70%)

Figure 3.7. Results of a principal components analysis (PCA) based on performances of 702 horses in different behavioural tests. Axis 1 corresponds to learning performances and axis 2, to responses to emotional situations (arena, novel object, bridge tests). For ease of reading, only two factors figure here (breed and type of work) but the PCA was performed on all factors. Camargue, Fjord and Merens appeared as the least emotive horses with average learning performances. Arab, Angloarab and French Saddlebreds appeared globally more emotional and had more difficulties in the learning task (chest opening). Dressage horses appeared as more emotional than most other horses. Of the nine factors tested (age, 2–20 years; sex, males, females, stallions; sire; site, 104 locations; housing, box/paddock; number of riders; type of work; food), only two did not appear as having a significant effect on variance (sex and age). The three most important factors appeared to be: sire, site and breed. From Hausberger *et al.* (2004).

complementary, but what is the most efficient way of assessing the characteristics that should be selected or rejected? The results of the existing studies make us aware of the multi-faceted nature of temperament. Young and older animals may require different types of evaluations. Horses, like other species (e.g. Budaev, 1997), do not show the same responses in different situations. Obviously, reactions to predators, novel objects, spaces or social separation certainly do not elicit the same emotional, that is physiological, internal states.

• Assessing large samples of progeny from given sires would be an important step forward.

• It is important to keep in mind the complexity of genetics, the interactions between genes and between genes and the environment. Breed differences may equally reflect differences in maternal influence (sometimes in utero) and may not have a genetic basis. It must also be borne in mind that selecting for certain behavioural traits may induce morphological characters or incidental behaviour changes.

Finally, it must be recognized from all the work so far, that genetic factors may induce a certain 'susceptibility' that may be reinforced or diminished according to environmental conditions. No selection program can promise 'perfect horses' in terms of behaviour without seriously considering how the horse is managed from an early age, its housing, food, human interaction, social environment and education (discussed elsewhere in this volume). All are involved in what can be seen as a horse's personality at a given age. Taking into account genetic susceptibilities should, however, enable us to adapt training procedures and thus to limit the risks of, for example, stereotypies in order to manage horses more effectively and more humanely.

Acknowledgements

We are grateful to H. Schuelke and C. Lunel for their help in preparing this manuscript. Our own studies have benefited from the financial support of 'Les Haras Nationaux'. We want also to thank particularly a number of breeders and riding teachers who have helped us both practically and in the thinking, among them the 'Haras de Roulefort', riding centres such as 'Fénicat', 'l'Ecole Nationale d'Equitation' and the different 'Haras Nationaux' who opened their doors several times to our studies and made them possible.

References

Alexander, S. L., Irvine, C. H. G., Livesey, J. H. & Donald, R. A. (1988). Effect of isolation stress on concentrations of arginine, vasopressin, α-melanocyte-stimulating hormone and ACTH in the pituitary venous effluent of normal horse. *Journal of Endocrinology*, 116: 325–34.

Anderson, M. K., Friend, T. H., Evans, J. W. & Bushong, D. M. (1999). Behavioral assessment of horses in therapeutic riding programs. *Applied Animal Behaviour Science*, 63: 11–24.

Bachmann, I., Bermasconi, P., Herrmann, R., Weishaupt, M. A. & Stauffacher, M. (2003). Behavioural and physiological responses to an acute stressor in crib-biting and control horses. *Applied Animal Behaviour Science*, 82: 297–311.

Bagshaw, C. S., Ralston, S. L. & Fisher, H. (1994). Behavioral and physiological effect of orally administered tryptophan on horses subjected to acute isolation stress. *Applied Animal Behaviour Science*, 40: 1–12.

Bates, J. E. (1989). Concepts and measures of temperament. In *Temperament in Childhood*, ed. G. A. Kohnstamm, J. E. Bates & M. K. Rothbart. New York: Wiley & Sons, pp. 3–26.

Belyaev, D. K. (1979). Destabilizing selection as a factor in domestication. *Journal of Heredity*, 70: 301–8.

Belyaev, D. K. & Trut, L. N. (1975). Some genetic and endocrine effects of selection for domestication in silver foxes. In *The Wild Canids*, ed. M. W. Fox. New York: Van Nostrand Reinhold, pp. 416–26.

Boissy, A. (1995). Fear and fearfulness in animals. *Quarterly Review of Biology*, 70: 165–91.

Boivin, X., Le Neindre, P., Chupin, J. M. & Garel, J. P. (1992). Influence of breed and early management on handling facility and open-field behavior of heifers. *Applied Animal Behaviour Science*, 32: 313–23.

Bowlby, J. (1978). *Attachement et Perte*, vol. 1, *L'Attachement*. Paris: Presses Universitaires de France.

Boyd, L. E. (1988). Time budgets of adult Przewalski horses: effects of sex, reproductive status and enclosure. *Applied Animal Behaviour Science*, 21: 19–39.

Brockmann, A. & Bruns, E. (2000). Schätzung genetischer Parameter für Merkmale aus Leistungsprüfungen für Pferde. *Zuchtungskunde*, 72: 4–16.

Budaev, S. V. (1997). 'Personality' in the guppy (*Poecilia reticulata*): a correlational study of exploratory behavior and social tendency. *Journal of Comparative Psychology*, 111: 399–411.

Budzynski, M., Soltys, L. & Wawiorko, J. (1992). Estimate of excitability of half-bred horses. In *43rd Annual Meeting of the EAAP, Madrid, 14–17 September 1992*.

Buss, A. (1989). Temperament as personality traits. In *Temperament in Childhood*, ed. J. A. Kohnstamm, J. E. Bates & M. K. Rothbart. Chichester: John Wiley & Sons, pp. 49–58.

Clément, F. & Barrey, E. (1995). Fluctuations de la fréquence cardiaque chez le cheval au repos. 2. Facteurs de variation biologiques liés au profil comportemental. *Comptes Rendus de l'Académie des Sciences, Sciences de la vie/Life Sciences*, 318: 867–72.

Clément, Y., Lepicard, E. & Chapouthier, G. (2001). An animal model for the study of the genetic bases of behaviour in men: the multiple marker strains (MMS). *European Psychiatry*, 16: 246–54.

Costa Jr., P. T. & McCrae, R. R. (1992). *Revised NEO Personality Inventory (NEO-PI-R) and NEO Five Factor Inventory (NEO-FFI) Professional Manual*. Odessa, FL: Psychological Assessment Resources.

Crabbe, J. C., Wahlsten, D. & Dudek, B. C. (1999). Genetics of mouse behavior: interactions with laboratory environment. *Science*, 284: 1670–2.

Crowell-Davis, S. L. (1986). Spatial relations between mares and foals of the Welsh pony (*Equus caballus*). *Animal Behaviour*, 34: 1007–15.

Dodman, N. H., Normile, J. A., Shuster, L. & Rand, W. (1994). Equine self-mutilation syndrome (57 cases). *Journal of the American Veterinary Medical Association*, 204: 1219–23.

Dossenbach, M. & Dossenbach, H. (1983). *The Noble Horse*. Sydney: Collins Pty, Ltd.

Erhard, H. W., Mendl, M. & Christiansen, S. B. (1999). Individual differences in tonic immobility may reflect behavioural strategies. *Applied Animal Behaviour Science*, 64: 31–46.

Fairbanks, L. A. (1996). Individual differences in maternal style. *Advances in the Study of Behavior*, 25: 579–611.

Faure, J. M. & Mills, A. D. (1998). Improving the adaptability of animals by selection. In *Genetics and the Behavior of Domestic Animals*, ed. T. Grandin. New York: Academic Press, pp. 235–64.

Feaver, J., Mendl, M. & Bateson, P. (1986). A method for rating the individual distinctiveness of domestic cats. *Animal Behaviour*, 34: 1016–25.

Fiske, J. C. & Potter, G. D. (1979). Discrimination reversal learning in yearling horses. *Journal of Animal Science*, 49: 583–8.

Franchini, M. (2001). *Les indiens d'Amérique et le cheval.* Paris: Zulma.

Francis, R. C. (1990). Temperament in a fish: a longitudinal study of the development of individual differences in aggression and social rank in the Midas cichlid. *Ethology,* **86**: 311–25.

French, J. M. (1993). Assessment of donkey temperament and the influence of home environment. *Applied Animal Behaviour Science,* **36**: 249–57.

Gardner, L. P. (1937). The responses of horses to the situation of a closed feed box. *Journal of Comparative and Physiological Psychology,* **23**: 13–36.

Gerlai, R. & Csányi, V. (1990). Genotype-environment interaction and the correlation structure of behavioral elements in paradise fish (*Macropodus opercularis*). *Physiology and Behavior,* **47**: 343–56.

Goldsmith, H. H., Buss, A. H., Plomin, R., Rothbart, M. K., Chers, T. A., Hinde, R. A. & McCall, R. B. (1987). Roundtable: what is temperament? Four approaches. *Child Development,* **58**: 505–29.

Gosling, S. D. & Bonnenburg, A. V. (1998). An integrative approach to personality research in anthropozoology: rating of six species of pets and their owners. *Anthrozoos,* **11**: 148–56.

Grandin, T. & Deesing, M. J. (1998). Genetics and behavior during handling, restraint, and herding. In *Genetics and the Behavior of Domestic Animals,* ed. T. Grandin. New York: Academic Press, pp. 113–44.

Haag, E. L., Rudman, R. & Houpt, K. A. (1980). Avoidance, maze learning and social dominance in ponies. *Journal of Animal Science,* **50**: 329–35.

Hall, C. S. (1941). Temperament: a survey of animal studies. *Psychological Bulletin,* **38**: 909–43.

Hansen, S. W. (1996). Selection for behavioural traits in farm mink. *Applied Animal Behaviour Science,* **49**: 137–48.

Hausberger, M. & Muller, C. (2002). A brief note on some possible factors involved in the reactions of horses to humans. *Applied Animal Behaviour Science,* **76**: 339–44.

Hausberger, M., Bruderer, C., Le Scolan, N. & Pierre, J.-S. (2004). The interplay of environmental and genetic factors in temperament/personality traits of horses. *Journal of Comparative Psychology* (in press).

Hausberger, M., Le Scolan, N., Bruderer, C. & Pierre J. S. (1998). Le tempérament du cheval: facteurs en jeu et implications pratiques. In *24ème Journée d'Etude de la Recherche Equine,* CEREOPA. Paris: Institut du Cheval, pp. 159–69.

Hausberger, M., Le Scolan, N., Muller, C., Gautier, E. & Wolff, A. (1996). Caractéristiques individuelles dans le comportement du cheval: prédictibilité, facteurs endogènes et environnementaux. In *22ème Journée d'Etude de la Recherche Equine,* CEREOPA. Paris: Institut du Cheval, pp. 113–23.

Hayes, K. E. N. (1998). Temperament tip-offs. *Horse and Rider,* **37**: 46–51.

Heird, J. C. & Grandin, M. J. (1998). Genetic effects on horse behavior. In *Genetics and the Behavior of Domestic Animals,* ed. T. Grandin, New York: Academic Press, pp. 203–34.

Heird, J. C., Lokey, C. E. & Cogan, D. C. (1986a). Repeatability and comparison of two maze tests to measure learning ability in horses. *Applied Animal Behaviour Science,* **16**: 103–19.

Heird, J. C., Whitaker, D. D., Bell, R. W., Ramsey, C. B. & Lokey, C. E. (1986b). The effects of handling at different ages on the subsequent learning ability of 2-year-old horses. *Applied Animal Behaviour Science,* **15**: 15–25.

Hosoda, T. (1950). On the heredity of susceptibility to windsucking in horses. *Japanese Journal of Zootechnical Science,* **21**: 25–8.

Houpt, K. A. & Kusunose, R. (2000). Genetics of behaviour. In *The Genetics of the Horse,* ed. A. T. Bowling & A. Ruvinsky. New York: CABI Publishing, pp. 281–306.

Houpt, K. A. & Lieb, S. (1994). A survey of foal rejecting mares. *Applied Animal Behaviour Science,* **39**: 188.

Houpt, K. A. & Rudman, R. (2002). Foreword to special issue on equine behavior. *Applied Animal Behaviour Science,* **78**: 83–5.

Houpt, K. A. & Wolski, T. R. (1980). Stability of equine hierarchies and the prevention of dominance related aggression. *Equine Veterinary Journal,* **12**: 18–24.

Jensen, P. (1995). Individual variation in the behaviour of pigs – noise or functional coping strategies? *Applied Animal Behaviour Science,* **44**: 245–55.

Jezierski, T., Jaworski, Z. & Gorecka, A. (1999). Effects of handling on behaviour and heart rate in Konik horses: comparison of stable and forest reared youngstock. *Applied Animal Behaviour Science,* **62**: 1–11.

Jones, R. B. (1987). The assessment of fear in the domestic fowl. In *Cognitive aspects of social behaviour in the domestic fowl,* ed. R. Zayan & I. J. H. Duncan. Amsterdam: Elsevier, pp. 40–81.

Juarbe-Diaz, S. V., Houpt, K. A. & Kusunose, R. (1998). Prevalence and characteristics of foal rejection in Arabian mares. *Equine Veterinary Journal,* **30**: 424–8.

Langlois, B. (1984). Cheval de loisir et de sport: aptitudes et sélection. In *Le Cheval. Reproduction, Sélection, Alimentation, Exploitation,* ed. R. Jarrige & W. Martin-Rosset. Paris: Editions INRA, pp. 423–35.

Lanier, J. L., Grandin, T., Green, R., Avery, D. & McGee, K. (2001). A note on hair whorl position and cattle temperament in the auction ring. *Applied Animal Behaviour Science,* **73**: 93–101.

Lankin, V. S. & Bouissou, M. F. (1998). Interspecific differences in fear of humans ('domestic behaviour') in farm animals. In *Proceedings of the 32nd Congress of the International Society of Applied Ethology,* p. 114.

Lebelt, D., Zanella, A. J. & Unshelm, J. (1998). Physiological correlates associated with cribbing behaviour in horses: changes in thermal threshold, heart rate, plasma β-endorphin and serotonin. *Equine Veterinary Journal,* **27** (Suppl.), 21–7.

Le Neindre, P. (1989). Influence of rearing conditions and breed on social behaviour and activity of cattle in novel environments. *Applied Animal Behaviour Science*, 23: 129–40.

Le Roy, I., Perez-Diaz, F., Cherfouk, A. & Roubertoux, P. L. (1999). Preweanling sensorial and motor development in laboratory mice: quantitative trait loci mapping. *Developmental Psychobiology*, 34: 139–58.

Le Scolan, N., Hausberger, M. & Wolff, A. (1997). Stability over situations in temperamental traits of horses as revealed by experimental and scoring approaches. *Behavioural Processes*, 41: 257–66.

Luescher, U. A., Mc Keown, D. B. & Dean, H. (1998). A cross sectional study on compulsive behaviour (stable vices) in horses. *Equine Veterinary Journal*, 27: 14–18.

Lyons, D. M. (1989). Individual differences in temperament of dairy goats and the inhibition of milk ejection. *Applied Animal Behaviour Science*, 22: 269–82.

Mackenzie, S. A. & Thiboutot, E. (1997). Stimulus reactivity tests for the domestic horse (*Equus caballus*). *Equine Practice*, 19: 21–2.

Mader, D. R. & Price, E. O. (1980). Discrimination learning in horses: effects of breed, age and social dominance. *Journal of Animal Science*, 50: 962–5.

Magnusson, D. & Cairns, R. B. (1995). Developmental Science: an integrated framework. In *Developmental Science*, ed. R. B. Cairns, G. H. Elder, Jr. & E. J. Costello. New York: Cambridge University Press, pp. 7–30.

Mal, M. E., Friend, T. H., Lay, D. C., Vogelsang, C. G. & Jenkins, O. C. (1991). Physiological responses of mares to short term confinement and isolation. *Journal of Equine Veterinary Science*, 11: 96–102.

Manteca, X. & Deag, J. M. (1993). Individual differences in temperament of domestic animals: a review of methodology. *Animal Welfare*, 2: 247–68.

Marinier, S. L. & Alexander, A. J. (1992). Use of field observations to measure individual grazing ability in horses. *Applied Animal Behaviour Science*, 33: 1–10.
 (1994). The use of a maze in testing learning and memory in horses. *Applied Animal Behaviour Science*, 39: 177–82.

Martin, P. & Bateson, P. (1988). Behavioural development in the cat. In *The Domestic Cat: The Biology of its Behaviour*, ed. D. C. Turner & P. Bateson. Cambridge: Cambridge University Press, pp. 9–22.

Martinez, R., Godoy, A., Naretto, E. & White, A. (1988). Neuroendocrine changes produced by competition stress on the Thoroughbred race horse. *Compendium of Biochemical Physiology*, 91A: 599–602.

McCall, C. A., Potter, G. D., Friend, T. H. & Ingram, R. S. (1981). Learning abilities in yearling horses using the Hebb-Williams closed field maze. *Journal of Animal Science*, 53: 928–33.

McCann, J. S., Heird, J. C., Bell, R. W. & Lutherer, L. O. (1988). Normal and more highly reactive horses. I. Heart rate, respiration rate and behavioral observations. *Applied Animal Behaviour Science*, 19: 201–14.

McCune, S. (1995). The impact of paternity and early socialisation on the development of cats' behaviour to people and novel objects. *Applied Animal Behaviour Science*, 45: 109–24.

McGreevy, P. D. & Nicol, C. J. (1995). Behavioural and physiological consequences associated with prevention of crib-biting. In *Proceedings of the 29th International Congress of the International Society for Applied Ethology, Universities Federation for Animal Welfare, Wheathampstead, UK*.

McGreevy, P. D., French, N. P. & Nicol, C. J. (1995a). The prevalence of abnormal behaviours in dressage, eventing and endurance horses in relation to stabling. *Veterinary Record*, 137: 36–7.

McGreevy, P. D., Cripps, P. J., French, N. P., Green, L. E. & Nicol, C. J. (1995b). Management factors associated with stereotypic and redirected behaviour in the Thoroughbred horse. *Equine Veterinary Journal*, 27: 86–91.

Meers, L., Van Avermaet, E., Moons, C. & Ödberg, F. O. (2002). Behavioural responses and heart rates of stereotypic and non-stereotypic geldings to four stimuli. In *Proceedings of the 4th Meeting on Behavioural Observations and Methods, Amsterdam 2002*.

Mendl, M. & Harcourt, R. (2000). Individuality in the domestic cat: origins, development and stability. In *The Domestic Cat: The Biology of its Behaviour*, 2nd edn, ed. D. C. Turner & P. Bateson. Cambridge: Cambridge University Press, pp. 47–64.

Mills, D. (1998). Personality and individual differences in the horse, their significance, use and measurement. *Equine Veterinary Journal*, 27 (Suppl.): 10–13.

Mills, P. C., Ng, J. C., Kramer, H. & Auer, D. E. (1997). Stress response to chronic inflammation in the horse. *Equine Veterinary Journal*, 29: 483–6.

Minero, M., Canali, E., Ferrante, V., Verga, M. & Ödberg, F. O. (1999). Heart rate and behavioural responses of crib-biting horse to two acute stressors. *Veterinary Record*, 145: 430–3.

Momozawa, Y., Ono, T., Sato, F. Kikusui, T., Takeuchi, Y., Mori, Y. & Kusunose, R. (2003). Assessment of equine temperament by a questionnaire survey to caretakers and evaluation of its reliability by simultaneous behavior test. *Applied Animal Behaviour Science*, 84: 127–38.

Morris, P. H., Gale, A. & Duffy, K. (2002). Can judges agree on the personality of horses? *Personality and Individual Differences*, 33: 67–81.

Moureaux, S., Verrier, E., Ricard, A. & Mériaux, J. C. (1995). Genetic variability within French race and riding horse breeds from genealogical data and blood marker polymorphisms. *Genetics Selection Evolution*, 28: 83–102.

Nicol, C. J. (2002). Equine learning: progress and suggestions for future research. *Applied Animal Behaviour Science*, 78: 193–208.

Pell, S. M. & McGreevy, P. D. (1999). A study of cortisol and beta-endorphin levels in stereotypic and normal Thoroughbreds. *Applied Animal Behaviour Science*, 64: 81–90.

Peng, Y., Zebrowitz, L. & Lee, H. K. (1993). The impact of cultural background and cross-cultural experience on impressions of American and Korean male speakers. *Journal of Cross-Cultural Psychology*, 24: 203–20.

Pontier, D., Rioux, N. & Heizmann, A. (1995). Evidence of selection on the orange allele in the domestic cat *Felis catus*: the role of social structure. *Oikos*, **73**: 299–308.

Porges, S. W. (1991). Vagal tone. In *The Development of Emotion Regulation and Dysregulation*, ed. J. Garber. Cambridge: Cambridge University Press, pp. 111–18.

Robinson, R. (1977). *Genetics for Cat Breeders*, 2nd edn. London: Pergamon Press.

Roubertoux, P. L. & Carlier, M. (2002). Invariants et variants génétiques: les apports de la génomique dans l'étude des processus cognitifs. In *Invariants et Variabilité dans les Sciences Cognitives*, ed. J. Lautrey, B. Mazoyer & P. Van Geert. Paris: Presses de la Maison des Sciences de l'Homme, pp. 25–39.

Roubertoux, P. L., Mortaud, S., Tordjman, S., Le Roy, I. & Degrelle, H. (1998). Behavior-genetic analysis and aggression: the mouse as prototype. In *Advances in Psychological Science. Vol. 2: Biological and Cognitive Aspects*, ed. M. Sabourin, F. Craik & M. Robert. London: Taylor and Co Group, pp. 1–30.

Seaman, S. C., Davidson, H. P. B. & Waran, N. K. (2002). How reliable is temperament assessment in the domestic horse (*Equus caballus*). *Applied Animal Behaviour Science*, **78**: 175–91.

Sgoifo, A., de Boer, S. F., Westenbroek, C., Maes, F. W., Beldhuis, H., Suzuki, T. & Koolhoss, J. M. (1997). Incidence of arrhythmias and heart rate variability in wild-type rats exposed to social stress. *American Journal of Physiology*, **273**: H1754–60.

Slater, P. J. B. (1981). Individual differences in animal behavior. *Perspectives in Ethology*, **4**: 35–49.

Stevenson-Hinde, J., Stillwell-Barnes, R. & Zunz, M. (1980). Subjective assessment of rhesus monkeys over four successive years. *Primates*, **21**: 66–82.

Tellington-Jones, L. (1996). *Comprendre et Influencer la Personnalité de son Cheval*. Paris: Vigot.

Vecchiotti, G. G. & Galanti, R. (1986). Evidence of heredity of cribbing, weaving and stall-walking in Thoroughbred horses. *Livestock Production Science*, **14**: 91–5.

Visser, E. K., Van Reenen, C. G., Hopster, H., Schilder, M. B. H., Knaap, J. H., Barneveld, A. & Blokhuis, H. J. (2001). Quantifying aspects of young horses'

temperament: consistency of behavioural variables. *Applied Animal Behaviour Science*, **74**: 241–58.

Visser, E. K., Van Reenen, C. G., Van der Werf, J. T. N., Schilder, M. B. H., Knaap, J. H. & Barneveld, A. E. A. (2002). Heart rate and heart rate variability during a novel object test and a handling test in young horses. *Physiology and Behavior*, **76**: 289–96.

Visser, E. K., Van Reenen, C. G., Rundgren, M., Zetterquist, M., Morgan, K. & Blokhuis, H. J. (2003a). Responses of horses in behavioural tests correlate with temperament assessed by riders. *Equine Veterinary Journal*, **35**: 176–83.

Visser, E. K., Van Reenen, C. G., Engel, B., Schilder, M. B. H., Barneveld, A. E. A. & Blokhuis, H. J. (2003b). The association between performance in show-jumping and personality traits earlier in life. *Applied Animal Behaviour Science*, **82**: 279–95.

Visser, K. (2002). Horsonality: a study on the personality of the horse. PhD thesis, Ponsen en Looijen BV, Netherlands.

Waring, G. H. (1983). *Horse Behavior. The Behavioral Traits and Adaptations of Domestic and Wild Horses, including Ponies*. Park Ridge, NJ: Noyes Publications.

Wolff, A. & Hausberger, M. (1992). Comparaison de caractéristiques comportementales chez les poulains: une étude quantitative. In *CEREOPA. 18ème Journée du Cheval*, Institut du Cheval, Paris, pp. 78–91.

Wolff, A. & Hausberger, M. (1994). Behaviour of foals before weaning may have some genetic basis. *Ethology*, **96**: 1–10.

(1996). Learning and memorisation of two different tasks in horses: the effects of age, sex and sire. *Applied Animal Behaviour Science*, **46**: 137–43.

Wolff, A., Hausberger, M. & Le Scolan, N. (1997). Experimental tests to assess emotivity in horses. *Behavioural Processes*, **40**: 209–21.

Xénophon (1995). *De l'art équestre*, Texte établi et traduit par E. Delebecque. Paris: Les Belles Lettres.

Zechner, P., Zohman, F., Sölkner, J., Bodo, I. ., Habe, F., Marti, E. & Brem, G. (2001). Morphological description of the Lipizzan horse population. *Livestock Production Science*, **69**: 163–77.

Zeiler, M. & Distl, O. (2000). Schätzung von genetischen Parametern für die Haflingerzucht. *Zuchtungskunde*, **72**: 241–57.

II

**The natural behaviour
of horses in the wild and
domestic environment**

4

Behavioural ecology of feral horses

Lee Boyd and Ronald Keiper

Introduction

Domestic horses are sometimes managed as free-ranging populations (Tyler, 1972; Gates, 1979). Over the centuries, some domestic horses escaped captivity to become feral or 'wild'. Some feral horse populations are unmanaged, while others are subject to periodic roundups, removals, and contraception, skewing age structure and sex ratios.

Feral horse populations are found worldwide in a variety of habitats, indicating a wide ecological tolerance (Figure 4.1). These habitats include deserts (Berger, 1977; Miller & Denniston, 1979; Stoffel-Willame & Stoffel-Willame, 1999), mountains (Feist & McCullough, 1976; Berger, 1986), temperate forests (Tyler, 1972; Salter & Hudson, 1982; Linklater *et al.*, 1999), barrier islands (Welsh, 1975; Rubenstein, 1981; Keiper, 1986; Stevens, 1988), swampy estuaries in the Camargue region of southern France (Duncan, 1980; Feh, 1990), at Toi Cape in Japan (Kaseda & Khalil, 1996), on the pampas of Argentina (Scorolli, 1999) and on the savannahs

(llanos) of Venezuela (Pacheco & Herrera, 1997). Feral horse populations show great variation in their behaviour and ecology as the result of differing environmental, social and management pressures.

Free-ranging domestic and feral populations provide insight into the natural behaviour of horses, which is valuable knowledge for improving the welfare of their captive relatives. (See Feh, Chapter 5, for related discussions of social interactions and behaviour of free-living equids.)

Social organization

Most feral horses live in small, stable groups called bands that inhabit large, overlapping home ranges. They display a female defence polygymous mating system, where one male forms a bond and breeds with several females. This type of mating system is considered to be an adaptation to seasonally changing ecological conditions (Klingel, 1975). Because the stallion defends his mares rather than a territory, the band is not restricted in its movements so it can make use of the best available food from season to season.

The Domestic Horse: The Origins, Development, and Management of its Behaviour, ed. D. S. Mills & S. M. McDonnell.
Cambridge University Press. © Cambridge University Press 2005.

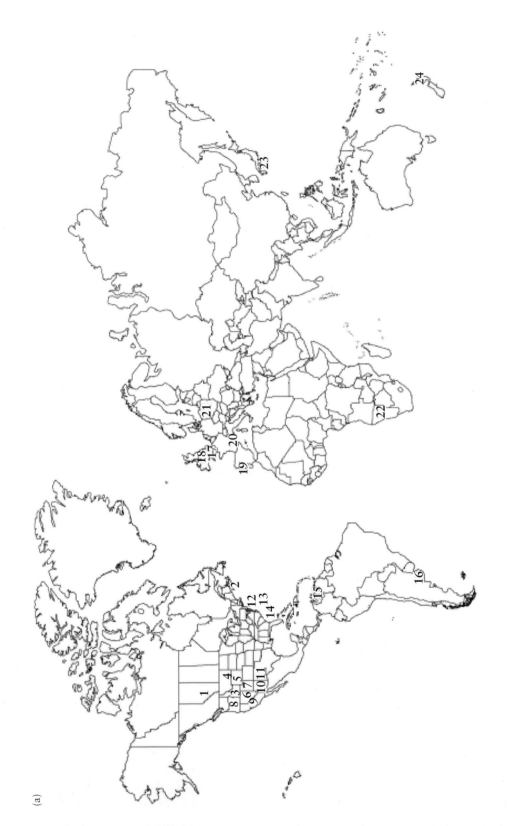

(a)

Figure 4.1(a). Map of free-ranging domestic and feral horse populations; see accompanying chart.

Figure 4.1(b). Key to Figure 4.1(a)

Country	Map	Population	Population type	Management during study	References
Canada	1	Alberta	Feral	None	Salter (1978), Salter & Hudson (1978, 1979, 1982), Salter & Pluth (1980)
	2	Nova Scotia: Sable Island	Feral	None	Welsh (1973, 1975), Lucas et al. (1991), Kimura (2001)
USA	3	Idaho	Free-ranging domestics	Castration of some males	Stebbins (1974)
	4	Wyoming/Montana: Pryor Mountains	Feral	Removals	Feist & McCullough (1975, 1976), Hall & Kirkpatrick (1975), Kirkpatrick et al. (1979), Perkins et al. (1979), Turner et al. (1981)
	5	Wyoming: Red Desert	Feral	Removals	Miller & Denniston (1979), Miller (1979, 1981, 1983), Boyd (1980)
	6	Nevada: Great Basin	Feral	Removals began in the 1980s	Pellegrini (1971), Green & Green (1977), Berger (1983a,b,c, 1986), Berger et al. (1983), Berger & Rudman (1985), Berger & Cunningham (1987), Roberts (1988), Bowling & Touchberry (1990), Eagle et al. (1993), Asa (1999), Ashley (2000), Ashley & Holcombe (2001)
	7	Utah: Cedar Mountain	Feral	Removals	Kirkpatrick et al. (1979)
	8	Oregon: Beaty Butte	Feral	Removals	Eagle et al. (1993), Asa (1999)
	9	California: Montgomery Pass Wild Horse Territory	Feral	None	Turner et al. (1992)
	10	Arizona: Grand Canyon	Feral	None	Berger (1977)
	11	New Mexico: Jicarilla Wild Horse Territory	Feral	None	Nelson (1980)
	12	Maryland/Virginia Assateague/Chincoteague Island	Feral	Removals on Chincoteague, contraception in the 1990s	Keiper (1976, 1980, 1985, 1986), Zervanos & Keiper (1979), Keiper & Keenan (1980), Keiper et al. (1980), Keiper & Berger (1982), Houpt & Keiper (1982), Keiper & Houpt (1984), Keiper & Sambraus (1986), Rutberg (1987, 1990), Rutberg & Greenberg (1990), Rudman & Keiper (1991), Kirkpatrick & Turner (1991a,b), Kirkpatrick et al. (1992), Rutberg & Keiper (1993), Powell (1999, 2000)
	13	North Carolina: Rachel Carson Estuarine Sanctuary Shackleford Island	Feral	None	Rubenstein (1981, 1982), Hoffmann (1985), Stevens (1988, 1990)
Venezuela	14	Georgia: Cumberland Island	Feral	None	Goodloe et al. (2000)
Argentina	15	Hato El Frfo	Feral	Removals	Pacheco & Harrera (1997)
England	16	Ernesto Tornquist State Park	Feral	None	Scorolli (1999)
	17	New Forest	Free-ranging domestics	Removals, skewed sex ratio	Tyler (1972), Pollock (1980)
Portugal	18	Exmoor	Free-ranging domestics	Removals, skewed sex ratio	Gates (1979)
France	19	Peneda-Gerés National Park	Feral	Removals, skewed sex ratio	Oom & Santos Reis (1986)
	20	Camargue	Free-ranging domestics	Removals, skewed sex ratio	Goldschmidt-Rothschild & Tschanz (1978), Wells & von Goldschmidt-Rothschild (1979), Boy & Duncan (1979), Duncan & Vigne (1979), Duncan (1980, 1982, 1983, 1992), Duncan & Cowtan (1980), Duncan et al. (1984a,b), Feh (1990, 1999), Monard & Duncan (1996), Monard et al. (1996)
Poland	21	Popielno Experimental Station	Free-ranging domestics	Removals	Jaworowska (1976, 1981)
Namibia	22	Namib Desert	Feral	Removals	Stoffel-Willame & Stoffel-Willame (1999)
Japan	23	Misaki, Toi Cape	Feral	Castration of some males	Kaseda (1981, 1982, 1983), Kaseda et al. (1982, 1984, 1995, 1997), Kaseda & Khalil (1996), Kahlil & Kaseda (1997, 1998), Kahlil & Murakami (1999a,b)
New Zealand	24	Kaimanawa	Feral	Removals	Linklater et al. (1999, 2000), Cameron & Linklater (2000), Linklater & Cameron (2000)

A number of bands living in the same geographically localized area form a population called a herd. While the herd has been considered simply a loose grouping of horse bands living in the same area, Miller and Denniston (1979) suggested that a herd may really be a structured social unit where bands follow similar movement patterns within their common home range and display an interband dominance hierarchy.

In unmanaged feral populations the most commonly observed type of band is the harem or family band, composed of a stable core of sexually mature mares three years or older, their immature offspring and one or more adult stallions (Feist & McCullough, 1975; Keiper, 1976; Miller, 1981; Berger, 1986). Other kinds of bands include all-male bachelor bands and bands of mares and their offspring.

Bachelor bands are formed by dispersal of young males from their natal band and by older males that have lost membership in a harem group. Group size varies from lone bachelors to groups of 16 horses (McCort, 1984). In contrast to family bands, bachelor bands tend to be unstable in composition, although some authors have reported long-term associations between several males (Miller, 1979; Salter & Hudson, 1982).

Mare–offspring bands may result from the death of a family band stallion, with the mares remaining together without a stallion for up to six months (Keiper, 1985) or from fission of the family band (Kaseda, 1981).

Although the social structure of different feral horse populations is similar, family band composition and size varies. On Cumberland Island (off the Atlantic coast, Georgia, USA), 85.9% of the bands were composed of mares, juveniles, foals and a single stallion, 10.7% of the bands had two or more stallions who shared dominance and 3.4% of the bands consisted only of mares and their offspring (Goodloe *et al.*, 2000). Similarly, 88% of the bands in the Granite Range in the USA (Berger, 1986) and 83% of the bands in Tornquist Park in Argentina (Scorolli, 1999) contained only one adult stallion.

By contrast, in some populations up to half the bands contain more than one and as many as five stallions (Linklater *et al.*, 1999). It may be, however, that in many multi-male bands the additional males were simply young animals, whose ages were not clearly known, who had not yet dispersed from their natal

band. Green and Green (1977), for example, admitted they had difficulty distinguishing between immature (two- to three-year-old) and adult horses and therefore included immature males in the adult male category, so the actual percentage of bands with more than one adult male might actually be lower. Multimale bands might also result when young male and female horses disperse from the same or different family bands and form a mixed-sex peer band.

Family band sizes ranged from 2 to 35 horses and averaged 3.4 to 18.1 animals for the various populations studied (Table 4.1). Band size may be influenced by a number of factors including differences in foaling rates and foal mortality, the age of the stallion, the adult sex ratio and the quality and quantity of food within the home range. Zervanos and Keiper (1979) reported a positive correlation between group size and available forage biomass.

Pacheco and Herrera (1997) found that bands living in tropical areas were larger than bands in temperate regions. They concluded that the larger size of the bands was related to the dry and rainy seasons of the tropical climate. During the dry season water sources decreased in number, so bands clustered around the remaining water. In the wet months the llanos flooded, leaving only patches of higher ground where several bands gathered. Because bands were closer together, stallions had a greater opportunity to compete for females and transfer of horses between bands was common, leading to larger band size.

With respect to age of the stallion, male Misaki horses began forming harem groups of one to four mares at the age of four years. As the stallion aged, the number of mares in the band increased, reaching a maximum size of about nine mares when the stallion was seven to nine years old. Harem size then gradually decreased as the stallion continued to age (Kaseda & Khalil, 1996).

The adult sex ratio of the population may also affect band size. When the ratio of males/females in the Misaki population was 1:3, family bands averaged only 3.8 mares whereas when the adult sex ratio was 1:6, bands averaged seven mares (Kaseda & Khalil, 1996). When there are more males in the population, there may be stronger competition between stallions for the smaller number of mares resulting in smaller band size (Keiper, 1986). Berger (1986) reported a negative correlation between average band size and the number of males in the population. On the other

Table 4.1. *Feral horse harem band structure*

Population	Family size		Mares per band		No. of bands	Reference
	Mean	Maximum	Mean	Maximum		
Sable Island, Canada	5.5	20	–	–	40–50	Welsh, 1975
Shackleford Banks, USA	12.3	–	–	–	–	Rubenstein, 1981
Assateague Island, USA	9.1	17	3.8	8	10	Keiper, 1986
Cumberland Island, USA	4.6	–	1.8	–	171	Goodloe *et al.*, 2000
Venezuela	18.1	35	–	22	–	Pacheco & Herrera, 1997
Tornquist Park, Argentina	8.4	–	3.7	–	30	Scorolli, 1999
Kaimanawa, New Zealand	4.5	7	2.2	11	36	Linklater *et al.*, 2000
Toi Cape, Japan	6	13	1.5	3	13	Kaseda, 1981
Alberta, Canada	7.7	17	–	–	23	Salter & Hudson, 1982
Pryor Mountains, USA	5	9	1.8	3	44	Feist & McCullough, 1975
Grand Canyon, USA	4.5	6	3	4	4	Berger, 1977
Granite Range, USA	–	13	3.1	7	13	Berger, 1986

hand, the presence of subordinate males in a band may contribute to increased band size. Stevens (1990) found that each of the 10 mares that changed bands in her study left single-male bands and concluded that the presence of additional males on the periphery of the band more effectively prevented mares from wandering off.

Home range and territoriality

In essentially all feral horse populations studied, bands occupy specific, undefended, non-exclusive geographic portions of their environment known as the home range. The home range is the area used during day-to-day activities and incorporates not only grazing sites and waterholes but also shade, wind breaks, mineral licks and refuges from insects. Within the home range are core areas where most of the animals' activities are concentrated. Home-range boundaries are well defined and remain constant over a period of months to years (McCort, 1984).

Because resources are usually not abundant or evenly distributed, they are shared among several bands. For example, in the Pryor Mountain Wild Horse Range, located on the Wyoming–Montana border in the USA, only 10% of the study area was used exclusively by one harem (Feist & McCullough, 1976) and Miller (1983) noted that some Red Desert horse home-ranges overlapped almost completely. In the Granite Range, Berger (1986) reported that in a given year 90% of low altitude home ranges and 87% of high altitude ranges overlapped. However, average

Table 4.2. *Size of home range of feral horse bands*

Home-range size (km²)	Location	Reference
0.9–6.6	Sable Island, Canada	Welsh, 1975
2.2–11.4	Assateague Island, USA	Zervanos & Keiper, 1979
6	Shackleford Banks, USA	Rubenstein, 1981
2.6–14.4	Alberta, Canada	Salter, 1978
0.8–10.2	New Forest, UK	Tyler, 1972
2.8–3.7	Exmoor, UK	Gates, 1979
0.6–17.7	Kaimanawa, New Zealand	Linklater *et al.*, 2000
1.8–8.0	Tornquist Park, Argentina	Scorolli, 1999
3–32	Pryor Mountains, USA	Feist & McCullough, 1976
6.7–35.6	Granite Range, USA	Berger, 1986
8–48	Grand Canyon, USA	Berger, 1977
10.3–78	Sone Cabin Valley, USA	Green & Green, 1977
5.2–31.1	Wassuk Range, USA	Pellegrini, 1971

core area overlap was only 43% for low altitude sites and even less (30%) for high altitude locations.

Feral horse home ranges vary in size from 0.6 to 78 km² (Table 4.2). Home ranges in mainland populations tend to be larger than home ranges of island populations and those in mesic habitats tend to be smaller than home ranges in xeric habitats, but there is considerable overlap in home-range size. In the New

Forest ponies in the UK (Tyler, 1972) and for the horses on Assateague Island in the USA (Zervanos & Keiper, 1979) there was no correlation between band size and home-range size. In contrast, there was a significant correlation between band size and home-range size in Kaimanawa wild horses in New Zealand (Linklater *et al.*, 2000). The home ranges of harem bands in the Granite Range in the USA were smaller than those of bachelor bands (Berger, 1986), while in New Zealand the home ranges of harem and bachelor bands were similar in size (Linklater *et al.*, 2000).

In some populations home-range size varied seasonally, although there was no consistent trend across populations. In the Granite Range home-range sizes increased during the fall, winter and spring apparently because the horses had to range more widely to secure adequate forage (Berger, 1986). Similarly the home ranges of Kaimanawa horses in winter were 21% larger than in summer (Linklater *et al.*, 2000). On the other hand, on Assateague Island average home-range size was larger in summer (6.48 km^2) than in winter (4.32 km^2). The increased size of the home range in summer resulted because bands travelled long distances from grazing areas to refuge sites on the beach and in the bay to reduce insect harassment (Zervanos & Keiper, 1979).

Territoriality in the classical sense, where bands defend areas of exclusive use against other bands, has only been reported on Shackleford Island off the coast of North Carolina, USA (Rubenstein, 1981) and in the New Forest (Pollock, 1980). On Shackleford Island, bands on the wider western end of the island occupied overlapping home ranges. Territories, whose boundaries ran the width of the island, only occurred on the narrow eastern part of the island where visibility was unrestricted and vegetation ran along the long axis of the island. Apparently here the problems of defending resources were reduced sufficiently to make territorial defence efficient. Today there is no longer any evidence that any Shackleford stallion defends a specific territory (S. Stuska, pers. comm., 2002).

While Tyler (1972) found no evidence of territorial behaviour in the New Forest ponies, Pollock (1980) observed 'ritualized territorial displays' between three stallions that featured repeated dunging and mutual withdrawal. Other scientists (Feist & McCullough, 1976; Salter & Hudson, 1982; Berger, 1986; Keiper, 1986) concluded that the stud piles created by repeated dunging did not mark territorial

boundaries. Instead they functioned to orientate horses to a familiar trail or waterhole location, to warn other stallions away from an area being used by another band or to provide socially relevant information such as identity and dominance status of the marker.

Habitat selection

Although home-range boundaries were generally constant throughout the year, habitat utilization varied seasonally, as factors such as quantity, quality and availability of forage changed. Habitat utilization by horses in the Camargue was primarily determined by the abundance of high quality food. In spring the horses fed in the marshes where there was an abundance of green biomass (540 g/m^2) with 17.5% crude protein. In contrast vegetation in the grasslands contained just 130 g/m^2 of green biomass and only 5.2% protein content. On the other hand, in winter the emergent marsh vegetation was dead with a protein content of only 4.1% so the horses fed more in the grasslands that were dominated by perennial grasses and forbs with 7.8% crude protein content (Duncan, 1983).

On Assateague Island, horses grazed during the summer primarily in the low saltmarsh (44.5%), in the shrub zone (22.1%) and on the dunes (15.3%), but in winter they increased their utilization of the dunes (32.8%) and the shrub zone (33.7%) where they were protected to some extent from the cold winter winds and could browse on shrubs like poison ivy, sumac and bayberry (Keiper & Berger, 1982). Misaki horses also demonstrated a seasonal shift in habitat utilization. In summer they grazed on grasslands but transferred to a winter range in forests or weedy grasslands where native grasses, forbs and trees were available (Kaseda, 1982).

Some horse populations living in mountainous regions display vertical migrations. In the Granite Range horse bands moved from low- to high-altitude sites in late spring/early summer, apparently in accordance with newly emerging, higher quality vegetation. On the other hand, some bands remained at low altitudes throughout the year, but likely because they failed to exploit the higher quality high-altitude grasses, exhibited lower reproductive success than bands that did migrate (Berger, 1986).

Kaimanawa horses also displayed seasonal altitudinal movements, although in a reverse direction.

The altitude at which bands were located increased through the autumn and peaked in winter, then showing an abrupt decline at the beginning of spring prior to foaling. Apparently the combined effect of low wind, air temperatures below freezing and consequent overnight frosts created frost inversion layers that resulted in colder air temperatures in river basins and valleys, whereas the inversion layers were prevented at high altitudes by wind blowing across the mountain tops. With warmer temperatures in the spring, the horses returned to the lush river valleys for foaling (Linklater *et al.*, 2000).

Availability of water may influence behaviour as well as seasonal habitat utilization. While horses living in the Namib Desert make use of ephemeral pools of water after rainfall, their only permanent source of water is a water hole 10 to 25 km from their home range. The mean interval between visits varies from 30 hours in summer to 72 hours in winter and is inversely correlated to the maximum daytime temperature (Stoffel-Willame & Stoffel-Willame, 1999). Other feral horses have been reported to drink one to three times a day (Keiper & Keenan, 1980).

Horses in the Grand Canyon of Arizona, USA grazed widely in late winter and early spring but concentrated in a small area around a water source in late summer (Berger, 1977). In the Stone Cabin Valley of Nevada, USA the southern end of the valley was little used by horses despite good forage apparently because water was not always available. However, following heavy rain or snow, large pools of water accumulated in shallow depressions and up to 150 horses migrated there and fed in the area until the standing water dried up. Then they returned to their primary feeding area to the north (Green & Green, 1977).

Wind and snow are other environmental variables that may influence habitat selection. In the Granite Range, 80% of high-altitude bands moved to low-altitude sites within four days after snowstorms. Here they spent their time in exposed areas of grassland and shrub, although during and after strong winds and snowstorms, the horses sought refuge in ravines and small clumps of juniper where they browsed on bitterbrush, rabbitbrush and mahogany. With severe wind chills, the animals sought shelter in forests. Bands that remained at high altitude were exposed to cold temperatures, extreme wind chill and deep snow that often resulted in the death of the entire band (Berger, 1986).

In Alberta, Canada where winter snow depth averaged 30 cm, feral horses utilized south-facing birch and mixed shrub meadows during periods of good weather but sought shelter in alder thickets and black spruce woods during storms. The horses took advantage of the reduced snow depth under the trees to paw through snow as deep as 60 cm to the underlying vegetation. With the onset of snow melt and the new growth of vegetation, the horses increased their utilization of meadows and reduced their use of forested habitats (Salter, 1978).

Utilization of natural salt licks by feral horses in the western foothills of Alberta also appears to vary seasonally. Although horses visited natural mineral licks throughout the year, utilization was highest from May to July. By analyzing vegetation and faecal samples, Salter and Pluth (1980) found that during the spring the herbaceous vegetation available to horses was low in sodium, while at the same time loss of this mineral in the faeces was high. They concluded that the peak use of natural mineral licks in spring coincided with a peak in sodium stress. Like other limited resources, mineral licks were shared by a number of bands (Salter & Hudson, 1979).

Feral horses on Sable Island, Canada were observed eating organic soils and lapping at saltwater foam on the beach (Welsh, 1975) and may have been trying to supplement their mineral requirements, although mineral levels in the soil and foam were not determined.

Finally tabanid flies may also influence what kinds of habitats feral horses utilize. On hot, humid summer days these biting insects disrupt normal activities and induce long-distance movements and habitat shifts. In response to insect harassment the feral horses of Assateague Island move from grazing areas in the marsh and on the dunes to other habitats that serve as anti-insect refuge sites. These include mudflats, the beach and the waters of Chincoteague Bay. These habitats were used significantly more in summer, when pest densities are higher, than in winter (Keiper & Berger, 1982). Similarly during hot summer days Camargue horses spent more of their diurnal resting time on bare or sparsely vegetated resting sites (Duncan & Cowtan, 1980), but showed no preference for these areas at night or in other seasons.

Despite living in a very different physical environment, horses in the Granite Range of Nevada, like those on Assateague Island and in the Camargue,

moved to specific habitats to reduce insect harassment. During summer they grazed in low meadows
early in the day when it was cooler and pest densities
were lower, but as insect activity increased they moved
to the higher, cooler, windy slopes and ridges where
they rested during the afternoon (Keiper & Berger,
1982).

While insects influence habitat selection, it is not
clear that they alter the social behaviour of horses.
Duncan and Vigne (1979) observed that group size in
Camargue horses increased during the months when
tabanids were active and that horses in larger groups
had fewer flies on them. They concluded that the
horses formed into larger social groups in summer
as a way of reducing insect harassment. In contrast,
Rutberg (1987) found that group sizes of Assateague
horses did not change in response to fly abundance
and concluded that the size, cohesiveness and internal structure of the groups was probably determined
more by food distribution and social factors than
by insects. It may be that the skewed sex ratio,
where mares greatly outnumbered stallions, was more
responsible for increasing band size in the Camargue
than the biting flies.

Time budgets

Horses divide their time between activities that allow
them to satisfy their basic requirements for food,
water, movement and rest. The time budget is the
amount of time invested by animals in each of these
activities and it varies with the characteristics of the
environment, the season of the year and the age and
gender of the animal.

Free-ranging horses spend between 90% and 98%
of their time feeding and resting (Salter & Hudson,
1979; Duncan, 1980; Keiper *et al.*, 1980; Kaseda,
1983), although, there is considerable variation in the
total time spent grazing. This variation seems to be
primarily related to the availability of forage. Horses
living in environments with sparse vegetation such as
deserts (Stoffel-Willame & Stoffel-Willame, 1999) or
alpine forests (Salter & Hudson, 1982) spend about
18 hours a day grazing, while horses living in marshes
graze as little as 13 hours a day (Duncan, 1980; Keiper
et al., 1980).

Time budgets vary over the course of the day and
from season to season. In summer horses on Shackleford Island grazed 61% of the time compared to 79%
in winter (Rubenstein, 1981). Similarly Assateague

horses graze only 65% of the time and rest 32%
compared to 81% grazing and 16% resting in winter (Keiper *et al.*, 1980). At night they spend 54% of
the time grazing and 40% resting (Keiper & Keenan,
1980). These activities follow a definite rhythm, with
periods of intense grazing alternating with periods of
rest and with all other behaviours (drinking, grooming, aggressive and reproductive behaviour) occurring
at low frequencies. Typically horses graze heavily just
after sunrise and around sunset. In the summer, a
major resting period occurs in the heat of the day
when the horses move to anti-insect refuges such
as mud flats, the beach (Figure 4.2) or in the bay
(Rubenstein, 1981; Keiper, 1985). A second major
resting period takes place between midnight and
04:00, when recumbency comprises almost one-third
of the total time (Keiper & Keenan, 1980). The
decrease in insect activity and the cooler temperatures
at night appear to be the primary factors responsible
for the increase in nocturnal resting.

Horses in Alberta showed little seasonal variation,
grazing about 77% of the time in both winter and
summer (Salter & Hudson, 1979). Misaki horses
exhibited a similar pattern, grazing 79% of the day
in summer and 71% in winter. However, resting at
night increased from only 8.7% in summer to 22.8%
in winter (Kaseda, 1983). Instead of increasing time
spent in searching for and harvesting food on low
quality winter rangeland, activities that obliged the
horses to expend more energy to maintain body temperature, Kaseda (1983) concluded that the horses
moved into forests and rested on cold nights as a way
of reducing energy expenditure.

Time budgets also vary depending on the age and
gender of the horse and on the quality of the home
range. Camargue stallions spent more time standing alert and trotting or galloping but less time foraging than did females (Duncan, 1980) and young
horses spent more time lying down than adults (Boy &
Duncan, 1979). Within two to four weeks of weaning,
however, foals developed time budgets closely similar to older horses (Boy & Duncan, 1979). In the
Granite Range, males spent less total time feeding
than females, and lactating mares spent more time
feeding than barren mares (Berger, 1986). In Alberta,
foals spend an average of 41% of their time foraging,
13% standing resting, 26% lying resting, 4% suckling and 15% in other activities (Salter & Hudson,
1979). Finally, stallions, barren mares and lactating

Figure 4.2. Assateague Island feral horses using a bay as an insect refuge. (Photo by Ron Keiper.)

mares living in low quality home ranges spent more time feeding than those in medium and high quality home ranges (Berger, 1986). Colts nursed longer than fillies (Duncan *et al.*, 1984a; Berger, 1986) and foals from high quality home ranges averaged more than a minute per hour longer in nursing duration than foals from low quality home ranges (Berger, 1986).

The frequency of other maintenance behaviours also varies daily or seasonally. On Assateague, auto- and allogrooming occurred more frequently from April through to June when the horses were shedding their winter coats and when courtship activities were increasing (Keiper, 1985). At night, allogrooming occurred significantly more often between 19:00 and 23:00, when the horses were more active, than it did between midnight and 04:00 when the horses were standing resting or recumbent (Keiper & Keenan, 1980).

Dominance hierarchies and band interactions

Individuals who encounter one another regularly on a long-term basis often establish dominance hierarchies that determine priority of access to resources, and horses are no exception. Dominant mares spend more time foraging (Powell, 2000) and have faster growing offspring (Duncan, 1992). Dominant mares may chase subordinate oestrous mares away from stallions (Tyler, 1972; Stebbins, 1974; Monard & Duncan, 1996; Powell, 2000). Stallions investigate more eliminations by dominant mares than subordinate mares

and dominant mares engage in more sexual behaviour (Powell, 2000).

Individual recognition

The presence of hierarchies implies individual recognition. Feist and McCullough (1976) presented anecdotal evidence that mustangs could identify one another individually using sight and sound, and suggested that olfaction was probably also involved. Kin recognition appears to be by prior association; if kin are unfamiliar with one another, consanguineous matings have been known to occur. Some evidence exists for how long recognition persists in the absence of contact: Berger (1986) found that if colts who had dispersed more than 18 months previously reencountered their sires they were treated as unfamiliar rivals, whereas those encountered within 18 months appeared to be recognized. Recognition of bands may occur as well. Miller and Denniston (1979) observed that some dominant bands appeared to be recognized at distances of at least 50 m. When these particular bands approached a water hole, subordinate bands moved away from the well immediately without contest.

Intraband hierarchies

Horses within a band form linear dominance hierarchies. Hierarchies remain reasonably stable over time, with injury, old age and death accounting for most changes (Tyler, 1972; Keiper & Sambraus, 1986).

L. Boyd and R. Keiper

Table 4.3. *Factors affecting intraband dominance ranks*

Factor	Populations	Gender	Effect	No effect	Reference
Gender	New Forest, UK		X		Tyler, 1972
	Pryor Mountains, USA		X		Feist & McCullough, 1976
	Camargue, France			X	Wells & Goldschmidt-Rothschild, 1979
				X	Houpt & Keiper, 1982
Age	New Forest, UK	M, F	X		Tyler, 1972
	Pryor Mountains, USA	B	X		Feist & McCullough, 1976
	Camargue, France	M, F	X		Wells & Goldschmidt-Rothschild, 1979
	Assateague Island, USA	M, F	X		Houpt & Keiper, 1982; Keiper & Sambraus, 1986
		F			Rutberg & Greenberg, 1990; Powell, 2000
Size	New Forest, UK	F	X		Tyler, 1972
	Grand Canyon, USA	M, F	X		Berger, 1977
	Assateague Island, USA	F	X		Rutberg & Greenberg, 1990
	Camargue, France	M		X	Feh, 1990
Tenure	New Forest, UK	F		X	Tyler, 1972
	Camargue, France	F	X		Monard & Duncan, 1996
Maternal rank	Assateague Island, USA	M, F		X	Keiper & Sambraus, 1986
	Camargue, France	M	X		Feh, 1999
Presence of foal	Red Desert, USA	F		X	Boyd, 1980
	Assateague Island, USA	F		X	Keiper & Sambraus, 1986

F, female; M, male; B, both sexes.

Rates of 0.25–2 aggressions per hour are used to maintain the hierarchy (Houpt & Keiper, 1982; Rutberg & Greenberg, 1990; Stevens, 1990; Powell, 1999). Rank is correlated with age (Table 4.3), enabling older high-ranking mares to reduce involvement in aggression (Rutberg & Greenberg, 1990). Size may also be an important determinant for females of similar age; larger mares in the New Forest population had a competitive edge (Tyler, 1972; Pollock, 1980). Presence or absence of a foal does not affect a mare's rank (Boyd, 1980). Offspring rank as an adult correlated with maternal rank for sons (Feh, 1999) but not for daughters (Keiper & Sambraus, 1986). Monard and Duncan (1996) found that the mares' order of arrival in the harem was the prime determinant of rank, which would usually be correlated with relative age. Newly arrived mares are at the bottom of the hierarchy except that they outrank any daughters of resident mares who have not yet dispersed. The exceptions were mares who immigrated into harems containing familiar subadults whose mothers had been low-ranking; in these cases the immigrants outranked such females.

The stallion was the dominant harem member in three of four feral horse bands studied by Berger (1977) and in all bands on Shackleford Banks (Rubenstein, 1981), and in Polish wild horses (Jaworowska, 1976), in New Forest ponies (Pollock, 1980) and feral horses in Canada and the American West (Welsh, 1973; Feist & McCullough, 1976; Salter, 1978). During her study of free-ranging Appaloosa horses, Stebbins (1974) found stallions to be dominant to mares and geldings. When water is limited, Grand Canyon and Shackleford Banks harem stallions drink first (Berger, 1977; Stevens, 1988) although Namib feral stallions do not (Stoffel-Willame & Stoffel-Willame, 1999). In contrast, Keiper and Sambraus (1986) did not find the stallion to be the dominant member of any band on Assateague Island. Wells and Goldschmidt-Rothschild (1979) found the Camargue stallion to be subordinate to his mares except when he was herding them. Stallions are most often farther from their nearest neighbour than are the mares or youngsters to each other and thus stallions are involved in fewer encounters of intraband aggression (Keiper & Sambraus, 1986) which makes determination of their rank difficult. Some of the conflicting conclusions about the position of the stallion may arise from the observer's classification of behaviour. Some researchers consider herding by males to be aggression (Miller, 1981) and include such acts in the determination of rank, whereas Keiper and Sambraus (1986) defined herding by the stallion as a sexual rather than an aggressive act (Figure 4.3). Likewise, inclusion of

Figure 4.3. Stallion herding his mares on Assateague Island. (Photo by Ron Keiper.)

aggression by mares toward stallions when refusing courtship or protecting newborns may, or may not be, included in determination of rank. Tyler (1972) observed New Forest stallions to be dominant in food contexts but regardless they received aggression from oestrous mares and those with newborn foals. The degree of intrasexual competition among males may also be a factor in where a stallion ranks within his harem. In populations with a 1:1 adult sex ratio, males have many rivals and only the most dominant males acquire females, increasing the likelihood that they are dominant also to their mares. The intensity of intrasexual selection may be so reduced in populations from which males are removed that even young or less competitive males can obtain a harem and may be outranked by some of their older mares. Keiper and Sambraus (1986) speculated that if a young male took over an already established harem he would be younger than many of the mares and often lower ranking, whereas males present at the formation of a new harem of dispersing females are likely to be older and higher ranking.

In multi-male bands, one stallion is clearly dominant and the others subordinate (Hall & Kirkpatrick, 1975; Green & Green, 1977; Miller, 1981; Salter & Hudson, 1982; Stevens, 1988, 1990; Linklater & Cameron, 2000).

Bachelor stallions also have a linear dominance hierarchy in spite of the more fluid membership of these groups (Feist & McCullough, 1975; Rubenstein,

1982). Dominant bachelors are the most likely to obtain mares (Welsh, 1975; Feist & McCullough, 1976).

No one horse consistently serves as leader (Welsh, 1975, Feist & McCullough, 1976; Berger, 1977). The stallion is frequently last (Pellegrini, 1971; Jaworowska, 1976) except in the context of final approach to a water hole when the leader was typically the stallion (Berger, 1977; Miller, 1979; Wells & Goldschmidt-Rothschild, 1979). The stallion typically positions himself between the mares and danger when they flee (Oom & Santos Reis, 1986).

Interband hierarchies

The home ranges of feral horses often overlap and encounters between bands frequently occur. Mares and juveniles from different bands rarely interact, instead continuing with their activity or watching as the stallions display or fight (Feist & McCullough, 1976; Salter & Hudson, 1982). Stallions typically leave to meet between the bands and display with arched necks, tails held high and prancing motions. Faecal marking and mutual sniffing may occur (Rubenstein, 1981; Salter & Hudson, 1982). Escalations include pushing, biting, striking with a foreleg (Figure 4.4), rearing, kicking and chasing behaviours. Many stallions receive cuts and scrapes (Rubenstein, 1981) (Figure 4.5). Feh (1999) counted scars on two 20-year-old Camargue stallions; one had

Figure 4.4. Fight between Red Desert stallions over water access. (Photo by Lee Boyd.)

Figure 4.5. Scarred Assateague Island stallion. (Photo by Ron Keiper.)

35 scars and the other had 146. Ninety-six per cent of the adult stallions in the Granite Range of Nevada were scarred from bites (Berger, 1986). More seriously injured stallions may lose their harem (Feist & McCullough, 1975) and even their life if their wounds become infected (Keiper, 1986). Three of approximately 35 adult males in the Granite Range population died from injuries related to fighting (Berger, 1986).

Many feral horses live in arid or coastal environments in which fresh water is a limited resource. Bands may compete with one another for access to water (Berger, 1977). Where water shortages are transient, the band currently occupying the water hole typically defeats intruders (Stevens, 1988). In arid environments with persistent shortages, larger bands have priority access to water (Berger, 1977; Miller & Denniston, 1979; but see Pellegrini, 1971). A nearly linear dominance hierarchy exists between bands (Goldschmidt-Rothschild & Tschanz, 1978; Miller & Denniston, 1979). Both stallions and mares participate in the aggression. In multi-male bands, any male or combination of males may be involved, but there is no correlation between a band's rank and the number of adult males it contains (Miller & Denniston, 1979). If only one of the males participates it is often the dominant stallion (Miller & Denniston, 1979).

Figure 4.6. Dung pile on Assateague Island along a roadway. (Photo by Ron Keiper.)

Marking

Stallions use their own urine to mark the urine and faeces of their mares. They also use faecal-marking to create 'dung piles' or 'studpiles'. By the time a juvenile male is three years old these elimination marking behaviours are performed in full as part of his behavioural repertoire (Hoffmann, 1985). Elimination marking behaviour is rarely exhibited by mares or immature horses (Turner *et al.*, 1981).

Marking the eliminations of the mares occurs more frequently during the breeding season than during the rest of the year (Feist & McCullough, 1976; Turner *et al.*, 1981; Hoffmann, 1985; Kimura, 2001) and is closely correlated with seasonal testosterone levels (Turner & Kirkpatrick, 1986). Typically, a stallion will detect an elimination, approach, sniff it and perhaps flehmen, defecate or urinate on top of the elimination, and possibly smell it again before moving away (Welsh, 1973; McCort, 1984). The response is more likely to contain all of these components during the breeding season (Turner *et al.*, 1981). Kimura (2001) found that concentrations of fatty acids were higher in the faeces of oestrous mares than non-oestrous mares. The urine of stallions contains high levels of long-lasting pungent cresols. When stallions use urine to mark the faeces of oestrous mares, the fatty acid concentration decreases to levels found in

the faeces of non-oestrous mares. Thus stallions may mask the difference between the odour of oestrous and non-oestrous faeces by urinating on them. This might reduce the attraction potential of the faeces to rival suitors.

Dung piles, also known as stud piles, (Figure 4.6) are located throughout a stallion's home range (Rubenstein, 1981) in areas devoid of vegetation (Hall & Kirkpatrick, 1975) often along trails (Salter & Hudson, 1982). The greatest density of faecal piles is along trails used by many bands, for example leading to water sources (Feist & McCullough, 1976). Use of stud piles occurs year round (Hoffmann, 1985). The stallion approaches a pile, sniffs it, steps over it, defecates, and then typically turns and sniffs his excretion. No other stallion need be present to elicit faecal marking, but the behaviour is common during encounters between stallions. In the Red Desert of Wyoming, 25% of all aggressive interactions among males involved faecal marking of stud piles, and marking occurred in 69% of all encounters around these piles (Miller, 1981). In Alberta, 62% of all interactions among males involved faecal deposition (Salter & Hudson, 1982). In comparison with urine, faeces contain a more complex mixture of volatile substances and may create a unique chemical signature reflecting an individual's age,

Figure 4.7. Play fight on Assateague Island. (Photo by Ron Keiper.)

gender and reproductive status (Kimura, 2001). Feist & McCullough (1975) reported that stallions who were members of a bachelor group defecated on dung piles in order of their rank, from lowest to most dominant, as was also observed in Camargue bachelors (Goldschmidt-Rothschild & Tschanz, 1978). Stallions in multi-male bands also typically defecate in order of rank (Feist & McCullough, 1976; Miller, 1981).

Alliances or coalitions

In the Camargue, high-ranking stallions held single-male harems, but low-ranking sons of low-ranking mothers formed long-term alliances leading to multi-male bands (Feh, 1999). These alliances formed between similar age males while they were in a bachelor group. Males within an alliance were not maternal half siblings, but might possibly have been paternal half siblings, otherwise they were unrelated. Alliance partners maintained close proximity to one another and frequently allogroomed, although this was not the case in the Kaimanawa population (Linklater & Cameron, 2000). Sixty per cent of the Camargue alliances lasted two to three years, after which the stallions separated permanently to establish single-male bands. The remaining alliances lasted at least 16 years, or up until the death of one of the stallions.

In other populations male alliances also formed prior to harem attainment. A pair of paternal half brothers dispersed together into the bachelor group on Shackleford Banks and achieved much higher status than young males who arrived singly (Rubenstein, 1982). They preferred each other as grooming partners, gained experience through play fights with one another (Figure 4.7) and acted cohesively to drive off other males.

In the Granite Range population, 17 alliances were observed but most were short-lived (Berger, 1986). Only two exceeded seven months, but one of these alliances lasted 2.5 years, and the other at least four years. In the 13 cases where genealogies were known, the stallions were unrelated. One stallion was dominant to the other and performed all of the observed matings in 15 of the alliances.

In the Rachel Carson Estuarine Sanctuary, only the dominant stallion was observed mating in multi-male bands (Stevens, 1990). If there were more than two stallions in Kaimanawa harems, only the two highest ranking males were observed mating (Linklater & Cameron, 2000). However, for multi-male bands in the Red Desert, 49% of the matings were by the dominant stallion, 42% by one of the subordinate stallions and 9% by stallions from other bands (Miller, 1981). Three mating systems were apparent in the Red Desert:

Figure 4.8. Red Desert multi-male band. The white stallion, grey stallion, and the bay stallion behind the grey, co-owned this band for more than three years. (Photo by Lee Boyd.)

- The dominant male did the majority of the breeding.
- The stallions mated the same mare serially during an oestrous period, with little aggression.
- In one harem each of the three males formed a consortship with one of three females who were in oestrus at the same time (Figure 4.8).

If the band flees in alarm, the dominant male leads the females while the subordinate male is at the rear between the mares and danger (Pacheco & Herrara, 1997). The subordinate stallion confronts approaching rivals while the dominant stallion drives the mares away (Miller, 1981; Rubenstein, 1982; Berger, 1986; Feh, 1999; Linklater & Cameron, 2000). This strategy is evidently effective as Linklater and Cameron (2000) observed copulations by males from outside the harem to occur only in single-male bands.

Why participate in alliances? Linklater and Cameron (2000) weigh the evidence and discount explanations involving mutualism or reciprocal altruism. They mention making the best-of-a-bad-job as a hypothesis, but clearly favour what they call the consort hypothesis. Under this hypothesis, multi-male bands result when mares solicit more than one stallion during harem formation, stimulating mate loyalty by more than one stallion. This hypothesis fails to explain why some bachelors form alliances prior to the acquisition of mares. Because it takes a developmental approach, the consort hypothesis is not mutually exclusive of the functional approaches proposed by other authors (Sherman, 1988; Feh, 2001).

Single-male Red Desert bands had higher natality in good years (Boyd, 1980). In the Camargue population, single-harem males have higher reproductive success than dominant stallions in multi-male bands, who in turn have a greater reproductive success than subordinate stallions. The latter have a higher reproductive success than bachelors who sneak copulations (Feh, 1999). Reproductive success was similar for single-male harem-holders and those stallions who switched from an alliance to single male status within a harem after a few years.

Granite Range stallions who hold single-male harems have longer tenure (Berger, 1986), perhaps indicating that they are good competitors from the outset. It would seem that if a stallion is capable of monopolizing mares he should do so. But if a stallion is not a top-notch competitor, an alliance provides a way to obtain mares and achieve matings more quickly than by remaining a bachelor.

The dominant male of an alliance may lose matings to subordinate males but this may be compensated for by the extra females that the males together can hold. There was a significant positive correlation between number of females and number of males in Venezuelan feral horse bands, and a greater number of females bred in larger bands (Pacheco & Herrara, 1997). Both Perkins et al. (1979) and Miller (1981) found that multi-male bands were larger, and in the Red Desert population these bands were more stable (Miller, 1981). Mares in stable bands produce more foals than mares from unstable bands (Berger, 1986; Kaseda *et al.*, 1995). Although Boyd (1980) recorded significantly higher natality in Red Desert single-male bands, the natality of multi-male bands was less affected by a severe winter, perhaps because the larger band size provided warmth and combined effort to dig through snow. The Sable Island population also experiences harsh winters and Welsh (1975) found lower natality and survival in smaller bands, which he attributed in part to less huddling protection and also to the fact that small bands were often newer and less stable.

Conversely, Linklater and Cameron (2000) found no difference between single- and multi-male harems of Kaimanawa feral horses regarding the size of the bands or the number of mares therein. There was also no difference in band stability or quality of home range. Mares in single-male bands were in significantly better condition throughout the year than mares in multi-male bands, and had higher foaling rates and lower rates of offspring mortality (Linklater *et al.*, 1999; Linklater & Cameron, 2000). But single males were sometimes cuckolded by rivals from outside the band whereas multiple males were not (Linklater & Cameron, 2000).

Boyd (1980) also found no difference in foal survival between small and large bands or single- versus multi-male bands in the Red Desert of Wyoming. But in the Camargue population, foal survival was higher in multi-male bands (Feh, 1999).

Subordinate stallions may eventually inherit the band (Feh, 1999), but Berger (1986) observed this to occur only once in the Granite Range horses. However, half of the methods of harem formation reported and described below involve some form of short- or long-term association among stallions, after which one male may ultimately 'inherit' the harem.

Alliances among females have not been widely observed (Monard & Duncan, 1996). However, Berger (1986) described two females who became separated from their original band and once incorporated into a new band they reciprocally defended each other from aggressive approaches by the stallion.

Harem stability

Harem groups are reasonably stable over time (Pellegrini, 1971; Feist & McCullough, 1975; Green

Table 4.4. *Percentage of adult mares voluntarily transferring between bands per year*

Population	Percentage	Reference
Camargue, France	2.0%	Monard *et al.*, 1996
Assateague Island, USA	7.5%	Rutberg, 1990
Granite Range, USA	10.0%	Berger, 1986
Bird Shoal-Carrot Island, USA	27.0%	Stevens, 1990
Cumberland Island, USA	29.6%	Goodloe *et al.*, 2000

& Green, 1977; Salter, 1978; Oom & Santos Reis, 1986), due to:

- herding of mares by the stallion;
- active defence by the stallion;
- mare fidelity to one another and to a lesser extent to the stallion (Salter, 1978; Klingel, 1982). This mare fidelity causes all-mare groups to persist in the absence of a stallion. Group loyalty of resident mares conversely leads to aggressive behaviour toward new arrivals (Goldschmidt-Rothschild & Tschanz, 1978; Rubenstein, 1982; Rutberg, 1990; Rutberg & Greenberg, 1990; Monard & Duncan, 1996).

Adult changes

Red Desert bands experienced an average of 0.75 adult changes per band per year (Miller, 1979). Over the three-year period during which the Kaimanawa horses of New Zealand were studied, 83% of the mares and 88% of the stallions remained in the same band (Linklater *et al.*, 2000). On Assateague Island approximately 20% of the sample mares transferred bands but 12.5% of these transfers were actually the result of take-over by a new harem stallion (Rutberg, 1990). Some populations exhibit higher rates of change (Table 4.4). Most band changes occur during late winter and early spring when nutritional stress is greatest and the horses are spread out while foraging extensively (Miller, 1979; Stevens, 1990; Asa, 1999; Goodloe *et al.*, 2000). The breeding season is also a common time of change (Goldschmidt-Rothschild & Tschanz, 1978; Salter, 1978; Nelson, 1980). Stevens (1990) found no significant differences in the ages of mares who changed bands, nor in the ages of their harem stallions, nor the size of the harems, but all mares who transferred came

Table 4.5. *Age at dispersal (years)* $\overline{X}$ = *mean*

Population	Both sexes	Male age	Female age	Reference
Bird Shoal-Carrot Island, USA	1–2			Hoffmann, 1985
Pryor Mountains, USA		1–3		Hall & Kirkpatrick, 1975
New Forest, UK			2–4	Tyler, 1972
Misaki, Japan		$\overline{X} = 2.00$	$\overline{X} = 1.98$	Kaseda *et al.*, 1997
Granite Range, USA		$\overline{X} = 2.30$	$\overline{X} = 2.10$	Berger & Cunningham, 1987
Cumberland Island, USA		$\overline{X} = 2.10$	$\overline{X} = 1.86$	Goodloe *et al.*, 2000
Assateague Island, USA		$\overline{X} = 1.73$	$\overline{X} = 2.05$	Rutberg & Keiper, 1993
Camargue, France			median = 2.00	Monard *et al.*, 1996

from single-rather than multi-male bands, presumably because the subordinate stallions helped guard and retain the mares. Band stability increased with stallion age and tenure.

In the Peneda-Gerês National Park of Portugal, mares and their foals sometimes left the harem for a few days and rejoined it or another harem later (Oom & Santos Reis, 1986). The sex ratio of this population was strongly skewed toward females and the lack of competition may have caused the stallions to be less assiduous about sequestering mares.

Misaki feral horses are unusual in that although they have very stable harems in summer, in winter some harems break up into smaller units that do not all include a stallion (Kaseda, 1981). Half the mares were in permanent consort relationships with a stallion, lasting up to 10 years, and often terminated only by the death of the mare or stallion. Another 30% left the stallion during the winter but returned to the same harem the next spring. The remaining mares rarely kept stable consort relationships with any stallion and changed partners at one- to two-year intervals. During the winter the horses move from grasslands into forests for shelter which Kaseda felt prevented larger groups from remaining together. In this population the majority of males are gelded so the ratio of stallions to mares ranged from 1:2 to 1:7 (Kaseda, 1981; Kaseda & Khalil, 1996). The gelding of rivals may have reduced male competition to the point that stallions no longer need to keep their harems together outside the breeding season.

Nelson (1980) reported many mare transfers and all-mare groups in revegetated areas of the Jicarilla District of New Mexico. A band in which two stallions alternated dominance lost and gained members (Asa, 1999).

Harem instability is known to affect reproductive rate. Females belonging to unstable bands have significantly lower reproductive success than those from stable bands (Berger *et al.*, 1983). Mares that changed bands on Cumberland Island had a 44% foaling rate as compared to a 71% foaling rate for mares in stable bands (Goodloe *et al.*, 2000).

Stallion tenure in the Granite Range of Nevada averages 3.1 years (Berger, 1986) and on Cumberland Island 2.8 years (Goodloe *et al.*, 2000). In short-term studies these figures are minimum estimates; in the Granite Range 33% of the stallions still held their harems at the end of the five-year study. Stallions on Sable Island (Welsh, 1975) and Assateague Island (Keiper, 1985) had tenures of 10 years or more, and Feh (1999) recorded tenures of up to 18 years in the Camargue population.

Emigration – juvenile dispersal

The majority of changes in harem membership reflect dispersal of immature horses from their natal bands (Feist & McCullough, 1976). Long-term association of the parents in a stable harem necessitates that both sexes of offspring disperse if inbreeding is to be avoided (Monard & Duncan, 1996; Monard *et al.*, 1996). Separation is characterized by active dispersal, but also passive separation may result if the mother dies (Khalil & Kaseda, 1997). Juveniles leave their natal harem at approximately two years of age (Table 4.5), but departure-age ranges from one to five years and a few individuals do not disperse, depending on circumstance (Table 4.6). For example, Monard *et al.* (1996) found that all Camargue fillies dispersed, at a median age of 24 months, with more than half dispersing between 1.5 and 2.5 years (range: 12–62 months).

Table 4.6. *Percentage of juveniles dispersing by age*

Population	Yearlings	2 years	3 years	4 years	5 years	Reference
Misaki						
– both sexes	59%	32%	5%	3%	1% (all)	Kaseda *et al.*, 1984
– males	28%	48%	10%	all		Kaseda *et al.*, 1997
– females	21%	31%		all		
Kaminawa						
– both sexes	–	–	93%			Linklater *et al.*, 2000
Granite Range						
– both sexes	–	–	–	97%		Berger, 1986
Assateague Island						
– males	–	–	–	–	97%	Rutberg & Keiper, 1993
– females	–	–	–	–	81%	

Where a single value appears in a row it is a cumulative percentage at that age with no information available from earlier ages.

Most dispersals in the Misaki population occurred during the winter when the quality and quantity of forage was reduced and some harems disbanded (Khalil & Kaseda, 1997). Dispersal was common during the breeding season in many other populations studied.

The factors affecting dispersal age include:

- Opportunity for continued maternal investment. Most Misaki feral horses whose mothers failed to produce additional offspring delayed their dispersal (Kaseda *et al.*, 1984). However, in the Camargue, dispersal age was not correlated with number of maternal siblings or the birth of the next sibling (Monard *et al.*, 1996). Young Misaki males born to primiparous mothers all departed and returned to their natal harems several times before permanently separating, whereas sons of multiparous mothers departed permanently, although the average age at departure was not significantly different between the two categories (Khalil & Kaseda, 1997).
- Puberty. Dispersal of Camargue fillies typically occurred during oestrus and no females dispersed prior to achieving puberty (Monard *et al.*, 1996). Rutberg and Keiper (1993) reported that Assateague fillies dispersed at the average age of 24.6 months, about the time of onset of first oestrus. Turner and Kirkpatrick (1986) examined 21 two-year-old male mustangs and found that the 19 who had not already dispersed had undescended testicles and low plasma testosterone levels, whereas the two that had already dispersed had descended testicles and testosterone levels typical of adult males.
- Presence of peers in the natal harem. Age of male departure from Assateague Island bands was strongly positively correlated with the number of peers in their natal group, as was also reported for Sable Island horses (Welsh, 1975; Rutberg & Keiper, 1993).
- Opportunity for reproduction. Dispersal age of Camargue fillies declined as the number of available harems for them to join increased (Monard *et al.*, 1996). Five Granite Range fillies did not depart at puberty but mated with males from outside the band, rather than their father (Berger, 1986). Fillies may not disperse if their father is deposed. Misaki juveniles whose mothers only associated with stallions during the breeding season left later than those whose mothers consorted with stallions year-round (Kaseda *et al.*, 1984). Two Assateague males who did not disperse took over some or all of the mares in their father's band (Rutberg & Keiper, 1993).

Monard *et al.* (1996) found that the age of female dispersal was not correlated with the number of mares or aggression by mares or stallions in the natal group, or with juvenile body weight or condition.

Active separation is most commonly voluntary, but involuntary dispersal can occur in response to aggression. Both stallions and mares have occasionally been observed to eject juvenile males from harems in the Pryor Mountains (Hall & Kirkpatrick, 1975).

Stallions have been known to harass sons resulting in their expulsion, but this is not commonplace (Green & Green, 1977; Boyd, 1980; Rubenstein, 1982). Only 11% of Granite Range colts were bitten by their sire in the week before their departure. However, although little overt aggression was seen, if a new stallion took over the band, 40% of the colts departed within two months (Berger, 1986). Khalil & Kaseda (1997) also observed the departure of a three-year-old male immediately after his mother began consorting with a new stallion.

Active expulsion of daughters by stallions is rare (Boyd, 1980; Rubenstein, 1982; Rutberg & Keiper, 1993; Monard *et al.*, 1996) except in a few Polish wild horse harems (Jaworowska, 1981), one case in Alberta (Salter, 1978), and one case in the Camargue (Monard *et al.*, 1996). More often Camargue stallions attempted to prevent daughters from interacting with stallions of other groups, herding them back to their natal harem (Monard *et al.*, 1996). However Sable Island stallions make no effort to keep daughters (Welsh, 1973). Rubenstein (1982) reported intense grazing competition on Shackleford Banks and observed mares driving daughters away and Welsh (1973) also observed maternal aggression toward daughters of dispersal age in the harsh conditions of Sable Island.

Both male and female juveniles may return to their natal bands briefly and are accepted (Tyler, 1972; Welsh, 1975; Rubenstein, 1982).

Departing fillies join other harem groups. No Camargue fillies became solitary; 80% entered existing harems and the others joined bachelors to form new harems (Monard *et al.*, 1996). In the Misaki population, 48% of dispersing two-year-old females joined existing harems with their mothers, 14% joined existing harems on their own, and 38% formed the nucleus of new harems (Kaseda *et al.*, 1995). Fillies may change bands several times (Tyler, 1972; Feist & McCullough, 1975; Salter & Hudson, 1982; Monard *et al.*, 1996) before settling into one harem long term.

Young males enter bachelor groups until they are five to six years of age (Hoffmann, 1985).

Harem formation
Misaki colts whose mothers died became independent at an average of 1.2 years of age and did not begin forming a harem for another 3.9 years (Khalil & Kaseda, 1998). Young males whose mothers formed unstable relationships with stallions departed at an average age of two years and associated with various groups for 1.8 years before starting to form harems. Colts left stable harems at a significantly older age of 3.3 years and were bachelors for an average of 1.7 years. Misaki bachelors typically become solitary for a time immediately before they attempt to acquire females (Khalil & Murakami, 1999a).

Stallions obtain mares in one of the following ways (Miller, 1979; Nelson, 1980; Klingel, 1982):

- Acquiring unguarded females such as fillies who have just dispersed from their natal band, or adult mares who have separated from their harem (for example during parturition), or whose stallion has died.
- Challenging a harem stallion and defeating him in combat, obtaining the entire harem.
- Raiding, resulting in the abduction of part of a harem.
- Attaching themselves peripherally to a harem for some time, perhaps eventually ousting the resident stallion or departing with a fraction of the mares (Welsh, 1975; Salter, 1978; Miller, 1979; Rubenstein, 1982; Khalil & Murakami, 1999b).
- Belonging to a bachelor group that obtains females in one of the ways above, typically first point above. One bachelor, or a coalition of bachelors, eventually oust the other males, or if several females were obtained the group breaks up to form several harems.
- Remaining with his natal band and inheriting it (Miller, 1979; Rutberg & Keiper, 1993).

In the Misaki population with its female-biased sex ratio, about 20% of the bachelors were able to obtain females as early as their fourth year (Kaseda *et al.*, 1984) and another 40% by their fifth year of life (Kaseda *et al.*, 1997). In a later study of the same population, as the sex ratio approached 1:2, 70% of the stallions formed harems at the age of five years (Khalil & Murakami, 1999a). The age range at the time of first acquisition was 3.8–7.6 years with a mean of 5.2 years (Khalil & Murakami, 1999b). Most harems were formed at the beginning of the breeding season. Bachelors who entered the pre-harem phase in the winter had ready access to wandering females and were able to form harems in less than four months. Males who sought to obtain mares after the breeding season had begun were more likely to have to steal them, and spent an average of 1.3 years in this phase before they were successful. In the Granite Range of Nevada, only bachelors that were at least six years old were able to win mares through combat (Berger, 1986).

Forty-five per cent of the Granite Range mares were acquired when they were wandering unguarded and 75% of these were dispersing fillies. Forty-eight per cent of the mares were acquired by combat, and 6% via male alliances (Berger, 1986). About 28% of the mares acquired by Misaki bachelors were dispersing fillies; the rest were older mares who separated from stallions during the winter in this population. Juvenile females dispersing from their natal band prefer to join harems with other subadult females (Monard & Duncan, 1996).

Newly formed harems are often unstable and young stallions may lose their mares after a few weeks.

Reproduction

Reproduction in feral horses is strongly seasonal. In the Northern Hemisphere the breeding season extends from March through to August, peaking in May and June (Tyler, 1972; Feist & McCullough, 1975; Hall & Kirkpatrick, 1975; Welsh, 1975; Salter, 1978; Boyd, 1980; Nelson, 1980; Berger, 1986; Kirkpatrick & Turner, 1986; Lucas et al., 1991; Goodloe et al., 2000). Mares are sexually mature between one and two years of age, and stallions at two to three years of age, although it is uncommon for feral mares under the age of three years to foal and stallions are not behaviourally mature until the age of five or six years (Tyler, 1972; Hall & Kirkpatrick, 1975; Feist & McCullough, 1975; Welsh, 1975; Salter, 1978; Boyd, 1980; Keiper & Houpt, 1984; Hoffmann, 1985; Kirkpatrick & Turner, 1986; Lucas et al., 1991; Kaseda et al., 1995; Monard & Duncan, 1996, Asa, 1999; Stoffel-Willame & Stoffel-Willame, 1999). Most births occur at night or just before dawn (Berger, 1986). It is common for mares to foal in alternate years or in two out of three years (Tyler, 1972; Green & Green, 1977; Welsh, 1975; Salter, 1978; Boyd, 1980; Keiper & Houpt, 1984; Lucas et al., 1991; Cameron & Linklater, 2000). Reproductive lifespan varies from 2 to 21 years, and maximum lifespan in the wild is 18–25 years (Turner & Kirkpatrick, 1986; Kaseda et al., 1995). Three Misaki stallions were observed throughout their reproductive lifespan: two sired 24 and 25 foals during

their 10 and 11 years of tenure, respectively. The third sired just five foals in six years (Kaseda & Khalil, 1996).

Mate choice

Feral mares in Alberta approached and initiated courtship with dominant stallions while rejecting all approaches of subordinate stallions (Salter, 1978). Older experienced mares reject the attention of young inexperienced males by biting and kicking them (Tyler, 1972; Berger, 1986). In Miller's (1981) study of Red Desert horses, when stallions did not achieve intromission it was almost always because of the mare's aggression toward the stallion, suggesting that she was exerting some choice. Mares who had been abducted sometimes attempted to return to their original band for as long as two days (Miller, 1979); whether this indicates a preference for the original stallion or for the other mares is not known. Females may use the ability of the stallion to obtain and retain mares, often by combat, as a measure of the stallion's ability (Rutberg, 1990). Female choice is also not obviated by despotic behaviour of stallions. Rutberg (1990) observed Assateague Island mares moving out of sight of the stallion in areas where dense cover could thwart the stallion's attempt to find them unless the mare cooperated. Mares were also observed alone or accompanied only by immature offspring, implying that they were able to leave harems and remain unincorporated if they so chose. Gates (1979) also reported that Exmoor pony mares sometimes broke away from their harem temporarily, and after a gathering brought harems into close proximity, some mares shifted their allegiance from one stallion to another. During the course of a year, some mares in the Kaimanawa population moved between several bands or away from their own band for several days at a time (Linklater et al., 1999). And a significant portion of the Misaki harems break up in winter and re-form during the breeding season, giving mares the opportunity to change their allegiance.

Harem stallions do not necessarily interfere when daughters who have not yet dispersed are courted and bred by other males (Feist & McCullough, 1976; Boyd, 1980; Berger & Cunningham, 1987; Asa, 1999)(Figure 4.9). Berger & Cunningham (1987) reported that half of dispersing fillies observed to copulate mated with the first male they met, but 33% mated with a subsequent male rather than the first

Figure 4.9. Red Desert yearling not yet dispersed from her natal band is bred by a stallion from another band as the bands aggregate at a water hole. (Photo by Lee Boyd.)

male, and the remaining 17% mated with multiple males.

Inbreeding

Granite Range stallions do not breed daughters with which they are familiar, but familiarity wanes if parent and offspring are separated for more than 18 months, so that 4% of copulations for which the genealogies of the participants were known were between fathers and daughters (Berger & Cunningham, 1987). Polish wild stallions usually do not mate their daughters nor do young males show interest in mounting their mothers (Jaworowska, 1981). Welsh (1975) reported only one case of father–daughter mating among the feral horses of Sable Island, Nova Scotia. Duncan *et al.* (1984b) found no evidence of mother–son matings in Camargue horses but did observe father–daughter and sibling matings although the level of sexual activity was less than expected when these relatives were familiar with one another. Harem stallions were less closely related to mares in their own band than to mares in other bands, suggesting negative assortment with regard to kinship. In this same population, Monard *et al.* (1996) observed young females accepting the attention of males from other groups but rejecting courtship by close kin males from their natal group. Additionally dams interposed

themselves when related males began to court their daughters but not when unrelated males approached. Monard and Duncan (1996) showed that dispersing fillies preferred to move to harems containing familiar females similar to them in age but lacking familiar males. Only 30% of dispersing fillies joined harems headed by a stallion who was a close relative. Usually the stallion was unknown to them. In the few cases where the filly joined a familiar stallion, there were no other harems with unrelated males available.

On Assateague, Rutberg and Keiper (1993) found that 19% of young females remained in their natal harem after sexual maturity and father–daughter matings resulted in live offspring. However, Keiper and Houpt (1984) noted that father–daughter matings resulted in a foaling percentage of only 22.7% compared with 36.8% for half-sibling matings and 61.8% for mating with unrelated individuals.

'Extra-pair' copulation

Half of the observed copulations by harem males in the Jicarilla feral horse population were with females from other bands (Nelson, 1980). In single-male bands in the Red Desert, 14% of the observed copulations were by outside stallions (Miller, 1981). Two or more males bred the same mare during the same oestrus in 2% of the copulations observed by Berger and Cunningham (1987). Seventeen to 33% of mares in bands headed by vasectomized stallions produced foals (Asa, 1999). Eighty per cent of these bands contained a subordinate stallion and sneak breeding by bachelor males was a lesser possibility. Paternity testing confirmed that 14–15% of Misaki foals were not sired by the harem stallion (Kaseda *et al.*, 1982; Kaseda & Khalil, 1996). Ashley (2000) was able to attribute a large amount of reproductive success to subordinate and bachelor stallions in Great Basin feral horses. Blood typing of feral horses in the Great Basin confirmed that approximately one-third of foals were not sired by the harem stallion (Bowling & Touchberry, 1990). In only about half of either the single- or multi-male bands did one stallion sire all of the foals; in only one additional multi-male band were some of the foals sired by the resident subordinate stallion. It is possible that some pregnant mares had changed bands prior to this study, especially as some of the populations sampled had been disturbed by previous round-ups.

Abortion

Berger (1983a) reported that stallions who took over a band forced the mares to copulate even if pregnant, resulting in abortion by most of these mares. Some females who were not forced to copulate still aborted after the take-over. However in this study, pregnancy and abortion were assumed based on copulation date relative to the presence or absence of oestrous behavior. In addition to behavioural observation, Kirkpatrick and Turner (1991a) used urinary and faecal steroid analysis to monitor eight pregnant mares taken over by new stallions on Assateague Island. Rates of stallion aggression toward the new mares were extremely low and seven of the eight mares delivered live foals; the eighth mare experienced faetal loss during pregnancy although no signs of harassment were observed.

Ashley and Holcombe (2001) observed several incidents of harassment of mares by bachelor stallions, but only one forced copulation performed by a harem stallion upon a yearling. A few forced copulations were observed among Sable Island horses (Welsh, 1975) and Tyler (1972) observed some New Forest stallions mounting females aggressively immediately after the stallions were turned loose, having being segregated from the mares during the winter. Goodloe *et al.* (2000) reported two forced copulations on Cumberland Island. Abortion subsequent to forced copulation was not reported in any of these populations.

Post-partum parental care

Mares are very protective of their foals for the first few days after birth, and will threaten more dominant harem members including the stallion (Boyd, 1980). Mares wait for lagging young foals, and approach foals to collect them if danger appears (Boyd, 1980). Only when the band spooks abruptly is the foal temporarily left to keep up as best it can. After the foal is two to three months old, the foal bears most of the responsibility for locating and staying with its mother, although she may answer its calls and look for it if the foal is distressed (Boyd, 1980).

Mares typically nurse their offspring until the birth of their next foal. Many mares wean their foals at approximately 35–40 weeks of age (Jaworowska, 1976; Duncan *et al.*, 1984a; Berger, 1986) but some nurse up until 24 hours before the birth of their next foal (Welsh, 1975). Mares who fail to conceive may continue to nurse their yearlings and a few may allow both their current foal and their yearling to suckle (Tyler, 1972; Feist & McCullough, 1975; Welsh, 1975; Jaworowska, 1976; Boyd, 1980; Duncan *et al.*, 1984a; Oom & Santos Reis, 1986; Roberts, 1988)(Figure 4.10). The timing of weaning is probably influenced not only by the mare's reproductive status but also by her level of nutrition and body condition.

Harem stallions protect offspring and herd them to the group if the foals become separated (Feist & McCullough, 1975; Boyd, 1980). Stallions herd their offspring even after foals are old enough that the dam is no longer waiting for them or retrieving them herself (Boyd, 1980).

Stallions also play with offspring, particularly young males (Stebbins, 1974; Berger, 1986) which may teach important sparring skills. Male foals and yearlings were three times more likely to seek close proximity to the band stallion than female offspring (Boyd, 1980).

Both sires and dams use neighs to contact lost foals (Feist & McCullough, 1976).

Infanticide

A New Forest stallion was observed to maul a foal, and at least five other foals were suspected to have

Figure 4.10. Two year-old female still suckling from her dam in the Red Desert. The dam had no other younger offspring at this time. (Photo by Lee Boyd.)

been killed by several different stallions (Tyler, 1972). In the New Forest, most stallions were removed for the winter and new males added yearly, therefore they were unlikely to have sired the foals they killed. Camargue stallions newly introduced to a harem of females may kill male foals younger than six weeks of age that were sired by the preceding stallion (Duncan, 1982). In primates and lions, males may benefit by killing unrelated juveniles as their mothers then come into oestrus sooner. However, this benefit is less relevant to horses, for whom lactation does not necessarily inhibit oestrus. However, one potential benefit of infanticide to stallions is freeing mares from the stress of lactation which may have significant effects on body condition and foaling rate. Free-ranging and feral lactating mares have lower body condition scores than non-lactating mares (Pollock, 1980; Rudman & Keiper, 1991). New Forest pony mares with foals were twice as often in poor condition as expected by chance (Pollock, 1980). However, when their foals were removed, the deterioration of the dam's body condition slowed or improved immediately. On the southern portion of Assateague Island where many foals are removed yearly, the foaling rate is much higher than for mares on the northern portion of the island whose foals are not removed (Keiper & Houpt, 1984; Kirkpatrick & Turner, 1991b.

Infanticide has not been reported in the Granite Range (Berger, 1983a) nor in Namib feral horses (Stoffel-Willame & Stoffel-Willame, 1999). Kirkpatrick and Turner (1991a) report no record of infanticide in six years of observation on Assateague Island. However, when an eastern equine encephalitis outbreak orphaned foals, all three fillies were allowed to join other harems, but only one of four orphaned colts was accepted; harem stallions reacted aggressively toward the colts' attempts to join. Berger (1986) also observed males to be aggressive toward non-resident colts but not toward fillies. Stallions may be more likely to kill or reject juvenile males, who would grow up to be competitors, whereas unrelated juvenile females represent future mates (Duncan, 1982).

Predation

Predation, especially by brown hyenas, is a common source of mortality for young feral Namib horses (Stoffel-Willame & Stoffel-Willame, 1999).

Mountain lions killed an average of one feral horse per year in the Great Basin of Nevada (Berger,

1983b). Turner *et al.* (1992) found no evidence of mountain lion predation on adult feral horses in the Montgomery Pass Wild Horse Territory along the California–Nevada border, but 82% of foal carcasses discovered during the summer showed evidence of having been killed by mountain lions. During this season, 62% of mountain lion scats contained horse hair. The foals were less than six months old when attacked and most were younger than three months.

Feral horses are vigilant against coyotes when newborn foals are present, but ignore them when the foals are older. Dams, sires and siblings approach and/or drive off coyotes. Coyotes do not seem to be a threat to healthy foals, but in one case a sick foal that had been abandoned by its harem was killed by coyotes (Berger & Rudman, 1985). Coyotes elicited little reaction from feral horses in Alberta, Canada but the appearance of a single wolf caused a harem to bunch up and move a short distance away before they resumed grazing (Salter, 1978). Predation by wolves and mountain lions may be a significant source of mortality during severe winters in this population (Salter & Hudson, 1978). Wolves killed several foals that were less than three months old in the Peneda-Gerês National Park of Portugal (Oom & Santos Reis, 1986). Attacks on adults were rare and usually aimed at ill or disabled animals. Isolated horses were particularly targeted. When in a group the adults put themselves in a circle around the juveniles with their rumps outward and defended themselves by kicking. Winter was the time of most attacks as the wolves moved down the mountain to the lower elevations where the horses sheltered. Feral dogs were also a possible predator in this region of Portugal.

Effects of management on behaviour

Feral horse populations in the USA have increased greatly in number following federal protection in 1971. In order to control populations, horses are periodically gathered and placed in captivity. A variety of contraceptive methods are being explored as alternative or adjunct methods to manage population growth. All forms of management have the potential to significantly affect behaviour.

Immobilization of horses in the field may result in difficulty re-uniting the individual with its band if the band moves out of sight or if another male seizes the opportunity to pick up a harem member (Berger *et al.*, 1983).

In order to track vasectomized stallions, Eagle *et al.* (1993) outfitted them with radiocollars. When stallions fought, the radiocollars became the focus of attacks, leading to loss or extensive damage of the collars. Eagle *et al.* (1993) recommend freeze brands or implantable radiotransmitters as an alternative.

Kirkpatrick *et al.* (1979) studied blood levels of corticosteroids in 137 captured feral horses as a measure of stress. Each band was housed in a separate corral with hay and water for a few days before being roped and restrained. Roping of horses had no significant effect on corticosteroid levels with the exception of one stallion, and the range and mean levels were comparable to those of unstressed domestic horses.

Hansen and Mosley (2000) compared the behaviour of a control group of mustangs with those chased by helicopter but not captured three weeks before and three weeks after the simulated roundup. They found no difference in behaviour. However, none of the horses in their study spent any time in corrals. Horses to be adopted are penned with other bands and forced into closer contact than would be true on the range. Interband aggression is seen and bands released after being corralled may be permanently disrupted (Ashley & Holcombe, 2001). Hansen and Mosley (2000) found no difference in the foaling rate of mares who were destined for adoption, chased in a simulated roundup, or part of the ungathered control group. However, in a study by Ashley and Holcombe (2001), ungathered mares had higher reproductive success than those who were released within a few days or those who were permanently removed. A positive correlation existed between duration of the stress of capture and confinement and subsequent reproductive failure (Ashley, 2000). Abortions are common in mares awaiting adoption (Boyd, 1980) and Ashley and Holcombe (2001) speculate that the proximate reason may be a decline in progesterone levels due to acute stress. Selective removal of juveniles for adoption results in skewed age structures, with greater numbers of adult mares per band (Keiper, 1980).

Three methods of contraception: hormone injections, immunocontraception and vasectomy, have been examined. The hormonal forms of contraception prevent ovulation and therefore associated oestrous, breeding and mare–foal behaviours (Garrott, 1995). If young mares do not have the opportunity to experience these behaviours, can they competently perform them at an older age once contraception has ended?

Porcine zona pellucida (PZP) is an effective immunocontraceptive for feral horses and is reversible if used for less than three consecutive years (Kirkpatrick *et al.*, 1992). Powell (1999, 2000) found no difference in time budget, aggression given or received, social rank, or spatial relationship to the stallion, between untreated mares and those treated with PZP, although the study may have been confounded because all but one of the untreated mares had been receiving PZP until a year before the study. This contraceptive blocks fertilization resulting in extended breeding seasons which create many mating opportunities. Long-term research is needed to ascertain whether mares who fail to become pregnant as a result of contraception seek other mates, thereby disrupting band stability and reducing male tenure time (Garrott, 1995; Powell, 1999). Powell (2000) found that foals born immediately prior to contraception remained in their natal band longer and continued to suckle infrequently for up to four years. This is not atypical in unmanaged populations when mares fail to foal, but the percentage of affected mares is higher here and prolonged nursing might ultimately affect the mare's body condition and survival rate. Nulliparous mares contracepted with PZP have no opportunity to develop maternal behaviour until contraception ends.

Vasectomy of stallions is another means of contraception. In one study, 78% of vasectomized males retained harem stallion status over the next two years (Asa, 1999), but subordinate stallions attached themselves to some harems, attracted by the prolonged breeding season created when mares did not conceive. This extended season could cause mares to conceive and foal late, and both the lactating mare and the foal may not have enough time to store adequate body reserves to carry them through the winter (Garrott, 1995). Small foals may be more likely to become bogged down in snow drifts, which can be a significant cause of mortality (Berger, 1983c).

Conclusion

It is evident that feral horses are extremely adaptable to a variety of habitats and conditions. As long-term studies and studies of additional populations have added to our knowledge, a clearer picture is emerging of the causes underlying the development of territoriality, multi-male bands, and all-mare bands in some populations. Additionally, these feral horse studies provide useful information for the management of endangered equids and their domesticated cousins.

References

Asa, C. S. (1999). Male reproductive success in free-ranging feral horses. *Behavioral Ecology and Sociobiology*, **47**: 89–93.

Ashley, M. C. (2000). Feral horses in the desert: Population genetics, demography, mating, and management. Ph.D. Thesis, University of Nevada, Reno.

Ashley, M. C. & Holcombe, D. W. (2001). Effect of stress induced by gathers and removals on reproductive success of feral horses. *Wildlife Society Bulletin*, **29**: 248–54.

Berger, J. (1977). Organizational systems and dominance in feral horses in the Grand Canyon. *Behavioral Ecology and Sociobiology*, **2**: 131–46.

(1983a). Induced abortion and social factors in wild horses. *Nature*, **303**: 59–61.

(1983b). Predation, sex ratios, and male competition in equids (Mammalia: Perissodactyla). *Journal of Zoology (London)*, **201**: 205–16.

(1983c). Ecology and catastrophic mortality in wild horses: Implications for sociality in fossil assemblages. *Science*, **220**: 1403–4.

(1986). *Wild Horses of the Great Basin: Social competition and population size*. Chicago: University of Chicago Press.

Berger, J. & Cunningham, C. (1987). Influence of familiarity on frequency of inbreeding in wild horses. *Evolution*, **41**: 229–31.

Berger, J. & Rudman, R. (1985). Predation and interactions between coyotes and feral horse foals. *Journal of Mammalogy*, **66**: 401–2.

Berger, J., Kock, M., Cunningham, C. & Dodson, N. (1983). Chemical restraint of wild horses: effects on reproduction and social structure. *Journal of Wildlife Diseases*, **19**: 265–8.

Bowling, A. T. & Touchberry, R. W. (1990). Parentage of Great Basin feral horses. *Journal of Wildlife Management*, **54**: 424–9.

Boy, V. & Duncan, P. (1979). Time-budgets of Camargue horses. I. Developmental changes in the time-budgets of foals. *Behaviour*, **71**: 187–202.

Boyd, L. E. (1980). The natality, foal survivorship and mare–foal behavior of feral horses in Wyoming's Red Desert. M.S. Thesis, University of Wyoming, Laramie.

Cameron, E. Z. & Linklater, W. L. (2000). Individual mares bias investment in sons and daughters in relation to their condition. *Animal Behaviour*, **60**: 359–67.

Duncan, P. (1980). Time budgets of Camargue horses. II Time-budgets of adult horses and weaned sub-adults. *Behaviour*, **72**: 26–49.

(1982). Foal killing by stallions. *Applied Animal Ethology*, **8**: 567–70.

(1983). Determinants of the use of habitat by horses in a Mediterranean wetland. *Journal of Animal Ecology*, 52: 93–109.

(1992). *Horses and Grasses*. New York: Springer-Verlag.

Duncan, P. & Cowtan, P. (1980). An unusual choice of habitat helps Camargue horses to avoid blood-sucking horse-flies. *Biology of Behaviour*, 5: 55–60.

Duncan, P. & Vigne, N. (1979). The effect of group size in horses on the rates of attack by blood-sucking flies. *Animal Behaviour*, 27: 623–5.

Duncan, P., Feh, C., Gleize, J. C., Malkas, P. & Scott, A. M. (1984b). Reduction of inbreeding in a natural herd of horses. *Animal Behaviour*, 32: 520–7.

Duncan, P., Harvey, P. H. & Wells, S. M. (1984a). On lactation and associated behaviour in a natural herd of horses. *Animal Behaviour*, 32: 255–63.

Eagle, T. C., Asa, C. S., Garrott, R. A., Plotka, E. D., Siniff, D. B. & Tester, J. R. (1993). Efficacy of dominant male sterilization to reduce reproduction in feral horses. *Wildlife Society Bulletin*, 21: 116–21.

Feh, C. (1990). Long-term paternity data in relation to different aspects of rank for Camargue stallions, *Equus caballus*. *Animal Behaviour*, 40: 995–6.

(1999). Alliances and reproductive success in Camargue stallions. *Animal Behaviour*, 57: 705–13.

(2001). Alliances between stallions are more than just multimale groups: reply to Linklater and Cameron. *Animal Behaviour*, 61: F27–30.

Feist, J. D. & McCullough, D. R. (1975). Reproduction in feral horses. *Journal of Reproduction and Fertility*, 23 (Suppl.): 13–18.

(1976). Behavior patterns and communication in feral horses. *Zeitschrift für Tierpsychologie*, 41: 337–71.

Garrott, R. A. (1995). Effective management of free-ranging ungulate populations using contraception. *Wildlife Society Bulletin*, 23: 445–52.

Gates, S. (1979). A study of the home ranges of free-ranging Exmoor ponies. *Mammalian Review*, 9: 3–18.

Goldschmidt-Rothschild, B. von & Tschanz, B. (1978). Soziale Organisation und Verhalten einer Jungtierherde beim Camargue-Pferd. *Zeitschrift für Tierpsychologie*, 46: 372–400.

Goodloe, R. B., Warren, R. J., Osborn, D. A. & Hall, C. (2000). Population characteristics of feral horses on Cumberland Island, Georgia and their management implications. *Journal of Wildlife Management*, 64: 114–21.

Green, N. F. & Green, H. D. (1977). The wild horse population of Stone Cabin Valley, Nevada: A preliminary report. *Proceedings of the National Wild Horse Forum*, 1: 59–65. Reno: Cooperative Extension Service, University of Nevada.

Hall, R. & Kirkpatrick, J. E. (1975). *Biology of the Pryor Mountain Wild Horse*. US Dept. of the Interior, Bureau of Land Management, Salt Lake City, Utah. (Unpublished paper.)

Hansen, K. V. & Mosley, J. C. (2000). Effects of roundups on behavior and reproduction of feral horses. *Journal of Range Management*, 53: 479–82.

Hoffmann, R. (1985). On the development of social behaviour in immature males of a feral horse population (*Equus przewalskii* f. caballus). *Zeitschrift für Säugetierkunde*, 50: 302–14.

Houpt, K. A. & Keiper, R. (1982). The position of the stallion in the equine dominance hierarchy of feral and domestic ponies. *Journal of Animal Science*, 54: 945–50.

Jaworowska, M. (1976). Verhaltens beobachtungen an primitiven polnischen Pferden, die in einem polnischen Wald-Schutzgebiet – in Freiheit lebend – erhalten werden. *Säugetierkundliche Mitteilung*, 24: 241–68.

(1981). Die Fortpflanzung primitiver polnischer Pferde, die frei in Waldschutzgebiet leben. *Säugetierkundliche Mitteilung*, 29: 46–71.

Khalil, A. M. & Kaseda, Y. (1997). Behavioral patterns and proximate reason of young male separation in Misaki feral horses. *Applied Animal Behaviour Science*, 54: 281–9.

(1998). Early experience affects developmental behaviour and timing of harem formation in Misaki feral horses. *Applied Animal Behaviour Science*, 59: 253–63.

Kahlil, A. M. & Murakami, N. (1999a). Effect of natal dispersal on the reproductive strategies of the young Misaki feral stallions. *Applied Animal Behaviour Science*, 62: 281–91.

(1999b). Factors affecting the harem formation process by young Misaki feral stallions. *Journal of Veterinary Medical Science*, 61(6): 667–71.

Kaseda, Y. (1981). The structure of the groups of Misaki horses in Toi Cape. *Japanese Journal of Zootechnical Science*, 52: 227–35.

(1982). Seasonal changes in the home range and the size of harem groups of Misaki horses. *Japanese Journal of Zootechnical Science*, 54: 254–62.

(1983). Seasonal changes in time spent grazing and resting in Misaki horses. *Japanese Journal of Zootechnical Science*, 54: 464–9.

Kaseda, Y. & Khalil, A. M. (1996). Harem size and reproductive success of stallions in Misaki feral horses. *Applied Animal Behaviour Science*, 47: 163–73.

Kaseda, Y., Kahlil, A. M. & Ogawa, H. (1995). Harem stability and reproductive success of Misaki feral mares. *Equine Veterinary Journal*, 27(5): 368–72.

Kaseda, Y., Nozawa, K. & Mogi, K. (1982). Sire-foal relationships between harem stallions and foals in Misaki horses. *Japanese Journal of Zootechnical Science*, 53: 822–30.

(1984). Separation and independence of offsprings from the harem groups in Misaki horses. *Japanese Journal of Zootechnical Science*, 55: 852–7.

Kaseda, Y., Ogawa, H. & Khalil, A. M. (1997). Causes of natal dispersal and emigration and their effects on harem formation in Misaki feral horses. *Equine Veterinary Journal*, 29: 262–6.

Keiper, R. (1976). Social organization of feral ponies. *Proceedings of the Pennsylvania Academy of Science*, 50: 69–70.

(1980). Effect of management on the behavior of feral Assateague Island ponies. *Proceedings of the Conference on Scientific Research in the National Parks*, 8: 382–93.

(1985). *The Assateague Ponies*. Cambridge, MD: Tidewater Press.

(1986). Social Structure. In *Behavior,* ed. S. L. Crowell-Davis & K. A. Houpt, *Veterinary Clinics of North America: Equine Practice*, vol. 2, no. 3. Philadelphia: W. B. Saunders Company, pp. 465–84.

Keiper, R. & Berger J. (1982). Refuge-seeking and pest avoidance by feral horses in desert and island environments. *Applied Animal Ethology*, 9: 111–20.

Keiper, R. & Houpt K. (1984). Reproduction in feral horses: an eight-year study. *American Journal of Veterinary Research*, 45: 991–5.

Keiper, R. R. & Keenan, M. A. (1980). Nocturnal activity patterns of feral ponies. *Journal of Mammalogy*, 61: 116–18.

Keiper, R. R. & Sambraus, H. H. (1986). The stability of equine dominance hierarchies and the effects of kinship, proximity and foaling status on hierarchy rank. *Applied Animal Behaviour Science*, 16: 121–30.

Keiper, R. R., Moss, M. B. & Zervanos, S. M. (1980). Daily and seasonal patterns of feral ponies on Assateague Island. *Conference on Scientific Research in National Parks*, 2(8): 369–81.

Kimura, R. (2001). Volatile substances in feces, urine and urine-marked feces of feral horses. *Canadian Journal of Animal Science*, 81: 411–20.

Kirkpatrick, J. F. & Turner, J. W. Jr. (1986). Comparative reproductive biology of North American feral horses. *Journal of Equine Veterinary Science*, 6: 224–30.

(1991a). Changes in herd stallions among feral horse bands and the absence of forced copulation and induced abortion. *Behavioral Ecology and Sociobiology*, 29: 217–19.

(1991b). Compensatory reproduction in feral horses. *Journal of Wildlife Management*, 55: 649–52.

Kirkpatrick, J. F., Baker, C. B., Turner, J. W. Jr., Kenney, R. M. & Ganjam, V. (1979). Plasma corticosteroids as an index of stress in captive feral horses. *Journal of Wildlife Management*, 43: 801–4.

Kirkpatrick, J. F., Liu, I. K. M., Turner, J. W. Jr., Naugle, R. & Keiper, R. (1992). Long-term effects of porcine zonae pellucidae immunization on ovarian function of feral horses (*Equus caballus*). *Journal of Reproduction and Fertility*, 94: 437–44.

Klingel, H. (1975). Social organization and reproduction in equids. *Journal of Reproduction and Fertility*, 23 (Suppl.): 7–11.

(1982). Social organization of feral horses. *Journal of Reproduction and Fertility*, 32 (Suppl.): 89–95.

Linklater, W. L. & Cameron, E. Z. (2000). Tests for cooperative behaviour between stallions. *Animal Behaviour*, 60: 731–43.

Linklater, W. L., Cameron, E. Z., Minot, E. O. & Stafford, K. J. (1999). Stallion harassment and the mating system of horses. *Animal Behaviour*, 58: 295–306.

Linklater, W. L., Cameron, E. Z., Stafford, K. J. & Veltman, C. J. (2000). Social and spatial structure and range use by Kaimanawa wild horses (*Equus caballus*: Equidae). *New Zealand Journal of Ecology*, 24: 139–52.

Lucas, Z., Raeside, J. I. & Betteridge, K. J. (1991). Non-invasive assessment of the incidences of pregnancy and pregnancy loss in the feral horses of Sable Island. *Journal of Reproduction and Fertility*, 44 (Suppl.): 479–88.

McCort, W. D. (1984). Behavior of feral horses and ponies. *Journal of Animal Science*, 58(2): 493–9.

Miller, R. (1979). Band organization and stability in Red Desert feral horses. In *Symposium on the Ecology and Behavior of Wild and Feral Equids*, ed. R. H. Denniston. Laramie: University of Wyoming, pp. 113–28.

(1981). Male aggression, dominance, and breeding behavior in Red Desert feral horses. *Zeitschrift für Tierpsychologie*, 57: 340–51.

(1983). Seasonal movements and home ranges of feral horse bands in Wyoming's Red Desert. *Journal of Range Management*, 36: 199–201.

Miller, R. & Denniston, R. H. (1979). Interband dominance in feral horses. *Zeitschrift für Tierpsychologie*, 51: 41–7.

Monard, A.-M. & Duncan, P. (1996). Consequences of natal dispersal in female horses. *Animal Behaviour*, 52: 565–79.

Monard, A.-M., Duncan, P. & Boy, V. (1996). The proximate mechanisms of natal dispersal in female horses. *Behaviour*, 133: 1095–124.

Nelson, K. J. (1980). Sterilization of dominant males will not limit feral horse populations. *USDA Forest Service Research Paper RM-226.*

Oom, M. do Mar & Santos Reis, M. (1986). Ecology, social organization and behaviour of the feral horses in the Peneda-Gerês National Park. *Arquivos do Museu Bocage,* 3(10): 169–95.

Pacheco, M. A. & Herrera, E. A. (1997). Social structure of feral horses in the llanos of Venezuela. *Journal of Mammalogy*, 78: 15–22.

Pellegrini, S. W. (1971). Home range, territoriality, and movement patterns of wild horses in the Wassuk Range of Western Nevada. M.S. Thesis, University of Nevada, Reno.

Perkins, A., Gevers, E., Turner, J. W. Jr. & Kirkpatrick, J. F. (1979). Age characteristics of feral horses in Montana. In *Symposium on the Ecology and Behavior of Wild and Feral Equids*, ed. R. H. Denniston. Laramie: University of Wyoming, pp. 51–55.

Pollock, J. I. (1980). *Behavioural Ecology and Body Condition Changes in New Forest Ponies*. Horsham: RSPCA Scientific Publications.

Powell, D. M. (1999). Preliminary evaluation of porcine zona pellucida (PZP) immunocontraception for behavioral effects in feral horses (Equus caballus). *Journal of Applied Animal Welfare Science*, 2(4): 321–35.

(2000). Evaluation of effects of contraceptive population control on behavior and the role of social dominance in female feral horses, *Equus caballus*. Ph.D. Thesis, University of Maryland, College Park.

Roberts, L. M. (1988). Sexual differences in ontogeny of social behavior and maternal investment in feral horse foals (*Equus caballus*). M.S. Thesis, Humboldt State University, California.

Rubenstein, D. I. (1981). Behavioral ecology of island feral horses. *Equine Veterinary Journal*, 13: 27–34.

(1982). Reproductive value and behavioral strategies: coming of age in monkeys and horses. In *Perspectives in Ethology*, vol. 5, ed. P. P. G. Bateson & P. H. Klopfer. New York: Plenum Press, pp. 469–87.

Rudman, R. & Keiper, R. R. (1991). The body condition of feral ponies on Assateague island. *Equine Veterinary Journal*, 23(6): 453–6.

Rutberg, A. T. (1987). Horse fly harassment and the social behavior of feral ponies. *Ethology*, 75: 145–54.

(1990). Inter-group transfer in Assateague pony mares. *Animal Behaviour*, 40: 945–52.

Rutberg, A. T. & Greenberg, S. A. (1990). Dominance, aggression frequencies and modes of aggressive competition in feral pony mares. *Animal Behaviour*, 40: 322–31.

Rutberg, A. T. & Keiper, R. R. (1993). Proximate causes of natal dispersal in feral ponies: some sex differences. *Animal Behaviour*, 46: 969–75.

Salter, R. E. (1978). Ecology of feral horses in western Alberta. M.S. Thesis, University of Alberta, Edmonton.

Salter, R. E. & Hudson, R. J. (1978). Distribution and management of feral horses in western Canada. *Rangeman's Journal*, 5(6): 190–2.

(1979). Feeding ecology of feral horses in western Alberta. *Journal of Range Management*, 32: 221–5.

(1982). Social organization of feral horses in western Canada. *Applied Animal Ethology*, 8: 207–23.

Salter, R. E. & Pluth, D. J. (1980). Determinants of mineral lick utilization by feral horses. *Northwest Science*, 54: 109–18.

Scorolli, A. (1999). Demography and home ranges of a feral horse population in Ernesto Tornquist State Park. M.S. Thesis, National University of the South, Bahia Blanca, Argentina.

Sherman, P. W. (1988). The levels of analysis. *Animal Behaviour*, 36: 616–19.

Stebbins, M. C. (1974). Social organization in free-ranging Appaloosa horses. M.S. Thesis, Idaho State University, Pocatello.

Stevens, E. F. (1988). Contests between bands of feral horses for access to fresh water: the resident wins. *Animal Behaviour*, 36: 1851–3.

(1990). Instability of harems of feral horses in relation to season and presence of subordinate stallions. *Behaviour*, 112: 149–61.

Stoffel-Willame, M. & Stoffel-Willame, Y. (1999). Horses of the Namib. *Africa, Environment & Wildlife*, 7(1): 58–67.

Turner, J. W. Jr. & Kirkpatrick, J. F. (1986). Hormones and reproduction in feral horses. *Journal of Equine Veterinary Science*, 6: 250–8.

Turner, J. W. Jr., Perkins, A. & Kirkpatrick, J. F. (1981). Elimination marking behavior in feral horses. *Canadian Journal of Zoology*, 59: 1561–6.

Turner, J. W., Jr., Wolfe, M. L. & Kirkpatrick, J. F. (1992). Seasonal mountain lion predation on a feral horse population. *Canadian Journal of Zoology*, 70: 929–34.

Tyler, S. J. (1972). The behaviour and social organization of the New Forest ponies. *Animal Behaviour Monographs*, 5: 85–196.

Wells, S. M. & von Goldschmidt-Rothschild, B. (1979). Social behaviour and relationships in a herd of Camargue horses. *Zeitschrift für Tierpsychologie*, 49: 363–80.

Welsh, D. A. (1973). The life of Sable Island's wild horses. *Nature Canada (Ottawa)*, 2(2): 7–14.

(1975). Population, behavioural and grazing ecology of the horses of Sable Island, Nova Scotia. Ph.D. Dissertation, Dalhousie Univ., Halifax, Nova Scotia.

Zervanos, S. & Keiper, R. (1979). Seasonal home ranges and activity patterns of feral Assateague Island ponies. In *Symposium on the Ecology and Behavior of Wild and Feral Equids*, ed. R. H. Denniston. Laramie: University of Wyoming, pp. 3–14.

5

Relationships and communication in socially natural horse herds

Claudia Feh

Introduction

Horses are quite unique among ungulates, and in fact among mammals, in that stallions and mares maintain long-term bonds. They have developed a complex social and communication system based on close relationships with a few partners who, once adult, remain with each other for many years. They have an elaborate and sophisticated parental care system. Mothers, fathers, siblings and the peer group all play an important role in the social development of foals of both sexes until they leave their family band at about puberty or later. Horses and plains zebras are rare among the mammal species in that they have long-term co-operative alliances between unrelated individuals.

Observation methods

This chapter is mainly based on the few studies of feral horses where the populations had a more or less equal sex ratio, and where individual social behaviour

was followed quantitatively over many years (mustangs: Granite Range (Berger, 1986), Red Desert (Boyd, 1980; Miller, 1979) and Pryor mountains (Feist & McCullough, 1976) in the USA; Assateague ponies (Keiper & Houpt, 1984; Keiper & Sambraus, 1986; Keiper & Rutberg, 1993); Shackleford ponies (Rubenstein, 1986; Rubenstein & Hack, 1992); and Camargue horses (Wells & von Goldschmidt, 1979; Duncan *et al.*, 1984; Feh, 1987, 1990; Duncan, 1992; Monard, 1992; Feh & Mazières, 1993; Monard *et al.*, 1996; Monard & Duncan, 1998). Most of the studies lasted for 10 to 20 years, covering the natural lifespan of free-living horses. Human intervention was kept to a minimum: for example no supplementary feeding, no parasite treatment, no reproductive control and no disruption of the natural social structure except for the removal of young animals of dispersal age or after weaning took place. When populations were subject to some human management, some individuals were also removed for humane reasons, because they were sick or in a bad nutritional

The Domestic Horse: The Origins, Development, and Management of its Behaviour, ed. D. S. Mills & S. M. McDonnell.
Cambridge University Press. © Cambridge University Press 2005.

state. Results from an ongoing study, started in 1993, of the last wild horse population in the world – the Przewalski's horse *Equus ferus przewalski* – are included. Even if Przewalski's horses are probably not the ancestors of all domestic horses (Vila *et al.*, 2001; Jansen *et al.*, 2002) the parallel between their social behaviour and the feral domestic populations seems important. A paradox of our times, feral horses exist all over the world, whereas the last wild horses only survived in zoos until 10 years ago, when efforts to re-introduce them to their historic range in Mongolia started.

Certain topics of social organization and communication addressed in this chapter are also discussed by Boyd and Keiper in Chapter 4, who include a discussion of semi-free-ranging domestic horses. See also McDonnell in Chapter 7, for details on sexual behaviour.

Social organization of horses and other equids
Male–female bonds in mammals and their consequences on the social system

Stallions and mares maintain stable long-term relationships in contrast to most mammals, where sexes segregate and maintain bonds only during the breeding season (Clutton-Brock, 1989). Some canids, a few rodents and primate species such as gorillas, hamadryas baboons and red howler monkeys are the exception, where the same males stay with the same females all year round and over many breeding seasons. Typically, both sexes disperse at puberty in these species. In horses, it was shown that the causes for female dispersal was incest avoidance and not intraspecific competition (Monard *et al.*, 1996). As a rule, this is confirmed for mammal species where tenure length by males exceeds the age at first reproduction in females (Clutton-Brock, 1989). When horses are allowed to choose their mating partner freely, the inbreeding coefficient of the offspring is lower than expected should they mate randomly (Duncan *et al.*, 1984).

Equid social systems

Two broad types of social systems have been defined in the seven extant equid species, but the plasticity of equids seems to be underestimated and needs re-evaluation.

Feral horses (Feist & McCullough, 1976; Miller, 1979; Berger, 1986; Keiper & Sambraus, 1986;

Rubenstein, 1986; Duncan, 1992), plains zebras *Equus burchelli* (Klingel, 1972) and mountain zebras *Equus zebra* (Penzhorn, 1984) typically live in small and permanent family groups, often called bands, consisting of one adult stallion, one to three mares and their common offspring, following the classical definition for families where young animals stay in their parents' group beyond physical dependence. Przewalski's horses *Equus ferus przewalski* adopt the same group type (C. Feh, unpub. data). Reproductive groups with more than one adult stallion were observed by most authors studying feral horses or plains zebras (Berger, 1986; Miller, 1979; Klingel, 1972). This also occurs among Przewalski horses. In three populations where detailed data on individual relationships were available, it was shown that some unrelated, mostly low-ranking, stallions form alliances based on cooperative coalitions against their rivals (Berger, 1986; Schilder, 1990; Feh, 1999, 2001). In all family-forming equids, young or sometimes old stallions who have lost their mares to a competitor join bachelor bands.

Moehlman (1998) summarized the detailed studies carried out on feral and wild donkeys *Equus asinus* and Ginsberg (1987) on Grévy's zebra *Equus grévyi*. In both species, males usually defend breeding territories, and lasting bonds are found between females and their recent offspring only. Less is known about the social groups in the two remaining species – the Kiang and the Asian asses. Kiang *Equus kiang* seem to live in a variety of group types: all-male groups, mixed-sex groups of various numbers, all-female groups and solitary, and probably territorial males (Schaller, 1998). Group types also seem to vary widely across different populations of Asian asses *Equus hemionus*, the most flexible equid species with regard to its social organization. In Indian khur *Equus hemionus khur*, males defend breeding territories and only females and offspring associate over longer time periods (Shah, 1993). Persion onagers *Equus hemionus onager* show mixed-sex and all-male groups (Tatin *et al.*, 2003), whereas Turkmenian khulan *Equus hemionus kulan* (Raschek, 1973; Solomatin, 1973) and Mongolian khulans *Equus hemionus hemionus/luteus* in the Gobi B National Park live in family and all-male groups (Zhirnov & Ilinski, 1986; Feh *et al.*, 2001).

Herds, where several reproductive units associate, engage in the same activities and move around together, sometimes temporarily, sometimes on a

Table 5.1. *Group types and composition in unmanaged populations with equal sex-ratio (median numbers)*

	Species or subspecies					
	Family	Alliances	All-male	Solitary males	All-female	Herds
Feral horses Przewalski's horses Plains zebras Mountain zebras	1 stallion, 1–3 mares and their offspring	2 stallions (co-operation) 1–3 mares	Bachelors: young, before reproduction or old, after take over	Old or wounded	Not observed	Fission–fusion herds consisting of several family groups, sometimes up to several hundred individuals
Mongolian khulan Turkmenian khulan Persian onager Indian khur		Not observed		?	Groups of various size, 3–20 mares, sometimes with foals relatedness?	
Kiang	Not observed			Dominant stallions, territorial	Small groups, 1–3 mares matrilines	Not observed
Grévy's zebra African wild ass						

permanent basis, are a common feature of equid societies. Temporary herds consisting of several hundred individuals were observed in plains zebras (Klingel, 1972) and kiangs (Schaller, 1998), and up to 1200 Mongolian khulans, composed mainly of family groups with young foals, roaming the Gobi together on numerous occasions have been observed (C. Feh, unpub. data). Permanent herds with up to 12 family groups exist in both Camargue (Duncan, 1992) and Przewalski's horses (C. Feh, unpub. data). The families spread out during winter and gather closely together when the frequency of insects increases in summer. This is in contrast with both the kiang (Schaller, 1998) and khulan populations (Feh *et al.*, 2001), where herd size increases in winter, when predator attacks by wolves become more frequent and intense.

In family-forming equids, stallions are known to actively defend all members of their band against predators, and the permanent bond between stallions and mares in these societies may well have evolved as a response to large and cooperatively hunting predators which traditionally do not occur in the habitats of the other equid species (Feh *et al.*, 1994). Phylogenetic inertia appears to play an important role in the social system of equid species (Berger, 1998). Regardless of habitat type and resource distribution, horses and plains zebras stick to their family groups, whereas for example in Grévy's zebras, stallions and mares

consistently segregate. A summary of group types is presented in Table 5.1.

Individual social relationships and their ontogeny
Individual recognition

The most obvious proof of individual recognition between horses arises from the observation of interactions between mothers and foals, as most mares reject suckling foals other than their own. Learning how to distinguish between olfactory cues seems to play a primary role, as mares separated from their foals shortly after birth will reject their own offspring when reunited a few days later. Visual clues are involved in recognition over long distances. Foals a few weeks old regularly become separated from their families when they fall asleep while the other members of their band move on grazing. Upon waking up, young foals often approach a mare that is not their mother, whereas older foals go straight to their dam. Familiarity may be important in individual recognition. Berger and Cunningham (1987) concluded that after an 18-month separation, stallions were unable to discriminate between unfamiliar and previously familiar horses from their natal band. A field-experiment with Camargue horses found that, after a one-year separation from their mares, stallions immediately distinguished and joined up with their former mates in a large herd with many other mares present (Feh, 1999).

Number of preferred social partners

Testing the frequency distribution of 'being closest' and 'total interactions exchanged' against the negative binomial distribution, it has been shown that throughout their lives, horses have only one or two, rarely three, preferred social partners, regardless of the size of the band in which they live (Feh, 1987; Monard, 1992). Principal Component Analysis conducted on 'nearest neighbour' data and 36 different interactions indicates that these 'companions' spend most of their time in close proximity, always rest in each other's company, approach and follow each other frequently, exchange confident body-contacts and allo-groom (Feh, 1987; Monard, 1992). Besides these basic characteristic interactions, the horses add others according to their sex and age. Young horses typically play with each other and address submissive gestures towards adults, adult horses establish a dominance hierarchy and adult stallions herd and drive their families.

Ontogeny of relationships from birth to weaning

The first month is the most critical period in a free-ranging horse's life. Mortality is at its highest in the days following birth (Berger, 1986; Duncan, 1992), and many deaths are due to accidents, partly because foals become separated from their mothers, despite the protectiveness of the latter (Boyd, 1980). Foals of primiparous mothers stand a greater risk of dying than foals of multiparous mares, presumably due to the former's inexperience in defending the foal against other herd members during the first hours and days which are decisive to bonding (Berger, 1986; Duncan, 1992). The arrival of a new herd member arouses great curiosity in all other horses, and fathers play an important protective role by consistently circling the mare with her newborn foal and to keep all other individuals at bay. Interestingly, Camargue foals born into families where two stallions formed alliances had a 20% higher chance of survival than foals born into one-male families (Feh, 1999). The first band member allowed to contact the newborn foal is its older maternal sibling, male or female alike, but only after two to three days, once the foal is following the mother consistently (Murbach, 1976). Mutual grooming is initiated by the mother on the day of birth, and first play-running by the foal is performed in circles with

its mother in the centre. By the age of two weeks, foals make the acquaintance of their siblings, as well as of their father. Fathers groom the foal and tolerate its playfulness, sometimes responding in a gentle way much different from the more intense bites exchanged during play with their older sons. But the preferred play partners of the foal are other foals. By the age of one month, both in Camargue (Monard, 1984) and Przewalski's horses (Carenton, 1997), male foals played more often and longer than female foals, and in different ways. Whereas colts have a tendency to rear, bite and chase each other frequently, females predominantly kick out with their hind legs. This behaviour mirrors adult fighting techniques. Stallion fights are characterized by bites, rearing and chases, while mares kick at each other when asserting their rank position (Wells, 1978).

Weaning is a conflict between mother and foal. The mother starts to reject the sucking attempts of her offspring in relation to the date of birth of her next foal. In Camargue horses, most mares gave birth once a year, and the foal is weaned 15 weeks before the arrival of its sibling, at the age of seven to eight months. The foals nevertheless continue to stay in their family band and maintain close relationships based on grooming and proximity with their mother, father and older siblings. Besides the family members, the peer group becomes more important. When no foals of the same age were present in its natal band, young males went to play with colts of other families, and young females initiated contact in a similar way. Once its mother had a new foal, the now yearlings took great interest, groomed and played with the newborn and often rested close to it, 'baby-sitting' when their mother grazed at a distance (Wells, 1978).

Adolescence and dispersal

Young mares gradually emigrate from their natal band between two and three years of age during an oestrous period (Berger, 1986; Monard et al., 1996). When other reproductive groups are in the vicinity, most of the young mares leave of their own accord. They are neither expelled nor does aggressive behaviour by resident females, including their mother or their father, increase prior to departure (Monard & Duncan, 1998). When no such groups are around, fathers may chase them out. Mating attempts by all stallions from their natal bands were always actively refused by the young mares, moreover, their mothers frequently intervened. During oestrous, the young females often left their natal bands for a few hours or days, and mating by unfamiliar males from these other groups were always encouraged through presenting. Most of the mares integrated an already existing reproductive unit where they ranked lowest in the dominance hierarchy. Others, particularly mares with their first foals, were abducted in the days following parturition by young stallions that started their own family.

Young males follow a different path from females. They too leave their natal band at between two to three years of age, but, usually after a short and unsuccessful attempt at monopolizing one or more mares, they end up in an all-stallion group for one or three years, a pattern observed in all feral (Berger, 1986; Feh, 1999) and Przewalski's horses (C. Feh, unpub. data). Similar to females, the young stallions are sometimes chased out by the adult stallion of their band, generally their father, when not leaving of their own accord. Inside the all-stallion band, the main activity is play fighting, presumably to measure their respective strengths and establish their rank position. At this time some of the stallions form close bonds. By the age of four to five years, they are often seen in the proximity of reproductive units, sometimes in pairs. The stallions investigate all faeces to detect mares in oestrous, and take up contact with young mares at the periphery of their family groups with whom they groom frequently. The high ranking stallions manage, after a while, to monopolize one or two females on their own, while some low ranking stallions team up in pairs to defend mares against their rivals, forming cooperative alliances (Feh, 1999, 2001).

Adult relationships and dominance hierarchy

Despite the fact that dispersing mares are attracted by familiar females when choosing their new band, they do not maintain close relationships with these females and prefer to associate with unfamiliar and unrelated mares (Monard, 1992). Relationships between adult mares are mainly characterized by the maintenance of their respective rank position through exchanges of low intensity threats, and by allo-grooming, but only with one or two other mares in their family.

Similar to the adult mares, stallions have preferential relationships with one or two mares of their band only. They allo-groom frequently and rest in their company, yet they mate and drive or herd together all of their females in defence against rivals. Even outside oestrous, stallions approach their mares regularly with arched necks and a special vocalization – between a nicker and a squeal. Stallion–stallion interactions are foremost agonistic, as described in detail in other sections. Exceptions are alliance-stallions, where the pair rests in close proximity and even allo-groom each other, despite fighting at other times (Feh, 1999, 2001). Adult stallions when visiting the bachelors often play with them. Play occurs regularly between family leaders and their adult sons of four or five years old still present in their natal band, in both feral and Przewalski's horses, and is occasionally observed between family leaders outside the breeding season.

Adult horses living in natural social societies where individuals know each other from birth all form stable linear dominance hierarchies with occasional reversals and triangles, giving priority of access to limited resources such as water, food or wind shelter. Remarkably for ungulates, size or weight are not important, whether in males (Feh, 1990) or in females (Duncan, 1992). Both age and order of arrival in the reproductive group are the key factors determining rank position in Camargue (Monard *et al.*, 1996) as well as Przewalski's mares (Lescureux, 2001), while a mother's rank influences the rank of her sons in the herd and correlates with the number of foals he produces (Feh, 1990). Adult stallions are usually dominant over the mares of their family with regard to access to limited resources. In the Camargue horses studied by the author and colleagues, there was a clear linear dominance hierarchy between the 11 bands, most easily seen when they came to drink. The dominant band was not always first to drink, but subordinate bands left the watering place when they arrived.

Priority of access to resources is not to be confounded with leadership. Most of the time, one mare initiates the movement and leads the band to drinking or new feeding grounds. Neither in Camargue (Wells, 1978) nor in Przewalski's horses (C. Feh, unpub. data) was a correlation found between this leading position and rank.

Social communication
Vocal

Horses are open grassland animals, and they almost never lose visual contact with other band or herd members. This may be one reason why their acoustic communication repertoire is remarkably poor for an animal with such an elaborate social system. Typically, the highest and longest vocalization, the whinny, is used when horses lose visual contact with their band members, a stressful and dangerous situation for a horse in the wild. Acoustic signals may be averted in order to avoid attracting predators. Kiley (1972) identified five discrete vocal categories: the whinny, the nicker, the squeal, the sigh and the roar.

The nicker is a low tonality call. The mouth is closed and only the nostrils can be seen moving. It is used between familiar animals only, for example from mother to her foal or stallion to his own mare, and was interpreted as having a greeting function. The squeal is a high-pitched sound used in situations of social excitement and tension, whether during stallion contests, mare fights or stallion–mare interactions. Only sub-adult or adult horses communicate in this way. With the help of play-back experiments, Rubenstein and Hack (1992) have shown that squeals provide direct information about status during stallion encounters. Squeals by dominant animals are longer than that of subordinates, and they have a higher pitch at the onset of the vocalization. Regardless of their dominance status, stallions were more likely to respond to the sound of a low-ranking male. Therefore, stallions seem to be able to communicate their fighting abilities, which may reduce the intensity of the physical confrontation. The sigh much resembles a human sigh, and is often used in a resting, immobile position. Free-living horses often sigh after a rest, before starting to graze again. When startled, horses 'freeze', direct their attention to the source of fright and snort, by rapidly exhaling air through the nostrils. Many antelopes respond in a similar way when they locate a predator.

Olfactory

There are few detailed studies on olfactory communication of horses. Horses spend a great amount of time sniffing objects or herdmates, but little is known about the information they gather or transmit. As other ungulates and carnivores, horses exhibit 'flehmen', retracting their nostrils while inhaling and exhaling, moving material into their vomeronasal organ which is packed with secondary olfactory nerve cells. This behaviour is present in foals a few hours old when suddenly they are confronted with a pungent scent, and is specially developed in stallions when they sniff the urine or faeces of mares at the onset of oestrous.

Stallions urinate over the defecations or urine of their own mares as well as the mares they are in the course of 'courting', and deposit small amounts of faeces on the dung of other stallions, whether or not the stallions are present. Kimura's (2001) study showed that by urinating over the faeces of oestrus mares, stallions are possibly covering the scent identifying oestrus.

Visual and tactile

Tactile and visual signals are by far the most elaborate gestures in horses. The retraction of a nostril, the twitch of an ear all appear to have meanings to their social partner, and from the point of their nose to the tip of their tails, horses seem to communicate continuously.

When trying to evaluate a social group, equal attention should be paid to the many different forms of social interaction that are obvious to the observer. Below are listed a range of important expressions. Observations are based on the following breeds and studies: mustangs (Feist & McCullough, 1976; Berger, 1986), Camargue (Wells & von Goldschmidt, 1979; Feh, 1987; Monard, 1992; Feh & de Mazières, 1993), Przewalski's horses (Tatin, 1995; C. Feh, unpub. data). Other ethograms of the horse include those of Waring (2003) and McDonnell (2003).

Staying close
Actions: Maintaining close proximity (frequency of being 'nearest neighbour'), approach or follow the other horse.

These are the basic expressions between preferred partners, regardless of their age or sex. Typically, preferred partners rest together, whereas they are less selective when grazing. Relationships are not always symmetrical. Mothers maintain close distance with their young foals, and yearlings maintain close distance with their mothers. The function appears to be individual bonding and group cohesion.

Social investigation

Actions: Nose–nose contact, nose–body contact. Often, there is no direct skin contact, but the nostrils, which can be seen moving, are at a distance of 1–2 cm from the skin of the other horse.

Nose–nose sniffing takes place during greeting between two individuals, whether or not they are familiar with one another. Nose–elbow and nose–flank contact are typical for stallion encounters and courtship. These regions of the horse's body are tightly packed with sweat glands. Nose–genital contacts occur during courtship and stallion encounters. The function of these interactions certainly is olfactory information and transmission, but little is known about what information horses gather or transmit.

Comfort behaviour and reduction of social tension

Actions: Rubbing. The flat side or the under part of the head is rubbed against any other body part of another horse.

Allo-grooming. Two horses rhythmically (*c.* 2/s) scratch each other with their incisors (Figure 5.1), for up to 3 min or more.

Foals start to rub and allo-groom other horses from the first days after birth. Horses rub each other when something irritates their skin, for example to get rid of insects or to shed the winter haircoat in spring. Grooming is typically exchanged between preferred partners, such as mothers and foals, foals and siblings, fathers and foals, peer groups at all ages, including adult mares and adult stallions when alliance partners. High-ranking individuals groom more and initiate more grooming. Horses groom each other at different places of the body, but the preferred site lies at the base of the neck (Figure 5.1). Experimental imitation of grooming at this site produced a reduction in the heart rate of the recipient, whereas grooming on a non-preferred area did not. This preferred site lies close to a major ganglion of the autonomic nervous system and is used in acupressure by veterinarians to calm nervous horses.

Dominance behaviour

Actions: Threat to bite (Figure 5.2) and bites, threat to kick and kicks, attacks, chases. During all these interactions, the ears of the horse

Figure 5.1. Allo-grooming.

are laid flat against the neck. The head (in threat to bite) or the rump (in threat to kick) is turned towards the other horse. Attacks often follow unsuccessful bites or kicks.

Foals start to flatten their ears towards herd mates when they are a few days old, regardless of the recipient's age or sex. Only later do they appear to discriminate. The rank position often changes in young horses, but reaches stability once the individuals are adult and integrated in their own reproductive unit. Rarely will a mare challenge the rank position of another mare. Fights, mostly exchanging kicks, can go on for months before the issue appears settled. The function of dominance hierarchies are priority and organization of access to limited resources, such as water, food, shelters against wind, heat or insects, or a social partner. Dominance hierarchies are thought to have evolved in order to reduce incessant physical contests.

Figure 5.2. Threat to bite.

Submissive behaviour

Actions: Avoid. A horse actively deviates from its apparent itinerary to avoid meeting another individual. A horse waits before approaching a limited resource, until its access is no longer controlled by dominant individuals.

'Snapping'. The lips are pulled back, and the horse rhythmically claps its teeth, about 1/s (Figure 5.3). Frequently, the back is arched and the tail tucked in.

'Snapping' starts on the day of birth, but again, without discrimination. Young foals may snap towards a tree as a large unfamiliar object. Usually, only young horses snap towards adults. Young mares usually no longer show the behaviour once they have their first foal, and stallions once they have formed their own family. 'Snapping' can be a reaction to agonistic interactions by adult herd members (threats, bites, attacks), or mating attempts. On its own, 'snapping' can be an active submissive behaviour. Typically, young horses of both sexes approach their family stallion (most often their father), and teeth-clap against his mouth.

Figure 5.3. Snapping.

The behaviour may appease adult animals, but this is difficult to prove.

Social play

Actions: Bite, kick, rear, circle, chase. The ears are never flat against the neck when horses play. The bites are most often directed against the throat, the base of the neck and the legs, and they frequently circle when attempting to bite at each other's legs. For protection, the recipient may sit on its hind legs or go down on its fore legs. They kick with their fore legs when rearing, and with their hind legs when being chased.

Foals start to play with each other when they are two to three weeks old. Colts play more often and longer than fillies, and during their play, they rear, bite and chase the partner more frequently, whereas fillies kick with their hind legs. In young stallions the frequency of play peaks at two to four years of age, and the intensity of play gradually increases. They may inflict superficial wounds on each other. Young females play as much as their male consorts when living in a male bachelor band, but all mares stop playing once they have their first foal. Adult stallions continue to play with their adult sons, and other adult stallions outside the breeding season.

As the performed actions are extremely similar to adult fights for both sexes, play probably has a training function. In stallions, it may further allow them to know each other's fighting tactic and establish a dominance hierarchy before forming their own reproductive group.

Play sexual behaviour Elements and sequences of sexual behaviour are common in foal play, both of males and females. See McDonnell Chapter 7, for details and illustrations.

Fights

Actions: Same as in play. All actions are more intense and follow each other at a higher speed. The ears are laid flat against the neck during fights.

Adult mares exchange a sequence of several kicks when fighting for their respective rank position. Stallions start to fight when they are four to five years old. Some stallions develop individual fighting techniques, systematically aiming to bite the throat, neck, ears or tail of the rival, or specializing in boxing. The issue and function is access to natural resources for mares, and access to females for stallions. Both issues may be confounded in individual reproductive success.

Stallion ritual

Actions: Common sniffing of dung, strike, prance, push. Stallion rituals are behavioural sequences starting with the sniffing of a dung pile, followed by tossing of the head while squealing, striking alternately with their fore legs. One stallion leaves while the other defecates, or both stallions leave, prancing, or both stallions defecate at a distance.

Young stallions start to engage in ritual interactions when about two years old. Dominant stallions defecate last in the sequence, as a demonstration of their higher rank (Feh, 1987). If the rival does not accept this demonstration, the ritual can escalate into a fight. Parallel defecation at a distance appears to signal mutual acceptance of respective mare ownership. Rituals are usually followed by each stallion joining his mares and driving them in the opposite direction from the rival. In alliance families, the dominant and subordinate stallions share the task when meeting a rival. Most of the time, the subordinate stallion performs rank demonstrations (or fights) while the dominant stallion tends to their common mares.

The function appears to be mare ownership.

Figure 5.4. Herding or driving.

Group cohesion by stallions, 'herding' or 'driving'

Action: The stallion, with ears flat back against his neck, his head close to the ground, in extension of his neck, walks, trots or canters (Figure 5.4).

Stray mares, isolated foals or other family members are routinely brought back to the band and away from rivals in this way. Sometimes, young stallions show this behaviour towards younger siblings, regardless of their sex. Quite rarely, adult mares have been observed herding their foals.

The function is to increase the distance of the family members with regard to rivals, and to maintain group cohesion in general. In other family forming equids, stallions round up their band members when predators were in sight.

Conclusion

An understanding of the natural social behaviour and communication of free-ranging domestic or wild horses is a valuable basic tool for establishing a co-operative relationship with a domestic horse partner. Especially when the co-operation fails, whether the task is to make the horse jump an obstacle or load into a trailer, knowledge of the mechanisms of the communication system and the character of the natural intra- and inter-species relationships of horses can be helpful in achieving the goal or understanding the failure.

As a final comment, it is worth noting that free-ranging horses are always occupied, alert, and interactive with their herd mates as together they forage, seek shelter, groom, rest and protect one another. As such, free-ranging horses growing up in a natural social environment have never been observed with stereotypic behaviours, self-aggression or other problems that occur in stabled horses. Stabled horses are provided with what human caretakers believe to be good food, shelter and health care, but typically lose the long-term social relationships and communication with members of their own species that seem so important in their free-ranging counterparts. Social contact and relationships with horses are now being recognized in horse-behaviour research as important to the well-being of domestic horses (see Mills and McNicholas Chapter 11 and Cooper and Albentosa Chapter 16).

References

Berger, J. (1986). *Wild Horses of the Great Basin*. Chicago: University of Chicago Press.

(1998). Social systems, resources and phylogenetic inertia: an experimental test and its limitations. In *The Ecology of Social Behaviour*, ed. C. N. Slobodchikoff. London: Academic Press, pp. 157–86.

Berger, J. & Cunningham, C. (1987) Influence of familiarity on frequency of inbreeding in wild horses. *Evolution*, 41(1): 229–31

Boyd, L. (1980). The natality, foal survivorship and mare–foal behavior of feral horses in Wyoming's Red Desert. M.S. Thesis, University of Wyoming, Laramie.

Carenton, J.-M. (1997). Le jeu social chez les poulains Camargue et Przewalski en fonction du rang maternel. Maîtrise, Paul Sabatier University, Toulouse.

Clutton-Brock, T. H. (1989). Mammalian mating systems. *Proceedings of the Royal Society of London*, Series B, 236: 339–72.

Duncan, P. (1992). *Horses and Grasses*. New York: Springer.

Duncan, P., Feh, C., Gleize, J. C., Malkas, P. & Scott, A. M. (1984). Reduction of inbreeding in a natural herd of horses. *Animal Behaviour*, 32: 520–37.

Feh, C. (1987). Etude du développement des relations sociales chez des étalons de race Camargue et de leur contribution à l'organisation sociale du groupe. Thèse d'université, University of Aix-Marseille.

(1990). Long-term paternity data in relation to different aspects of rank for Camargue stallions. *Animal Behaviour*, 40: 995–6.

(1999). Alliances and reproductive success in Camargue stallions. *Animal Behaviour*, 57: 705–13.

(2001). Alliances between stallions are more than just multimale groups: reply to Linklater & Cameron. *Animal Behaviour*, 61: F27–F30.

Feh, C., Boldsukh, Ts. & Tourenq, C. (1994). Are family groups in equids a response to cooperatively hunting predators? The case of Mongolian khulans. *Revue d'écologie (Terre et Vie)*, 49: 11–20.

Feh, C. & de Mazières, J. (1993). Grooming at a preferred site reduces heart rate in horses. *Animal Behaviour*, 46: 1191–4.

Feh, C., Munkhtuya, B., Enkhbold, S. & Sukhbaatar, T. (2001). Ecology and social structure of the Gobi khulan in the Gobi B National Park. *Biological Conservation*, 101: 51–61.

Feist, J. D. & McCullough, D. R. (1976). Behaviour patterns and communication in feral horses. *Zeitschrift für Tierpsychologie (Ethology)*, 41: 337–71.

Ginsberg, J. (1987). Social organisation and mating strategies of an arid adapted equid: the Grévy zebra. Ph.D. Thesis, Princeton University, Princeton NJ.

Jansen, T., Forster, P., Levine, M. A., Oelke, H., Hurles, M., Renfrew, C., Weber, J. & Olek, K. (2002). Mitochondrial DNA and the origins of the domestic horse. *PNAS*, 99(16): 10905–10.

Keiper, R. & Houpt, K. (1984). Reproduction in feral horses: an eight year study. *American Journal of Veterinary Research*, 45: 991–5.

Keiper, R. & Rutberg, A. (1993). Proximate causes of natal dispersal in feral ponies: some sex differences. *Animal Behaviour*, 46: 969–75.

Keiper, R. & Sambraus, H. (1986). The stability of equine dominance hierarchies and the effects of kinship, proximity and foaling status on hierarchy rank. *Applied Animal Behaviour Science*, 16: 121–30.

Kiley, M. (1972). The vocalization of ungulates, their causation and function. *Zeitschrift für Tierpsychologie (Ethology)*, 31: 171–222.

Kimura, R., (2001). Volatile substances in feces, urine and urine-marked feces of feral horses. *Canadian Journal of Animal Science*, 81: 411–20.

Klingel, H. (1972). Das Verhalten der Pferde. In *Handbuch der Zoologie*, ed. H. G. Helmcke, D. Starck & H. Wermuth. Berlin: Walter de Gruyter, pp. 1–68.

Lescureux, N. (2001). Etude de la hiérarchie de dominance d'un troupeau de chevaux de Przewalski en semi-liberté. Maîtrise, University of Lille.

McDonnell, S. M. (2003). *The Equid Ethogram: A Practical Field Guide to Horse Behavior*. Lexington, KY: Eclipse Press.

Miller, R. (1979). Social organization and movements of feral horses in Wyoming's Red Desert. In *Proceedings of the Symposium on Ecology and Behavior of Wild and Feral Equids, University of Wyoming, Laramie*.

Moehlman, P. (1998). Behavioural studies of the donkey. *Applied Animal Behaviour Science* (Special Issue), **60**.

Monard, A. (1984). La tétée et le jeu: une étude des différences quantitatives entre poulains mâles et femelles de race Camargue. DEA (diploma) University of Rennes I, France.

(1992). Causes et conséquences du départ des jeunes femelles de leur groupe natal dans un troupeau de chevaux Camargue en semi-liberté. Ph.D. Thesis, Rennes University.

Monard, A. & Duncan, P. (1998). Consequences of natal dispersal in female horses. *Animal Behaviour*, **52**: 565–79.

Monard, A., Duncan, P. & Boy, V. (1996). The proximate mechanisms of natal dispersal in female horses. *Behaviour*, **133**: 1095–124.

Murbach, E. (1976). Die Entwicklung der sozialen Beziehungen vom Fohlen zu Artgenossen beim Camarguepferd. Diploma, Psychologisches Institut, University of Bern.

Penzhorn, P. (1984). A long-term study of social organisation and behaviour of Cape mountain zebras. *Zeitschrift für Tierpsychologie (Ethology)*, **64**: 97–146.

Raschek, V. (1973). Reproduction of the wild ass and its behaviour during rutting time on Barsa Kelmes Island. *Bulletin Moscow Prirodi*, **54**: 9–18.

Rubenstein, D. (1986). Ecology and sociality in horses and zebras. In *Ecological Aspects of Social Evolution*, ed. D. Rubenstein & R. Wrangham. Princeton: Princeton University Press, pp. 282–302.

Rubenstein, D. & Hack, M. (1992). Horse signals: the sounds and scents of fury. *Evolutionary Ecology*, **6**: 254–60.

Schaller, G. (1998). *Wildlife of the Tibetan steppe*. Chicago: Chicago Press.

Schilder, M. (1990). Interventions in a herd of semi-captive plains zebra. *Behaviour*, **112**: 53–83.

Shah, N. (1993). Ecology of wild ass in Little Rann of Kutch. PhD thesis. Baroda University.

Solomatin, A. (1973). *The Wild Ass*. Moscow: Academyia Nauk.

Tatin, L. (1995). Le rôle de diminution de la tension social du toilettage mutuel chez le cheval: étude faite sur le cheval de Przewalski. Maîtrise, Université de Provence, Marseille.

Tatin, T., Dareshoori, B. F., Tourenq, C., Tatin, D. & Azmayesh, B. (2003). The last population of the critically endangered onager *Equus hemionus onager* in Iran: urgent requirements for protection and study. *Oryx*, **37**: 4.

Waring, G. H. (2003). *Horse Behavior*, 2nd edn. New York: Noyes.

Wells, S. (1978). The behaviour and social structure of a herd of Camargue horses. M.Sc. Thesis, University of Cambridge.

Wells, S. & von Goldschmidt-Rotschild, B. (1979). Social behaviour and relationships in a herd of Camargue horses. *Zeitschrift für Tierpsychologie (Ethology)*, **49**: 363–80.

Vila, C., Leonard, J., Götherström, A., Marklund, S., Sandberg, K., Liden, K., Wayne, R. & Ellergreen, H. (2001) Widespread origins of domestic horse lineages. *Science*, **291**: 474–7.

Zhirnov, L. & Ilinski, V. (1986). *The Great Gobi National Park, a Refuge for Rare Animals of the Central Asian Deserts*. Moscow: UNEP.

6

Maintenance behaviours

Katherine A. Houpt

Introduction

The maintenance behaviour of horses falls into five general areas: ingestion/elimination, coat care, thermoregulation, locomotion and rest. Ingestive behaviours include eating and drinking. Elimination includes defecation and urination. Behavioural thermoregulation includes both means of conserving heat in cold weather and means of dissipating heat in hot weather. Autogrooming includes such actions as tail swishing, rolling, nipping, rubbing on inanimate objects or rubbing one part of the body on another. Resting behaviour takes the form of stand resting during which the horse can be drowsing or in slow wave sleep, and recumbent rest in which the horse is usually in either slow wave sleep or rapid eye movement (REM) sleep. Locomotion in the undisturbed horse is mostly walking in the course of grazing.

Feeding

The horse is a non-ruminant herbivore. The major fermentation site is the hindgut, in particular, the caecum and large colon. Food moves relatively quickly from the horse's small stomach through the small intestine to the large intestine. Here, bacteria digest cellulose to short chain fatty acids. Some of these fatty acids are absorbed and can be used for energy by the horse (Argenzio, 1992). The process is not as efficient as it is in ruminants because the major absorptive and digestive site – the small intestine – is proximal to the fermentation chamber. Horses can absorb some non-essential amino acids, but their ability to re-use urea to form essential amino acids is limited (Houpt & Houpt, 1971). Adult horses are not coprophagic, so they do not take advantage of bacterial products in their faeces as do lagomorphs. The nutritional strategy of the horse is to eat large amounts of forage, deriving only a small percentage of the total energy available (Janis, 1976; Demment & Van Soest, 1985).

Horses can thrive in areas where forage is high in fibre and low in quality because they have high rates of intake. In order to have high intake rates, they must devote the major portion of their time to eating. As fibre concentration increases, digestibility decreases. High gut fermentors can extract more from plants with low (<30%) cell wall content than from those with high cell wall content.

Having evolved to consume large quantities of high fibre, the horse now must live, under current management systems, on small amounts of fibre and calorically dense concentrate feeds. These feeds lead to a different population of microorganisms in the large intestine. These microorganisms produce larger amounts of lactic acid and other products that may lead to stereotypic behaviours, as well as laminitis, colic, etc. (see Mills, Chapter 15).

Grazing

Feeding behaviour by feral (Salter & Hudson, 1979; Keiper & Keenan, 1980; Rubenstein, 1981; Duncan, 1985), wild (Boyd, 2002), and pastured horses (Crowell-Davis *et al.*, 1985; Houpt *et al.*, 1986) is primarily grazing. Grazing is not just ingesting grass, but consists of appetitive and consumptive phases. The appetitive phase is the seeking of food in general and a specific type of food in particular. The consummatory phase is the process of ingesting the food. Grazing, therefore, is not a homogenous activity. It consists of a number of components that enable the horse to adjust to the forage availability and individual energy needs. The initial step is selection of a feeding patch.

The next step in feeding is prehension of the food. Horses use their prehensile upper lip to gather food, then bite off a clump with their incisors. The bite rate while grazing is about 30 000 per day (Mayes & Duncan, 1986). Once the food has been prehended, the horse chews, insalivates and swallows it. Horses seem able to chew with their molars while biting a new mouthful with their incisors. The upper lip is used to brush a thin layer of snow from the grass, but when the snow is thicker than 20 cm, the horse paws to uncover it (Salter & Hudson, 1979).

Grazing involves walking, but the horse does not move forward at a uniform rate. Rather, it eats several bites then takes a step or two forward or sideways. The area where a horse grazes without moving is called a feeding station. In our studies (Cosenza, 1999), the horse remained at a feeding station for

Table 6.1. *Grazing preferences based on observation*

Perennial rye grass *Lolium perenne*
Cocksfoot *Dactylis glomerata*
Timothy *Phleum bertolonii*
Tall fescue *Festuca arundinacea*
Wild white clover# *Trifolium pratense*
Canadian creeping red fescue *Festuca rubra*
Crested dogstail *Cynosurus cristatus*
Meadow foxtail *Alopecurus pratensis*
Creeping red fescue *Festuca rubra*
Rough stalked meadow grass *Poa trivialis*
Dandelion *Taraxacum officinale**
Smooth stalked meadow grass *Poa pratensis*
Ribgrass *Plantago lanceolata**
Chicory *Cichorium intybus**
Red fescue *Festuca rubra*
Yarrow *Achillea millefolium**
Late flowering red clover# *Trifolium pratense*
Brown top *Agrostis tenuis*

(Archer, 1973).
The results are based on observation of horses grazing on strips of pure stands of each plant. The species are given in decreasing order of preference. *, herb; #, legume.

about 12 seconds, then took several steps. The horse took about eight bites at each feeding station or 51 bites/minute (Cosenza, 1999).

Factors influencing grazing Selectivity. The horse uses both avoidance and preference (Ödberg & Francis-Smith, 1977) and also chooses selectively. They do not necessarily eat the most abundant plants, but rather choose certain plants. Young, growing grasses that are high in protein and digestibility are more likely to be selected. Almost the only studies of grass preference in horses are those of Archer (1973, 1978) who sowed pure plants and recorded what the horses ate. See Table 6.1.

Horses differ individually in their selectivity. Marinier and Alexander (1992) used two choice preference tests of all possible combinations of five different plants to determine if horses were consistent in their choice. They also determined whether the horse could sort a preferred plant from other plants and how much their choice was affected by bitterness. They found that horses could be grouped into efficient, semi-efficient and inefficient grazers. Inefficient grazers may be at greater risk of plant poisoning.

Horses can learn to avoid foods that make them ill. This is termed a conditioned taste aversion. Even when they have eaten three different types of feed and then become ill, they will avoid the novel food rather than the foods they have learned are safe to eat (Houpt *et al.*, 1990). In order for the horse to learn to avoid the food, it must produce nausea or gastrointestinal illness. Practical use can be made of that to teach horses to avoid poisonous plants. Locoweed *Oxytropis sericea* poisoning is a major cause of death in horses in many parts of the western USA. Plants such as locoweed do not produce rapid gastrointestinal signs; instead the toxins accumulate and cause central nervous system signs after some time. By pairing the taste of the plants with something that does make the horse ill immediately, a conditioned taste aversion can be formed. When horses were offered cut and dried locoweed paired with an intra-gastric dose of lithium chloride to produce illness, the horses would not eat the plant either cut or, to a lesser extent, on pasture (Pfister *et al.*, 2002).

Seasonal and temperature effects

During winter in the New Forest in England, ponies spent nearly all of the daylight hours grazing, interrupted for only 40 minutes at midday, whereas in the summer, more time was spent resting (Tyler, 1972). Duncan (1983) also found that more grazing occurs during summer than winter nights, but the total time spent grazing is relatively constant. The amount of time spent grazing varies with the season, with the age and sex of the horse, and less than would be expected with herbage availability (Duncan, 1985). As, for example, in Western Australia summer grazing activity is lowest during the warmest hours (08:00–16:00), but in winter the lowest point occurs in the early morning (03:00–06:00; low 3°C) (Arnold 1984; see Table 6.2). One seasonal effect on both free-ranging and pastured domestic horses is the dormancy of grass during the winter. At that time of year, when grass is scarce, horses will browse on branches of trees and shrubs and may become coprophagic. In pastures, this can lead to stripping of the bark from trees and destruction of wooden fences.

Eating hay

Horses are kept in three general management systems: (1) on pasture; (2) on pasture during the day, but in a box (stall) at night; and (3) box confinement most of the time with some hours outside in a grassless paddock. What the horse eats and how long it spends eating depends on its environment and the types of feeds provided to which it has access. Most stabled horses are provided with some hay and variable amounts of concentrates. Short-term (5 minutes) tests recently found that stalled horses showed a preference for multiple forages over single forage and displayed more 'positive' behaviours, perhaps reflecting their need for their stabled diets to provide more variation (Goodwin *et al.*, 2002).

When the feeding time of stalled horses is compared with the grazing time of free-ranging horses, there can be large differences or almost none depending on the amount of hay available to the horse. When fed hay ad libitum, horses in boxes or tie (straight) stalls spend approximately 60% of their time eating, the same percentage as that for grazing horses (see Figures 6.1 and 6.2; Tables 6.3 and 6.4). Eating hay differs from grazing in several ways: the horse does not move while eating hay; it does not have to bite to prehend the forage; depending on the manner in which the hay is provided, the horse may have its head up (hay rack or hay bag), at chest height (manger) or down in the natural position (floor). When given the opportunity, horses will remove the hay from a manger and place it on the floor (Sweeting *et al.*, 1985), indicating their preference.

The greatest difference between hay and natural forage is that hay is dry. The prevalence of respiratory disease in stabled horses is one consequence of this environmental and dietary change (Derksen, 1987). The horse must obtain more fluid as water

Table 6.2. *Grazing preferences based on amount consumed*

Creeping red fescue *Festuca rubra*
Tall fescue *Festuca arundinacea*
Crested dogstail *Cynosurus cristatus*
Cocksfoot *Dactylis glomerata*
Brown top *Agrostis tenuis*
Timothy *Phleum bertolonii*
Perennial rye grass *Lolium perenne*

(Archer, 1978).
The species are given in decreasing order of preference. Preferences were calculated from measurement of the height of the plants after grazing.

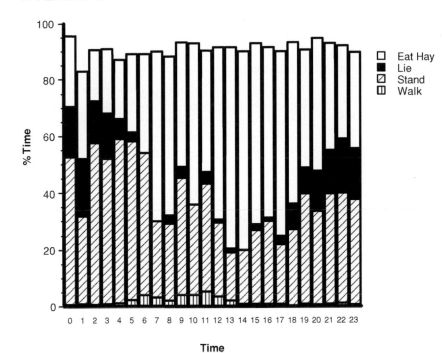

Figure 6.1. Time budgets (24-hour) of mares in a tie stall fed ad libitum hay.

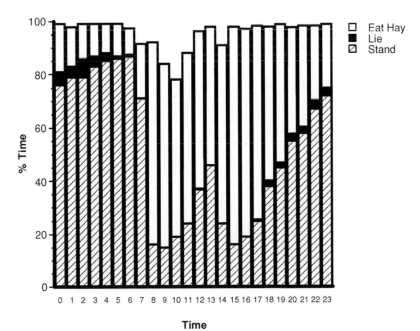

Figure 6.2. Time budgets (24-hour) of mares in a box stall fed ad libitum hay.

when eating hay rather than when grazing. Some horses will dunk their hay in their water which may be an attempt to rehydrate it. The closer the hay is to the water source, the more likely the horse is to place the hay in the water.

As prey animals, horses have evolved to be aware of predators. This behaviour is reflected in their vigilance while eating or drinking. Horses will often stop and lift their head from their feed. The presence of another horse inhibits that behaviour. Social facilitation of feeding (Sweeting *et al.*, 1985) may actually be a relaxation of vigilance behaviour. Holmes *et al.* (1987) found that dominant horses ate less if a solid partition separated that horse from a subordinate, but both dominant and subordinate horses ate more if a wire partition separated them.

Concentrates

The natural horse diet is high in long-stem roughage, but many contemporary equine diets are not. Grains are fed because the high performance horse needs more calories than hay alone can supply. Ponies consume grain at a rate of 60 g/minute, Thoroughbreds eat grain at a rate of 129 g/minute, but mixing the grain with chopped timothy can slow intake to 46 g/minute. Horses eat faster when grain is first presented – 140 g/minute, then 10 minutes later their intake drops to 100 g/minute. Peppermint flavouring does not seem to change the rate of grain intake, but apple, caramel and, especially, anise slows the rate of intake (Hintz *et al.*, 1989).

Complete pelleted diets have been devised for horses because they are convenient to the manager/owner, easily stored and may not trigger inhalant allergies in horses or owners (Derksen, 1987). The pellets can be fed in addition to hay or as a complete diet. Complete pelleted diets contain fibre, but it is ground. Therefore, the horse can consume its caloric requirement in 2 hours/day. For example, Hintz and Loy (1966) found that horses consumed a pelleted complete ration of barley bran and ground alfalfa more quickly (76 min/6 kg) than a non-pelleted

Table 6.3. *Grazing standing and moving on pasture*

Population	Time	Graze %	Stand %	Move %	Author
Pony	Night	55	33	3	Houpt *et al.*, 1986
Pony	Day	70	15–30		Crowell-Davis *et al.*, 1985
Przewalski	24 h	46	34	7	Boyd *et al.*, 1988
Camargue horse	24 h	59	27	10	Duncan, 1980
Misaki horse	24 h	76	18		Kaseda, 1983

Table 6.4. *Feeding standing and moving in stalls and grassless enclosures*

Population	Stall	Time	Diet	Feed %	Stand %	Move %	Author
Horse	Metabolism cage	day	Limited hay and concentrate	50	45	0	Willard *et al.*, 1977
Pony	Box stall	day		76	79	3	Sweeting *et al.*, 1985
Horse	Corral	day	Limited hay and grain	43	27	6	Houpt *et al.*, 1988
Pony	Box stall	night	Limited hay and grain	15	17	1	Houpt *et al.*, 1986
Horse	Box stall	night	Limited hay and grain	27	67	0.3	Shaw *et al.*, 1988
Pony	Pen	24 h	Ad lib grain	17			Laut *et al.*, 1984
Pony	Pen	24 h	Ad lib pellets	31			Ralston *et al.*, 1979
Horse	Tie stall	24 h	Ad lib hay	32	25		McDonnell *et al.*, 1999
Horse	Tie stall	24 h	Ad lib hay				Houpt *et al.*, 2001
Horse	Box stall	day	Ad lib hay and restricted oats	45			Doreau, 1978
Przewalski	Small corral	day		68	18	12	Boyd, 1988
	Large corral	day		44	45	8.5	Boyd, 1988

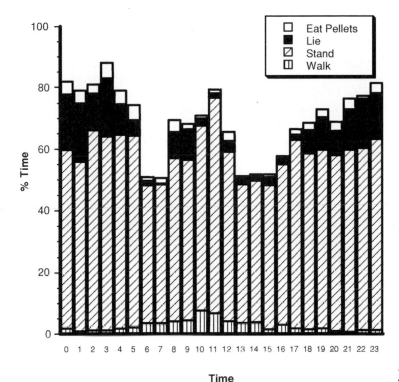

Figure 6.3. Time budgets (24-hour) of the same mare, in a box stall, fed pellets.

version of the same ration (107 min/6 kg). Horses eat pellets more quickly than extruded feed (Hintz *et al.*, 1989). When offered alfalfa hay ad libitum as wafers, pellets or loose hay, horses ate less loose hay than pellets or wafers. They exhibited wood-chewing only when fed the pellets. The questions is: what does the horse do when it is not grazing or eating hay? To a certain extent, the horse continues to forage, that is, it searches the bedding ingesting some wood shavings and more if the bedding is straw, finding stray pellets or consuming faeces. Coprophagia is a common response to any dietary insufficiency – calories, protein, fibre, etc. In our studies (i.e. author's laboratory), searching time increased from 1% when hay was available ad libitum to over 10% of the horses' time when pellets alone were fed (Elia, 2002). (See Figure 6.3.) Although the horse forages in the stall, it does not spend as much time as it would spend eating hay. Standing time, the default behaviour, also increases in the horse fed pellets from 34 to 58% of the time (Elia, 2002). The pelleted diets also stimulated wood chewing in the horses.

Chewing rate is higher per minute when pellets are consumed, but because eating time is so short, the total chews per day is much lower (Elia, 2002). This may have implications for gastrointestinal function because, in contrast to many species, horses have no psychic phase of salivation. Instead, they salivate only when they chew. Saliva is released by the parotid salivary gland on one side with each jaw movement (Alexander, 1966). This indicates that the horse consuming most of his calories as hard feed grain adds less saliva to his upper gastrointestinal tract. Nicol *et al.* (2002) have hypothesized that the functional basis of cribbing is to add saliva with its buffering capacity to the stomach.

One way to assess whether the horse 'cares' about roughage is to see how much effort it is willing to expend. Horses were taught to push a panel switch to obtain access to a bucket containing 100 g (a handful) of hay. To determine the strength of their motivation, the horses had to push the switch more times for each 100 g reward. A progressive ratio was used so that the horse pressed one, two, four, eight and 16 times for each successive reward. If they refused to complete the required number of presses or would not eat the reward, they were considered to have extinguished. If the horses had free choice hay, they would not press the switch to obtain hay. If they were being fed only a complete pelleted diet, they would press up to 14 times for 100 g of hay indicating that horses do have a motivation for hay (Elia, 2002). In the absence of hay or adequate hay, the horse may 'browse' on wood. Wood chewing is a common horse behaviour problem that can be caused by lack of roughage (Krzak et al., 1991). (See Mills, Chapter 15.)

Physiological controls of feeding

The taste of food is particularly important to the horse. They have a preference for sucrose in solution (Randall et al., 1978) or as a solid (Hawkes et al., 1985). Because horses have a sweet taste preference, many feeds are sweetened, usually with molasses. Molasses also has the advantage of preventing sorting of grains by the horse using its prehensile lips. Sweetened mixed grain is preferred to any of the grains alone. When plain grains are fed, oats are preferred to corn and cracked corn to whole corn, but wheat, barley, rye or soybean meal are not very palatable (Hawkes et al., 1985). When a series of two choice preference tests of a basal diet (corn, oats, wheat bran, soybean meal, molasses and limestone) versus the basal diet plus 5–10% of various by-products were performed, horses avoided higher concentrations of non-grain feeds such as blood meal, a mixture of bone and meat meal, or beet pulp. They would accept 20% alfalfa meal and up to 10% meat and bone meal. After initially avoiding distillers grain, horses would accept it and preferred it to the basal diet (Hawkes et al., 1985). Rapeseed meal is a good source of protein and is accepted by horses at concentrations up to 15% (Sutton & Stredwick, 1979). Substituting fat calories for carbohydrate calories has been advocated to reduce the risk of myopathies as well as cribbing. Metabolic advantages also have been found with high oil diets (Hoffman et al., 2001), as well as behavioural advantages (Holland et al., 1996; Kronfeld et al., 1999), so it is important to determine which form of lipid horses will eat. The order of preference for pure oils is corn, peanut, safflower and cottonseed. Soy and corn lecithins were accepted similarly and the more corn oil that was added, the more acceptable

the horse found it. Of all the forms of tallow, only hydrolyzed tallow chips were palatable (Holland et al., 1998).

Horses seem to meter their intake orally in that they do not continue to eat when sham fed as do other species (dogs and rats; Ralston, 1984). Horses will eat more kilograms of a calorically diluted diet and keep their caloric intake constant (Laut et al., 1984). Gastrointestinal factors are important. Clinically, the suppression of appetite due to gastrointestinal pain is well recognized. Physiologically, glucose levels in the blood appear to be more important in predicting meals, that is, horses eat when their plasma glucose levels fall below 90 mg/dl (Ralston et al., 1979). Fat appears to be important, not in controlling meal size, but in intra-meal interval (Ralston & Baile, 1983). Cairns et al. (2002) have found that horses will show an increased preference for a flavour paired with high energy. Volatile fatty acid concentration in the caecum may be important as satiety factors (Ralston et al., 1983). Sedatives, especially benzodiazepine sedatives, stimulate intake (Brown et al., 1976) so care must be taken that following tranquillization, horses do not ingest food that they are too sedated to swallow. Food intake is also affected by environmental temperature (Cymbaluk & Christison, 1990).

Elimination

Urination by mares occurs once every three to four hours and defecation once every two to five hours (Boyd, 1988; Tyler, 1972). (See Table 6.5.) When grazing, the horse may pause to urinate, in most cases, but will rarely pause to defecate. When walking, the horse is likely to stop to both urinate and defecate (Waring, 1983). Horses urinate more frequently after exercise (Caanitz et al., 1991). The stallion behaviour of urinating on urine and defecating on faeces is discussed in Chapter 6. Elimination patterns vary from study to study. Ödberg and Francis-Smith (1977) found that horses avoided faeces and used the roughs to defecate, but Arnold (1984), Redman and Goodwin (1999) and Putnam et al. (1987) reported that faeces were concentrated in the areas where the horses grazed and rested; i.e., they did not move far from their grazing or resting spot to defecate. Sweeting et al. (1985) found that stalled ponies moved away from their feeding area before eliminating on only half of the occasions.

Table 6.5. *Drinking, elimination and self-grooming*

Population	Environment	Drinking %	Defecation %	Urination %	Self-grooming %	Author
Przewalski's	Pasture	0.47	0.3	0.6	1.7	Boyd *et al.*, 1988
Pony	Box stall	0.4–0.8	0.1	0.1	3–4	Houpt *et al.*, 1988
Horse	Tie stall	0.4–1.4			0.05–0.20	McDonnell *et al.*, 1999

Drinking

Free-ranging horses in arid environments may drink only once every day or two. They must travel long distances from the grazing location to the water location. Sneddon (1993) and Sneddon *et al.* (1993) compared the responses of Namibian feral horses free ranging in the desert for 100 years with Boerpferd, a breed acclimated to a subtropical environment for the same period. The Namibian horses had higher plasma osmolality and vasopressin and lower aldosterone than the non-arid adapted horse. Apparently horses adapted to the arid conditions of the Namibian desert had become more efficient at water conservation in 100 years. Both breeds could rapidly rehydrate when water was made available.

Most domestic horses have water available at all times. In those circumstances, they drink frequently and tend to be prandial drinkers, consuming water shortly before, during or shortly after eating (Sufit *et al.*, 1985). Ponies consuming a pelleted diet drank 30 l/day. In general, horses consume daily 2–7 l/100 kg body weight (Cymbaluk, 1989; Houpt *et al.*, 2000, 2001; Sweeting & Houpt, 1987; Mueller & Houpt, 1991). Horses drank more when fed hay and less when eating pellets. Ponies offered water four times a day drank 3.2 l/kg dry matter when fed hay, but 2.0 l/kg dry matter when fed pellets. Fonnesbeck (1968) found similar results. Water intake is often expressed as l/kg dry matter, but there is a fallacy in expressing water intake on a dry matter of intake basis unless both are available free choice because if water is limited, food intake will decrease and vice versa. When horses' schedules of exercise and management are changed, water intake changes despite constant feeding schedules and amount (Welford *et al.*, 1999).

The amount of water and frequency of drinking also varies with the environmental temperature in confined and in pastured horses (Crowell-Davis *et al.*, 1985; Cymbaluk & Christison, 1989; Scheibe *et al.*,

1998). Scheibe, using a transponder-operated watering stall to measure individual intake of Przewalski's horses in a semi-reserve, found that intake increased from 4 l to 20 l/day as the temperature rose from below 0 to 24 °C. The warmer the weather, the more frequently horses drink. Drinking rate increased from every 8 hours to every hour as environmental temperature increased from 5 to 35 °C.

Horses spend more time drinking after treadmill exercise or turn-out in a paddock (Caanitz *et al.*, 1991; Houpt *et al.*, 2001). Lactation also stimulates water intake. Thoroughbred mares drank about 26 kg/day the day before foaling and 41 kg/day the day after foaling (Houpt *et al.*, 1995). Milk production in these mares is probably about 15 kg /day (Oftedal *et al.*, 1983) which would indicate that lactating mares drink enough to replace the fluid lost in milk quite accurately.

Horses drink more from buckets than from automatic waterers (Nyman & Dahlborn, 2001). Horses may also underdrink during or after transport. Masking the taste of water with apple or clover flavor did not influence consumption of water after shipping; but the horses preferred the apple to the clover flavouring. A novel source of water is better accepted in a familiar rather than a novel environment (Mars *et al.*, 1992). Unfortunately, a new water source is most likely to be offered at a new location. When an apple, blackcurrant, honey, orange, mint or vanilla solution was offered to horses, none were preferred to plain water (Murphy *et al.*, 1999).

The identified stimuli for thirst are an increase of plasma osmolality or a decrease in plasma volume. Ponies are not able to compensate for water deprivation precisely. When deprived for 19 hours, they over compensate (Sufit *et al.*, 1985) and when deprived for 36 hours, they under compensate (Mueller & Houpt, 1991). Although water deprivation will result in both stimuli, there are some circumstances where only one stimulus occurs. For example, water restriction leads to hyperosmolality (Houpt et al., 2000) and admin-istration of the diuretic frusemide results in hypo-volaemia (Sufit et al., 1985). This drug is given to horses in some circumstances before they race with a view to reducing the risk of pulmonary haemorrhage. However, it may give the horse an advantage because it may lose weight due to the diuretic effect (Gross et al., 1999).

Salt intake

Horses, like other herbivores, consume a sodium defi-cient diet, so must seek another source of that min-eral. Salt licks attract free-ranging horses. Stalled horses consume 19–143 g of salt a day, but there is marked individual variation (Schryver et al., 1987). Horses will increase their salt intake if sodium defi-cient (Houpt et al., 1991), but will not eat calcium supplements when deficient in calcium (Schryver et al., 1978). Sweating increases sodium loss in the horse, so one would expect more salt intake after exer-cise. Administration of frusemide causes sodium loss in the urine because the mechanism of action of the diuretic is to block chloride and sodium re-uptake in the renal tubules so the horse excretes a high sodium urine (Houpt et al., 1991) in comparison to the usual virtually sodium-free urine.

Coat care

Rolling is a behaviour that may be related to coat care or general comfort behaviour. It is the only way a single horse can rub the dorsal surface of the body. Before rolling, the horse will first lower its head, sniff the ground and often paw. Then it will flex all four limbs and sink to the ground to one side or the other. It may roll from side to side or remain on one side. If it does not roll to a second side, it may arise and lie down again with the second side down. When a normal, healthy horse has rolled, it will shake, dislodging some of the soil that clings to its coat. Rolling may also have a marking function because stallions roll more than mares. Although foals perform other forms of self-grooming more than adults, they roll less.

Horses swish their tails for two reasons: insect removal and aggression. Both insects and other horses can be considered irritants. Tactile stimuli produced by insects, especially biting flies, result in the fly swish-ing response. Horse tails are rarely docked now in Europe or North America except for some breeds of show horse. In some countries, docking or cutting the

tail so it is less than 20 cm long is common, especially for harness horses or horses whose saddles are stabi-lized by a crupper. Flies can also be dislodged if they stimulate the panniculus response, in which the skin is shaken. This might be a more frequent response in docked horses. Stamping is a response to insects on the legs. Horses can actually damage their legs due to the concussion of repeated stamping. A final response to external irritation is a simple swipe of the muzzle across the skin of the side or chest, presumably to brush off the irritating stimulus.

Horses can scratch themselves with their hind hooves, but this is more common in young horses and ponies. Horses rub their heads against their own front limbs or other objects. This is frequently seen when a horse, sweating after strenuous exercise, rubs his head on a person. Some rubbing is normal, but an increase in frequency or duration of rubbing may indicate a health problem. External parasites, especially lice or dermatitis, can produce pruritus to which the horse responds by rubbing. Culicoides fly bites often initi-ate an allergic response in horses called 'sweet itch' which is very pruritic. The intense rubbing can result in hair loss, especially of the mane. Horses will rub their heads or their rumps against solid inanimate objects. Rump or tail–head rubbing may indicate the presence of the pinworm Oxyuris.

In general, mares self-groom (2% of their time) more than stallions (1%) (Boyd, 1988). Grooming increases after exercise (Caanitz et al., 1991). See Table 6.5 for time spent grooming, drinking and elim-inating, and Figure 6.4 for the activities of a single horse.

Behavioural thermoregulation

Horses are fairly efficient at physiological thermo-regulation. They cool themselves by sweating and insulate themselves against the cold by growing a thick coat. There are breed and individual differences in ability to grow a thick coat. Cold-blooded horses and ponies grow thicker coats than Arabians and Thoroughbreds reflecting their origins. The Arabians are believed to have come from the desert areas of North Africa and the Arabian peninsula and the cold-blooded horse from northern Europe, although all horses appear to have evolved from a common ances-tor (Groves, 1994).

Behavioural thermoregulation involves selecting an environment and an activity that will minimize

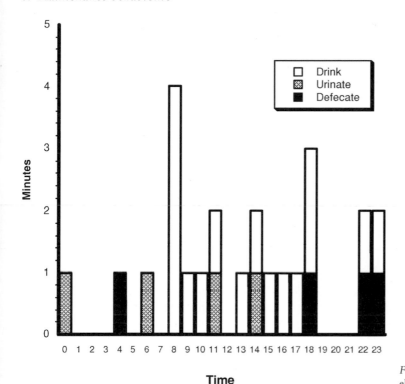

Figure 6.4. Twenty-four hour patterns of elimination and drinking in a mare.

thermal stress. This is an evolutionarily stable behavioural strategy because it is less costly to the animal than losing water by sweating or burning calories to keep warm.

Many mammalian species, such as swine, tend to clump in cold weather; in contrast, horses stand closer together in the summer, presumably to take advantage of one another's tails for protection from flies. Another strategy is to stand in water or on barren patches of ground (Duncan & Cowtan, 1980). In some climates, patches of snow remain even in the summer and horses will stand there to avoid insects (Keiper & Berger, 1982). When a water source is nearby, the number of drinking bouts per day is proportional to the environmental temperature (Crowell-Davis *et al.*, 1985).

Hot weather

In hot weather, horses will seek means to avoid hyperthermia. To avoid direct solar radiation, they will stand in the shade during the warmest parts of the day (Crowell-Davis, 1994) and possibly stand near rocks that retain a cooler temperature (L. Boyd,

pers. comm.). Horses do not drink more cool water in warm weather (McDonnell & Kristula, 1996), although they will drink more warm water in cold weather. Horses are least likely to be active during the hottest part of the day; instead they stand-rest (Tyler, 1972).

Cold weather

Cold, dry weather is well tolerated by most horses, but cold, wet weather is not. When the temperature drops below their critical temperature, horses shiver. The behavioural response to cold is to minimize exposure to the cold by seeking shelter and by reducing heat loss. To reduce heat loss, they turn so that the smallest surface area, the hindquarters, is facing the wind (Boyd & Houpt, 1994). As noted above, grazing horses usually spend more time grazing in the winter both because herbage is scarce and because their caloric needs are greater in the cold (Tyler, 1972). Horses on pasture graze on less exposed areas of the field when temperatures are low and wind speed is high. Horses exposed to the cold with no adaptation, (i.e. no heavy coat) can be taught to turn on heat lamps in cold weather (Houpt, 1991). Horses also avoid increasing the cold load by reducing intake of cold water. Reduction of intake can result in gastrointestinal problems because cold weather also stimulates food intake so that the horse is taking in more solids but less liquid. The fact that the horses are responding to the temperature of the water was shown by Kristula and McDonnell (1994) who found that, in cold weather, horses consumed more water when it was heated.

Horses stand rather than lie down when it rains, and standing increased 20 minutes per day for every degree (Celsius) drop in temperature during the winter when mean temperature was −2 °C (Duncan, 1985).

Locomotion

Horses travel great distances when necessary to a source of water. For example, in the Great Basin of the USA, water is scarce and horses have been observed to travel 16 km to water from a feeding area (Feist & McCullough, 1976). See Table 6.3 for percentage of time moving on pasture and Table 6.4 for percentage of time moving in stalls and pens. Isolation increases locomotion. The size of the pasture does not determine the distance a horse travels, but does influence the gait at which the horse travels. Yearlings canter less often in fields under 1.5 ha (Kusunose et al., 1985). Yearlings travel more when alone. Yearlings traveled 8–15 km during 7 hours of turn-out in a pasture when alone, but only 3–6 km when in a group of three (Kusunose et al., 1986).

Horses living in a small enclosure spent 7% of their time walking when other horses lived in adjacent paddocks, but 25% when alone (Houpt & Houpt, 1988). Horses on a large paddock spend less than half a minute per hour walking (Caanitz et al., 1991). Stallions walk more than mares. Perhaps the most interesting effect on locomotion is that of release into the wild of Przewalski's stallions. From the day of release, after a year of conditioning in a large enclosure, the stallions spent more time walking, and the increase in walking has persisted for several years (Boyd, 1998, 2002). The reason for this is not clear. Perhaps the stallion is patrolling to detect rival stallions or predators. The mares are not more active, so foraging for food or water is probably not the reason.

Because feral or pastured horses are not very active, the exercise requirement of the horse is difficult to

determine. Mares used in oestrogen production may be kept in tie stalls for two or more weeks without release. When pregnant mares were kept in similar conditions, and their behaviour recorded when they were released after two weeks, they spent more time in motion than horses released daily. This indicates compensatory exercise after exercise restriction (Houpt et al., 2001). If an activity is valuable to the animal, it will attempt to make up for deprivation by a compensatory increase in the activity. This is true of locomotion as well as of water, salt and food.

When given a choice, horses spent only 30–45 minutes outside in a barren paddock in the company of other horses. They chose to spend only 15 minutes there when alone. If the horses had not been released for two days, they chose to stay outside for an hour with other horses, but still chose to spend no more than 15 minutes when alone in the paddock (Lee et al., 2001). This study indicates that horses do want to leave their stalls, but only for a short period unless they are in the same enclosure as other horses. Similar results were found using operant conditioning techniques. The motivation for release from a stall was equal to the motivation to be with another horse, but the motivation for either was much less than for food (Lee, 2000).

Sleep and resting behaviour

Resting time in horses can be divided into stand-resting and lying. Lying occupies a small percentage of the horse's time, about an hour a day, but this time does not vary. If grazing time increases, stand-resting time, but not lying, will decrease. Resting behaviour varies with the season in free-ranging horses. In the winter, New Forest ponies spent less than an hour resting during the day. Most of the resting takes place between 09:00 and 13:00. Much more time, up to five hours, is spent resting during the day in the summer (Tyler, 1972); Camargue horses show a similar pattern: stand-resting during the day in summer, but at night in winter. Stand-resting time is inversely related to foraging time (Duncan, 1985). Pony mares on pasture stand-rest 13–32% of the day (Crowell-Davis, 1994). The smaller the paddock size, the less time horses spend stand-resting (Boyd, 1988). It is surprising that stalled horses rest less immediately after exercise (Caanitz et al., 1991) (See Table 6.6).

The stages of sleep are identified by their electrical characteristics. Wakefulness is characterized by

Table 6.6. *Lying*

Population	Time	Environment	Diet	Sternal % lying	Lateral % lying	Total % lying	Author
Horse	day	Metabolism cage	limited hay				Willard *et al.*, 1977
Horse	day		limited concentrate				
Pony	night		pasture	6	1.4	7.4	Houpt *et al.*, 1986
Pony	night	Box stall	limited hay and grain	12.3	0.5	12.8	Houpt *et al.*, 1986
		Box stall		2–4	0.3	2–4.3	Houpt *et al.*, 1986
Horse	night	Box stall	limited hay and grain			6	
Horse	24 h	Tie stall	ad lib hay			6.7	McDonnell *et al.*, 1999
		Tie stall	ad lib hay				Houpt *et al.*, 2001
Pony	day		pasture			0	Crowell-Davis, 1994
Przewalski's	day	Corral	limited hay and grain			0	Boyd, 1988
		Corral	limited hay and grain				Boyd, 1988
Przewalski's	24 h	Pasture		4.1	1.2		Boyd *et al.*, 1988
Horse	24 h	Pasture		4.7	0.6	5.3	Duncan, 1980

small amplitude, high frequency electroencephalograms. Slow wave sleep, as the name implies, is characterized by large amplitude, low frequency electrical activity. Paradoxical or REM is characterized by small amplitude, high frequency electroencephalograms and very low electrical activity in the muscles of the body. Sleep in horses has not been studied extensively, because of the difficulty of obtaining electroencephalograms from the head of an animal with large jaw and facial muscles (Dallaire & Ruckebusch, 1974). The few physiological studies have been supplemented with more numerous observational ones. Observational studies reveal that horses stand with a hindlimb flexed, their head lowered and their eyes half closed for two to four hours per day (Ruckebusch, 1972). This posture is believed to indicate that the horse is in slow wave sleep.

Horses sleep mostly at night. They spend 30–40% of the dark period sleeping, but most of the sleep is slow wave sleep which the horse is able to do standing. Rapid eye movement sleep has only been observed in the recumbent horse. The horse cannot support its weight during the complete muscle relaxation characteristic of REM sleep; therefore, it lies in lateral recumbency or in sternal recumbency with the muzzle supported on the ground. Each bout of REM sleep lasts only 3–10 minutes and is usually preceded and followed by slow wave sleep. Horses may dream during REM sleep because they vocalize and may kick or make running motions. Rapid eye movement sleep

is followed by periods of masturbation in stallions (Wilcox *et al.*, 1991).

Recumbent rest (presumably REM sleep) occupies more of the foal's time budget than that of the adult. There is a gradual decrease with age, so that even two-year-olds lie down more than adults (Boy & Duncan, 1979).

Horses do not lie down indiscriminately. For example, they may be reluctant to lie down in a new environment. Ruckebusch (1975) noted that ponies would not lie down their first night on pasture and lay down 30% fewer minutes for three nights. Reluctance to lie down may be one component of transport stress if transport lasts for more than one day, because horses may not lie down when they can feel motion or when in a narrow stall. Horses may be reluctant to lie down in confined spaces. Light horses restrained in tie stalls would not lie down and would lean against rails or walls. Some eventually fell to their knees, apparently when they entered REM sleep and lost muscle tone (Houpt *et al.*, 2001). Experience, particularly early experience, may be necessary before a horse is willing to lie down in a narrow place.

Confinement may not be the only reason a horse won't lie down; horses can be selective in choosing a bedding. An uncomfortable substrate, such as mud, may cause the horse to refuse to lie down. When given a choice, a horse will lie down on a more resilient surface. For example, when ponies were given a choice between a stall bedded with wood shavings and a bare concrete floor, none of the ponies lay down on the bare floor (Hunter & Houpt, 1989). The type of bedding, however, seems less important. Straw was not preferred to wood shavings in the American study (Hunter & Houpt, 1989), but was in a British study (Mills *et al.*, 2000). The British horses had to walk along a passage from one stall to another to find the bedding, whereas the American ones had only to step over a 6 cm barrier. Whether the difference in results is due to differences in the straw, the shavings, the test design or the horses tested remains unknown.

Conclusion

One means of assaying the welfare of an animal is to compare its behaviour patterns and time budget under domestic conditions with those of wild members of the same species. Horses spend the majority of their time grazing. This is true of free-ranging horses, whether they are living in the American West (Salter & Hudson, 1979; Berger, 1986), the Camargue region of France (Mayes & Duncan, 1986), the New Forest of England (Tyler, 1972) or the steppes of Mongolia (Boyd, 2002). From observation it would appear that pastured horses spend a similar amount of time grazing as do free ranging horses, i.e. about 60%. In the author's opinion, the optimum environment for the domestic horse should include ad libitum forage, ad libitum water, a source of salt, a soft, dry place to lie, close contact with other horses and opportunity for daily release for 30–60 minutes in the company of other horses. A challenge for the future is to determine objectively conditions for equine welfare.

References

Alexander, F. (1966). A study of parotid salivation in the horse. *Journal of Physiology*, 184: 646–56.

Archer, M. (1973). The species preferences of grazing horses. *Journal of the British Grassland Society*, 28: 123–8.

(1978). Further studies on palatability of grasses to horses. *Journal of the British Grassland Society*, 33: 239–43.

Argenzio, R. A. (1992). Pathophysiology of diarrhea. In *Veterinary Gastroenterology*, 2nd edn., ed. N. V. Anderson, Philadelphia: Lee and Febiger, pp. 163–72.

Arnold, G. W. (1984). Comparison of the time budgets and circadian patterns of maintenance activities in sheep, cattle and horses grouped together. *Applied Animal Behaviour Science*, 13: 19–30.

Berger, J. (1986). *Wild Horses of the Great Basin*. Chicago: University of Chicago Press.

Boy, V. & Duncan, P. (1979). Time budgets of Camargue horses. I. Developmental changes in the time budgets of foals. *Behaviour*, 71: 187–202.

Boyd, L. E., (1988). Time budgets of adult Przewalski horses: effects of sex, reproductive status and enclosure. *Applied Animal Behaviour Science*, 21: 19–39.

(1998). The 24-h time budget of a takh harem stallion (*Equus ferus przewalskii*) pre- and post-reintroduction. *Applied Animal Behaviour Science*, 60: 291–9.

(2002). Reintroduction of takhi (*Equus ferus przewalski*) to Mongolia: time budget and synchrony of pre- and post-release. *Applied Animal Behaviour Science*, 78: 87–102.

Boyd, L. E., Carbonaro, D. A. & Houpt, K. A. (1988). The 24-hour time budget of Przewalski horses. *Applied Animal Behaviour Science*, 21: 5–17.

Boyd, L. & Houpt, K. A. (Eds) (1994). *Przewalski's Horse: The History and Biology of An Endangered Species*. Albany, NY: State University of New York Press.

Brown, R. F., Houpt, K. A. & Schryver, H. F. (1976). Stimulation of food intake in horses by diazepam and promazine. *Pharmacology, Biochemistry and Behavior*, 5: 495–7.

Caanitz, H., O'Leary, L., Houpt, K. A., Petersson, K. & Hintz, H. (1991). Effect of exercise on equine behavior. *Applied Animal Behaviour, Science*, 31: 1–12.

Cairns, M. C., Cooper, J. J., Davidson, H. P. B. & Mills, D. S. (2002). Association in horses of orosensory characteristics of foods with their post-ingestive consequences. *Animal Science*, 75: 257–65.

Cosenza, M. (1999). Foraging behavior of the domestic horse (*Equus caballus*). Thesis, presented to The Division of Biological Sciences, Cornell University.

Crowell-Davis, S. L. (1994). Daytime rest behaviour of the Welsh pony (*Equus caballus*) mare and foal. *Applied Animal Behaviour Science*, 40: 197–210.

Crowell-Davis, S. L., Houpt, K. A. & Carnevale, J. (1985). Feeding and drinking behavior of mares and foals with free access to pasture and water. *Journal of Animal Science*, 60: 883–9.

Cymbaluk, N. F. (1989). Water balance of horses fed various diets. *Equine Practice*, 11(1): 19–24.

Cymbaluk, N. F. & Christison, G. I. (1989). Effects of diet and climate on growing horses. *Journal of Animal Science*, 67: 48–59.

(1990). Environmental effects on thermoregulation and nutrition of horses. *Veterinary Clinics of North America: Equine Practice*, 6: 355–72.

Dallaire, A. & Ruckebusch, Y. (1974). Sleep and wakefulness in the housed pony under different dietary conditions. *Canadian Journal Comparative Medicine*, 38: 65–71.

Demment, M. W. & Van Soest, P. J. (1985). A nutritional explanation for body-size patterns of ruminant and nonruminant herbivores. *American Naturalist*, 125: 641–72.

Derksen, F. J. 1987. Chronic obstructive pulmonary disease. In *Current Therapy in Equine Medicine*, 2nd edn. ed. N. E. Robinson, Philadelphia: WB Saunders, pp. 596–602.

Doreau, M. (1978). Comportement alimentaire du cheval à l'écurie. *Zootech*, 3: 291–302.

Duncan, P. (1980). Time-budgets of Camargue horses. II. Time-budgets of adult horses and weaned sub-adults. *Behaviour*, 72: 26–49.

(1983). Determinants of the use of habitat by horses in a Mediterranean wetland. *Journal Animal Ecology*, 52: 93–109.

(1985). Time-budgets of Camargue horses. III. Environmental influences. *Behaviour*, 92: 188–208.

Duncan, P. & Cowtan, P. (1980). An unusual choice of habitat helps Camargue horses to avoid blood-sucking horse-flies. *Biology of Behaviour*, 5: 55–60.

Elia, J. B. (2002). The effects of diet differing in fiber content on equine behaviour and motivation for fiber. Thesis, presented to the Faculty, Graduate School of Cornell University as partial fulfillment of Master of Science Degree.

Feist, J. D. & McCullough, D. R. (1976). Behavior patterns and communication in feral horses. *Zeitschrift für Tierpsychologie*, 41: 337–71.

Fonnesbeck, P. V. (1968). Consumption and excretion of water by horses receiving all hay and hay-grain diets. *Journal of Animal Science*, 27: 1350–6.

Goodwin, D., Davidson, H. P. B. & Harris, P. (2002). Foraging enrichment for stabled horses: effects on behaviour and selection. *Equine Veterinary Journal*, 34(7): 686–91.

Gross, D. K., Morley, P. S., Hinchcliff, K. & Wittum, T. E. (1999). Effect of frusemide on performance of Thoroughbreds racing in the United States and Canada. *Journal of the American Veterinary Medical Association*, 215(5): 670–5.

Groves, C. P. (1994). Morphology, habitat, and taxonomy. In *Przewalski's Horse: The History and Biology of An Endangered Species*, ed. L. Boyd, & K. A. Houpt. Albany: State University of New York Press, pp. 39–59.

Hawkes, J., Hedges, M., Daniluk, P., Hintz, H. F. & Schryver, H. F. (1985). Feed preferences of ponies. *Equine Veterinary Journal*, 17(1): 20–2.

Hintz, H. F. & Loy, R. G. (1966). Effects of pelleting on the nutritive value of horse rations. *Journal of Animal Science*, 25: 1059–62.

Hintz, H. F., Schryver, H. F., Mallette, J. & Houpt, K. (1989). Factors affecting rate of grain intake by horses. *Equine Practice*, 11(4): 35–42.

Hoffman, R. M., Wilson, J. A., Kronfeld, D. S., Cooper, W. L., Lawrence, L. A., Sklan, D. & Harris, P. A. (2001). Hydrolyzable carbohydrates in pasture, hay, and horse feeds: direct assay and seasonal variation. *Journal of Animal Science*, 79(2): 500–6.

Holland, J. L., Kronfeld, D. S. & Meacham, T. N. (1996). Behavior of horses is affected by soy lecithin and corn oil in the diet. *Journal of Animal Science*, 74: 1252–5.

Holland, J. L., Kronfeld, D. S., Rich, G. A., Kline, K. A., Fonetnot, J. P., Meacham, T. N. & Harris, P. A. (1998). Acceptance of fat and lecithin containing diets by horses. *Applied Animal Behaviour Science*, 56(2–4): 91–6.

Holmes, L. N., Song, G. K. & Price, E. O. (1987). Head partitions facilitate feeding by subordinate horses in the presence of dominant pen-mates. *Applied Animal Behaviour Science*, 19: 179–82.

Houpt, K. A. (1991). Animal behavior and animal welfare. *Journal American Veterinary Medical Association*, 198(8): 1355–60.

Houpt, K. A., Eggleston, A., Kunkle, K. & Houpt, T. R. (2000). Effect of water restriction on equine behaviour and physiolgoy. *Equine Veterinary Journal*, 32(4): 341–4.

Houpt, K. A. & Houpt, T. R. (1988). Social and illumination preferences of mares. *Journal of Animal Science*, 66: 2159–64.

Houpt, K. A., Houpt, T. R., Johnson, J. L., Erb, H. N. & Yeon, S. C. (2001). The effect of exercise deprivation on the behaviour and physiology of straight stall confined pregnant mares. *Animal Welfare*, 10: 257–67.

Houpt, K. A., Hunt, P. & Monier, S. (1995). The relation of maternal fluid balance to offspring well-being. *Eastern Psychological Association, Boston, MA, March 31– April 2.*

Houpt, K. A., Northrup, N., Wheatley, T. & Houpt, T. R. (1991). Thirst and salt appetite in horses treated with frusemide. *Journal Applied Physiology*, 71(6): 2380–6.

Houpt, K. A., O'Connell, M. F., Houpt, T. A. & Carbonaro, D. A. (1986). Night-time behavior of stabled and pastured peri-parturient ponies. *Applied Animal Behaviour Science*, 15: 103–11.

Houpt, K. A., Perry, P. J., Hintz, H. F. & Houpt, T. R. (1988). Effect of meal frequency on fluid balance and behavior of ponies. *Physiology and Behavior*, 42: 401–7.

Houpt, K. A., Zahorik, D. M. & Swartzman-Andert, J. A. (1990). Taste aversion learning in horses. *Journal of Animal Science*, 68: 2340–4.

Houpt, T. R. & Houpt, K. A. (1971). Nitrogen conservation by ponies fed a low-protein ration. *American Journal Veterinary Research*, 32: 579–88.

Hunter, L. & Houpt, K. A. (1989). Bedding material preferences of ponies. *Journal of Animal Science*, 67: 1986–91.

Janis, C. M. (1976). The evolutionary strategy of the Equidae and the origins of rumen and cecal digestion. *Evolution*, 30: 757–74.

Kaseda, Y. (1983). Seasonal changes in time spent grazing and resting of Misaki horses. *Japanese Journal of Zootechnical Science*, 54(7): 464–9.

Keiper, R. R. & Berger, J. (1982). Refuge-seeking and pest avoidance by feral horses in desert and island environments. *Applied Animal Ethology*, 9: 111–20.

Keiper, R. R. & Keenan, M. A. (1980). Nocturnal activity patterns of feral horses. *Journal of Mammalogy*, 61: 116–18.

Kristula, M. A. & McDonnell, S. M. (1994). Drinking water temperature affects consumption of water during cold weather in ponies. *Applied Animal Behaviour Science*, 41: 155–60.

Kronfeld, S., Holland, J., Hoffman, R. & Harris, P. (1999). Dietary influences on behaviour and stress. *Proceedings Equine Veterinary Journal*, **28** (Suppl.): 64.

Krzak, W. E., Gonyou, H. W. & Lawrence, L. M. (1991). Wood chewing by stabled horses: diurnal pattern and effects of exercise. *Journal of Animal Science*, **69**: 1053–8.

Kusunose, R., Hatakeyama, H., Ichikawa, F. *et al.* (1986). Behavioural studies on yearling horses in field environments. 2. Effects of the group size on the behaviour of horses. *Bulletin of the Equine Research Institute*, **23**: 1–6.

Kusunose, R., Hatakeyama, H. & Kubo, K. *et al.* (1985). Behavioural studies on yearling horses in field environments. I. Effects on the field size on the behaviour of horses. *Bulletin of the Equine Research Institute*, **22**: 1–7.

Laut, J. E., Houpt, K. A., Hintz, H. F. & Houpt, T. R. (1984). The effects of caloric dilution on meal patterns and food intake of ponies. *Physiology and Behavior*, **35**: 549–54.

Lee, J. Y. (2000). Motivation of horses for release from straight stall confinement and for exercise. Thesis: Presented to the Faculty of the Graduate School of Cornell University in partial fulfillment of the requirements for the Degree of Master of Science.

Lee, J., Floyd, T. & Houpt, K. (2001). Operant and two-choice preference applied to equine welfare. In *Proceedings of the International Society of Applied Ethology*: 110.

Marinier, S. L. & Alexander, A. J. (1992). Use of field observations to measure individual grazing ability in horses. *Applied Animal Behaviour Science*, **33**: 1–10.

Mars, L. A., Kiesling, H. E., Ross, T. T., Armstrong, J. B. & Murray, L. (1992). Water acceptance and intake in horses under shipping stress. *Equine Veterinary Science*, **12**(1): 17–20.

Mayes, E. & Duncan, P. (1986). Temporal patterns of feeding in free-ranging horses. *Behaviour*, **96**: 105–29.

McDonnell, S. M. & Kristula, M. A. (1996). No effect of drinking water temperature (ambient vs. chilled) on consumption of water during hot summer weather in ponies. *Applied Animal Behaviour Science*, **49**: 159–63.

McDonnell, S. M., Freeman, D. A., Cymbaluk, N. F., Schott II, H. C., Hinchcliff, K. & Kyle, B. (1999). Behaviour of stabled horses provided continuous or intermittent access to drinking water. *American Journal of Veterinary Research*, **60**(11): 1451–6.

Mills, D. S., Eckley, S. & Cooper, J. J. (2000). Thoroughbred bedding preferences, associated behaviour differences, and their implications for equine welfare. *Animal Science*, **70**: 95–106.

Mueller, P. J. & Houpt, K. A. (1991). A comparison of the responses of donkeys (*Equus asinus*) and ponies (*Equus caballus*) to 36 hours of water deprivation. In *Donkeys, Mules and Horses in Tropical Agricultural Development*, ed. D. Fielding & R. A. Pearson. Edinburgh: University of Edinburgh Press, pp. 86–95.

Murphy, K., Wishart, S. & Mills, D. (1999). The acceptability of various flavoured solutions by Thoroughbred horses.

Proceedings of the Equine Veterinary Journal **28** (Suppl.): 67.

Nicol, C. J., Davidson, H. P. D., Harris, P. A., Waters, A. J. & Wilson, A. D. (2002). Study of crib-biting and gastric inflammation and ulceration in young horses. *Veterinary Record*, **151**: 658–62.

Nyman, S. & Dahlborn, K. (2001). Effect of water supply method and flow rate on drinking behavior and fluid balance in horses. *Physiology and Behavior*, **73**: 1–8.

Ödberg, F. O. & Francis-Smith, K. (1977). Studies on the formation of ungrazed eliminative areas in fields used by horses. *Applied Animal Ethology*, **3**: 27–34.

Oftedal, O. T., Hintz, H. F. & Schryver, H. F. (1983). Lactation in the horse: milk composition and intake by foals. *Journal of Nutrition*, **113**: 2096–106.

Pfister, J. A., Stegelmeier, B. L., Cheney, C. D., Ralphs, M. H. & Gardner, D. R. (2002). Conditioning taste aversions to locoweed (*Oxytropis sericea)* in horses. *Journal of Animal Science*, **80**: 79–83.

Putnam, R. J., Pratt, R. M., Ekins, J. R. & Edwards, P. J. (1987). Food and feeding behaviour of cattle and ponies in the New Forest. *Journal of Applied Ecology*, **24**(2): 369–80.

Ralston, S. L. (1984). Controls of feeding horses. *Journal of Animal Science*, **59**: 1354–61.

Ralston, S. L. & Baile, C. A. (1983). Effects of intragastric loads of xylose, sodium chloride and corn oil on feeding behaviour of ponies. *Journal of Animal Science*, **56**: 302–8.

Ralston, S. L., Freeman, D. E. & Baile, C. A. (1983). Volatile fatty acids and the role of the large intestine in the control of feed intake of ponies. *Journal of Animal Science*, **57**: 815–25.

Ralston, S. L., Van den Broek, G. & Baile, C. A. (1979). Feed intake patterns and associated blood glucose, free fatty acid and insulin changes in ponies. *Journal of Animal Science*, **49**: 838–45.

Randall, R. P., Schrug, W. A. & Church, D. C. (1978). Response of horses to sweet, salty, sour and bitter solutions. *Journal of Animal Science*, **47**: 51–5.

Redman, P. & Goodwin, D. (1999). Grazing and defaecation behaviour of a bachelor group of Przewalski horses (*Equus prezewalskii*) under free-ranging and enclosed conditions. In *Proceedings of the Equine Veterinary Journal*, **28** (Suppl.): 68.

Rubenstein, D. I. (1981). Behavioural ecology of island feral horses. *Equine Veterinary Journal*, **13**: 27–34.

Ruckebusch, Y. (1972). The relevance of drowsiness in the circadian cycle of farm animals. *Animal Behavior*, **20**: 637–43.

  (1975). The hypnogram as an index of adaptation of farm animals to change their environment. *Applied Animal Ethology*, **2**: 3–18.

Salter, R. E. & Hudson, R. J. (1979). Feeding ecology of feral horses in western Alberta. *Journal of Range Management*, **32**: 221–5.

Scheibe, K. M., Eichhorn, K., Kalz, B., Streich, W. J. & Scheibe, A. (1998). Water consumption and watering behaviour of Przewalski horses (*Equus ferus przewalskii*) in a semireserve. *Zoo Biology*, **17**: 181–92.

Schryver, H. F., Parker, M. T., Daniluk, P. D., Pagan, K. I., Williams, J., Soderholm, L. V. & Hintz, H. F. (1987). Salt consumption and the effect of salt on mineral metabolism in horses. *Cornell Veterinarian*, **77**: 122–31.

Schryver, H. F., VanWie, S., Daniluk, P. & Hintz, H. F. (1978). The voluntary intake of calcium by horses and ponies fed a calcium deficient diet. *Journal Equine Medicine and Surgery*, **2**: 337–40.

Shaw, E. B., Houpt, K. A. & Holmes, D. F. (1988). Body temperature and behaviour of mares during the last two weeks of pregnancy. *Equine Veterinary Journal*, **20**: 199–202.

Sneddon, J. C. (1993). Physiological effects of hypertonic dehydration on body fluid pools in arid-adapted mammals. How do Arab-based horses compare? *Comparative Biochemistry and Physiology*, **104**(2): 201–13.

Sneddon, J. C., Van derWalt, J., Mitchell, G. & Hammer, S. (1993). Effects of dehydration and rehydration on plasma vasopressin and aldosterone in horses. *Physiology and Behavior*, **54**: 223–8.

Sufit, E., Houpt, K. A. & Sweeting, M. (1985). Physiological stimuli of thirst and drinking patterns in ponies. *Equine Veterinary Journal*, **17**: 12–16.

Sutton, E. I. & Stredwick, R. V. (1979). Acceptance of rapeseed meal (cv. Candle) by horses. *Canadian Journal of Animal Science*, **59**: 819–20.

Sweeting, M. P., Houpt, C. E. & Houpt, K. A. (1985). Social facilitation of feeding and time budgets in stabled ponies. *Journal of Animal Science*, **160**: 369–74.

Sweeting, M. P. & Houpt, K. A. (1987). Water consumption and time budgets of stabled pony (*Equus caballus*) geldings. *Applied Animal Behaviour Science*, **17**: 1–7.

Tyler, S. J. (1972). The behaviour and social organization of the New Forest ponies. *Animal Behaviour Monographs*, **5**: 85–196.

Waring, G. H. (1983). *Horse Behavior: The Behavioral Traits and Adaptations of Domestic and Wild Horses, Including Ponies*. Park Ridge, NJ: Noyes Publications.

Welford, D., Mills, D., Murphy, K. & Marlin, D. (1999). The effect of changes in management scheduling on water intake by the Thoroughbred horse. *Proceedings of the Equine Veterinary Journal*, **28** (Suppl.): 71–2.

Wilcox, S., Dusza, K. & Houpt, K. A. (1991). The relationship between recumbent rest and masturbation in stallions. *Equine Veterinary Science*, **11**(1): 23–6.

Willard, J. G., Willard, J. C., Wolfram, S. A. & Baker, J. P. (1977). Effect of diet on cecal pH and feeding behaviour of horses. *Journal of Animal Science*, **45**: 87–93.

7

Sexual behaviour

Sue M. McDonnell

Introduction

The general social organization and reproductive patterns of equids under natural herd conditions are discussed by Feh in Chapter 5. In that chapter, reproductive behaviour is presented from the perspective of relationships and social communication, with a discussion of the various equid species. This chapter will focus on sexual behaviour of the horse. It begins with a brief description of each of the various circumstances under which breeding occurs under natural herd conditions and descriptions of the specific precopulatory and copulatory behaviour sequences of horses. This is followed by a description of the sexual behaviour of horses under the most popular methods of domestic horse breeding, along with a summary of what is known about factors influencing sexual behaviour in domestically bred horses. Finally, some of the common problems of sexual behaviour and reproduction in domestic stallions and mares are discussed.

Breeding under natural social conditions

The body of literature describing the behaviour and ecology of the wild, feral and semi-feral equid populations throughout the world includes considerable information on sexual behaviour. Several representative observational studies have described sexual behaviour of horses in varying depths (Feist, 1971; Tyler, 1972; Miller, 1981; Salter & Hudson, 1982; Keiper & Houpt, 1984; McCort, 1984; Keiper, 1985; Stevens, 1987, 1990). Variations in social organization among the various horse populations studied have been addressed in Chapters 4 and 5. Among the horse populations studied, there is fairly general agreement of the observations on sexual behaviour. The following summary is based both on the information available from these landmark studies of free-ranging populations as well as the author's observations of breeding within feral and semi-feral herds. Many of the observations are based on a semi-feral herd of Shetland-type ponies that is maintained by the

The Domestic Horse: The Origins, Development, and Management of its Behaviour, ed. D. S. Mills & S. M. McDonnell.
Cambridge University Press. © Cambridge University Press 2005.

author at the University of Pennsylvania in Chester County, Pennsylvania, USA. Accessibility for continuous year-round daily observations over a 10-year period of this population so far, together with DNA-confirmed parentage, and whole-life histories of all individuals, has made this herd an extremely useful model for filling in gaps or confirming observations from many elegant more challenging studies over the decades of truly feral populations. While there are obvious limitations on questions of inter-herd interaction, migration, and extreme environmental stress, for the details of sexual or other categories of behaviour within a herd, observations of this herd appear to be consistent with reports of free-ranging horse herds.

Breeding within established harems

As described by Feh in Chapter 5, under natural social conditions most of the breeding involving mature mares is done within relatively stable harem groups that are typically comprised of one mature breeding stallion and one or more mature mares and their young offspring. A harem stallion is with his mares year-round, devoting considerable effort to keeping mares and young of the harem together and away from other stallions and threats. The harem stallion also appears to monitor continually the reproductive status of the harem mares year-round with periodic olfactory investigation of the mare and of voided urine and faeces, known as elimination marking behaviour (Turner *et al.*, 1981).

In feral and semi-feral horse herds breeding under free-ranging conditions, the annual reproductive cycle includes a peak in foaling and breeding in late winter, spring and early summer, when weather and nutritional conditions are usually favourable for the lactating mare and growing foal. Although the harems are together and the harem stallions work to maintain and protect their band year-round, the level of vigilance and response to threats, as well as the elimination marking behaviour and attentiveness to mares, are conspicuously greater during the foaling and breeding season months. For one to several days before a mare foals, stallions seem sexually attracted to the mare, with herding and increased sexual approaches of the mare (S. McDonnell, pers. obs.). The mare is generally non-receptive and moves off. In some cases, the stallion's sexual 'pestering' continues during parturition, when the harem stallion may approach in a sexual posture and even attempt to mount the

mare. During parturition, the harem stallion typically appears interested in the foetal fluids and responds not only with olfactory investigation, but also with erection and increased sexual arousal. Yearling and older young harem males that have not yet left their natal band may participate in investigation of foetal fluids and teasing of the foaling mare. These young band males are more or less tolerated by the harem stallion. During and immediately after parturition, the stallion may appear to take advantage of the mare's inability to effectively resist mating attempts. The basis of this increased sexual attractivity of the periparturient mare is not known. Simple endocrine changes associated with late gestation and parturition are a plausible component. This is supported by observations of similar apparent sexual attractivity of domestically managed mares undergoing early and mid-gestation pharmacologically-induced pregnancy termination (S. McDonnell, pers. obs.). Associated with this periparturient increase in sexual attractiveness of the mare is the stallion's increase in vigilant protectiveness of his harem as well as attentiveness and protectiveness of the particular foaling mare. This often results in the harem band moving off to the edge of the herd during parturition.

Mares generally return to oestrus cycling and ovulation within one to three weeks after parturition. With reasonable nutritional conditions, most will conceive on the first post-parturient ovulation (Figure 7.1). This occurs at about 10–15 days after foaling. With a gestation of about 340 days, the foaling and breeding season, therefore, maintain approximately an annual cycle. If a mare does not conceive, subsequent periods of oestrus begin at approximately three-week intervals and last an average of approximately five days each, with ovulation near the end of that period.

When a mature harem mare is in oestrus, the harem stallion tends to linger close to the mare, with obviously increased attentiveness and general reactivity to her movements, postures and elimination of urine or faeces. There is an obvious change in character and frequency of the approaches to the mare, from general dutiful 'housekeeping' or 'monitoring' to more frequent enthusiastic sexual approach with the typical prancing arched neck, tail high posture and nicker vocalization. Elimination marking behaviour similarly increases in frequency and character, as does herding of the harem away from other harems and from bachelors. A harem stallion tends to be especially

Figure 7.1. Breeding at first post-parturient oestrus with young foal at side. (Photo by Elkanah Grogan, Equine Behavior Laboratory, University of Pennsylvania.)

vigilant to potential threats by intruding males whenever one or more of his harem mares are in oestrus.

The age of puberty and first adult sexual behaviour appears to vary somewhat among individuals and populations. Differences in nutrition and the age when first exposed to seasonal increasing photoperiod may be factors known to affect age of maturation in domestic horses. Most fillies begin ovarian cycling in their yearling or two-year-old year. When these young females come into oestrus the harem stallion relaxes his guard of them and allows them to wander outside the harem for breeding, allowing them to return (see further description below). As described by Feh in Chapter 5, most young females leave their natal (or their dam's) band on their own, are expelled by their harem stallion or are forced away by other males. Colts begin heterosexual breeding also either in their yearling or second year, usually in transitional gangs described below. For most harems, young males over the age of two years appear to be at liberty to come and go from the natal band during the transitional period. Among the various populations studied, there appears to be some variation in how well and how long young males appear welcome to stay, or to come and go from their natal band. Most eventually leave permanently of what appears to be their own accord. Some appear to be expelled, usually by their harem stallion.

Some harems have more than one mature stallion sharing roles of harem maintenance (see Chapters 4 and 5 for brief reviews, and Stevens 1987, 1990 for detailed review). Most reports of two-stallion harems describe one stallion being the main breeding stallion and the second stallion an attached 'lieutenant' or 'assistant'. The assistant harem stallion appears to be delegated the role of 'outer' or 'first' defence against intruders as well as other harem maintenance activities such as retrieving young. The main harem stallion guards and does the breeding of the mature mares. In the author's 10-year study of the semi-feral herd of ponies, in which systematic observations were done twice daily year-round, the author observed the development of three assistant harem stallions as well as one apparent failed attempt. This herd is maintained with 50–75 animals, typically consisting of seven to nine harems and a bachelor band at any one time. At most times over the 10-year period, there have been zero to two harems with assistants and the remainder, single-stallion harems. Two assistants emerged over a period of months each lingering in the vicinity of one of two harems. Each of these assistants was a mature bachelor stallion that was unrelated and had no previous affiliation with the harem stallion that they assisted. In each case the assistant stallion neither tried to control nor steal mature or young harem mares, nor did it respond sexually to the mares when in oestrus, but rather maintained proximity and appeared to be volunteering outer defence of the harem band. In each case, initially the main harem stallion attempted to repel the lingering stallion. However, within one to two weeks the fighting between the main harem stallion and the assistant gradually diminished as a

co-operative defence relationship and mutual tolerance developed. After about one year, the relationships appeared stable, with only posturing, and no
frank fighting. During the second year, the assistant's
duties expanded to 'babysitting' and retrieval of the
young as they developed and started wandering off
in inter-harem playgroups. The two harem stallions
that took on assistants were aged – one was 20 years
of age and the other in his early 30s. In the first case,
the assistant was a 15-year-old long-term bachelor
and in the other case the assistant was a five-year-
old solitary bachelor. In another similar, but failed
instance of an emerging assistant harem stallion, a
four-year-old male in transition from his natal band
lingered near and alternately battled and offered assistance with defence for five months through the breeding season with a 14-year-old stallion of 10 years'
successful maintenance of a large harem. In a spectacular battle at the end of the breeding season, the
younger stallion was finally expelled. The remaining
assistant harem stallion relationship that has developed in this herd involved a four-year-old stallion
that had left his natal band and joined a bachelor
band for a period of about 18 months when he and
two other similar-aged bachelor 'pals' co-operated to
sequester a mature mare and her foal during post-
parturient oestrus from a large harem. They fought
together and received assistance in the battle from
several other harem and bachelor stallions. This continued for a period of 10 days, at which time they
clearly had established control of the mare. At that
time, one of the pair obviously took the role of harem
stallion and the other two appeared to willingly defer.
They 'tagged along' as if continuing a bachelor relationship with their former associate, without signs of
interest in sexual interaction or control of his mare.
The main harem stallion continued to interact with
the 'tag along' males in a sparring manner typical of
bachelor-band stallions, rather than to engage in serious fights as if to repel these cohorts. Over the subsequent three to four weeks, one of the tag-along pals
appeared to be assuming responsibility for first line
of defence for the new small band, and co-operating
with the main harem stallion during major aggressive encounters. Over a period of a year he became an
assistant. The second tag-along stallion did not appear
to collaborate with the others in defence, and during
the following spring, left to join a co-operative effort
with another same-aged previous bachelor associate

to completely depose a mature harem and split up
its mares to form two new harems. No matings have
been observed of these assistants with harem mares
and DNA paternity test results indicate that none of
these stallions has sired offspring with mature harem
mares or their young female offspring during their
tenure as assistants (S. McDonnell, pers. obs.). One
assistant was observed breeding with a yearling filly
from another harem band, but did not sire that filly's
foal.

While apparently it is extremely rare in most populations studied for an assistant harem stallion to be
allowed breeding access to a mature harem mare,
assistants have been reported to breed young mares
within the band or young fillies visiting on the perimeter of the band. If the harem is threatened by intruders or challenged by competing males, the assistant
and main harem stallion typically co-operate in the
defence of the harem. During periods of heightened
vigilance, for example during foaling of harem mares,
the two stallions appear to co-operate in the defence
of the band. During stage two of parturition of a
harem mare, the main and assistant harem stallions,
as well as older male offspring who have not yet left
the natal band, may appear to work together in protecting the band. These males have been observed circling the band, systematically marking the perimeter
with faecal piles, and may strut and posture among
the stud/dung piles. They work together to aggressively repel any nearby bands, keeping seemingly curi-
ous onlookers (often youngsters from neighbouring
bands) at bay. Another interesting example of co-
operation among a main harem stallion and an assistant has been observed repeatedly in the semi-feral
herd, maintained by the author, when under challenging conditions of deep snow in winter. Under normal environmental conditions, a harem band trekking
from one area to another within their habitat typically is led by a mature mare, followed by other
mares and their young, and followed at the rear by
the harem stallion. If the band has an assistant harem
stallion, he typically follows some distance behind
the main harem stallion. One interesting exception
to this trekking order has been observed in winter
when the herd has been trekking a path through fresh
deep snow. An assistant harem stallion typically leads
the way laboriously 'punching' a path through the ice
crust, followed by the adult mares that further clear
the path for the juveniles following behind the mares,

with the main the harem stallion bringing up the rear.

Breeding outside of established harems

In addition to the breeding of the harem stallions within their established harems, a few specific types of sexual interaction and breeding occur outside of the established harem bands. One of these involves more or less temporary assemblages of mostly young animals. Typically one or more young fillies that are still in their natal band, or that are in transition from their natal band, attract gangs of young males that similarly are either in a stage of coming and going from their natal bands or have recently left their natal bands. In most cases, these assemblages are momentary and repeated periodically throughout the young female's period of oestrus. Some or all of the young males and females may return to their natal bands between periodic assemblages. The copulatory encounters are typically brief, with little or no tending or protection of the female before or after copulation as would be the case for harem stallions. In these groups, the young females are typically bred many times in rapid succession in what appears to be a 'line up' of yearling, two-year-old and older males. In some circumstances, the young female appears to join the assemblage for the duration of her oestrus, and in other cases, the female seems to go on periodic excursions (as frequently as hourly) back and forth from the natal band, and often on a circuit of visits to one or more transitional gangs, bachelor bands or harems before returning again to the natal band. When oestrus ends, the young mare stays with her natal band and is herded and protected by the harem stallion of her natal band.

Another commonly reported circumstance in which breeding occurs outside an established harem band involves bachelor stallions that appear to be attempting to form a new harem. This may involve one or more established bachelor stallions attempting to sequester a mare from an established harem. This may involve either a mature mare or more often a young mare. The most common time for bachelors to show interest in stealing a harem mare appears to be around the time of parturition or when the female is in post-partum oestrus. In the case of the mature mare, there is typically violently aggressive resistance from the established harem stallion. In the case of a young mare, her established band stallion (often her sire) offers little resistance.

The term 'sneak breeding' is used to describe breeding within or at the edge of an established harem either by an opportunistic assistant harem stallion with a mature harem mare or by an established bachelor with a young or mature mare within an established harem band. In contrast to the situation in which a bachelor stallion appears to be trying to steal a mare and form his own harem, these sneak breedings clearly appear to be a 'breed and run', rather than an attempt to gain long-term control of the mare. They often occur when the harem stallion is distracted by or addressing other threats.

Precopulatory and copulatory sequences

A conspicuous feature of precopulatory and copulatory behaviour under natural social conditions is the strong role the female plays in soliciting attention of the stallion, seemingly determining the time of breeding, either by actively facilitating successful mounting, insertion and dismount or by rejecting the stallion's attempts. Oestrus behaviour (Figure 7.2) includes increased locomotor activity, frequent approaches to the stallion, swinging of the hindquarters toward the stallion, lifting the tail, flexing a foreleg while turning the head back and frequent urinations. In addition, rhythmic eversion of the vulva exposing the clitoris and lighter internal membranes, called 'winking' or 'flashing', is a characteristic behavioural sign of readiness for breeding. The female mating stance includes a 'sawhorse squat', with the tail lifted off to the side of the perineum, the head turned back toward the stallion, typically with one foreleg flexed. This posture appears to assure the stallion that she will not resist mounting. The typical precopulatory or 'teasing' sequence of the stallion begins with nose-to-nose contact and an associated soft nicker vocalization. It includes oronasal contact (sniffing, nuzzling, licking) proceeding from the shoulder, elbow, ventral abdomen and udder, to the hind legs and perineum. Stallions exhibit a conspicuous flehmen response following oronasal contact with urovaginal fluids (see Stahlbaum & Houpt, 1989, for detailed discussion of this behaviour). Other elements of precopulatory behaviour of the stallion include nipping or grasping the mare, usually at the flank, or hocks, leaning into the mare, resting the chin on the rump as if ready to mount, and mounting without an erection or over the back of the mare as if testing her willingness to stand for breeding.

Figure 7.2. Elements of precopulatory and copulatory interactive sequences (adapted from McDonnell, 2003).

Copulatory behaviour (Figure 7.2) may follow closely to precopulatory behaviour, or it may be interspersed with bouts of precopulatory behaviour. It includes mounting, insertion by the stallion, facilitation of insertion by the mare, organized rhythmic thrusting, and ejaculation. Insertion and organized thrusting usually span 20–25 s before ejaculation commences. The mare may move forward during breeding as if to accommodate or in some instances escape vigorous thrusting. The stallion typically remains mounted for as long as 20–30 additional seconds following ejaculation. The mare usually accommodates dismount by slowly stepping forward. Postcopulatory behaviour includes attention by the male to spilled urovaginal fluids and ejaculate on the mare or ground, including sniffing, flehmen response and sometimes urine covering. The stallion and mare often linger near one another for a few minutes following each copulation. Most copulations occur within less than a minute, and are a relatively quiet event.

Across the period of apparent oestrus, behaviour of the mare tends to vary from day to day in the intensity and combination non-receptive, proceptive and receptive elements. During the first two or three days, receptivity generally increases and non-receptivity decreases reaching a period of one or two days of greatest receptivity, followed by a period of declining proceptive and receptive behaviour and an increase with non-receptive behaviour. Work mostly with domestic mares indicates that this pattern is related to ovulation, with the maximum proceptivity and receptivity corresponding to the day before and the day of ovulation, with a noticeable decline in proceptive and receptive behaviour and increase in non-receptive elements following ovulation (Asa et al., 1979; Asa, 1986). Most mares have days during their oestrus period during which they alternately solicit and rebuff the advances of a stallion, from encounter to encounter, or even within an encounter. While most mares follow a similar pattern of their oestrus period from cycle to cycle, there is considerable variation between mares in the duration and character of the oestrus period. While the mare may be attractive and/or interactive with the stallion for many days, for most mares there are usually only one or two days of peak receptivity. On those days, the mare is typically bred many times, at intervals of a few minutes to one to two hours. Stallions at liberty with multiple mares in oestrus have been observed to breed

as often as one or more times per hour throughout the daylight and dark hours (Bristol, 1982). Sexual interaction during pregnancy is apparently uncommon under natural social conditions, however there are few populations in which accurate pregnancy data are available.

Breeding of domestic horses

Housing and handling systems for domestic breeding horses generally limit natural sociosexual interaction and mate selection. To the extent allowed, actual precopulatory and copulatory behaviour of domestic horses are relatively similar to that of horses under natural social conditions. However, restraint is often imposed that limits natural behaviour. Another difference that can have important implications for domestic breeding horses is that breeding animals first must be successful in an athletic or show career during which it is required that they refrain from natural social and sexual behaviour, deferring to direction from humans.

General breeding farm husbandry

Other than in pasture breeding systems, domestic stallions are generally housed so as to avoid direct contact and unsupervised sexual interaction with mares. Further, to avoid inter-male aggressive interaction that in the typical domestic environment would be dangerous to animals, people and property, stallions are generally housed in individual stalls or paddocks. Colts and fillies, if kept in mixed sex groups as weanlings, are separated into same sex groups before puberty to avoid unintentional breeding. Although stallions may no have direct contact with mares or other stallions, they are often kept together in buildings or adjacent enclosures that allow social communication. Although a variety of housing systems is used, the traditional method at most large breeding farms in much of the world is to house the stallions in barns with other stallions and with no mares, and to house mares together in barns with no stallions. There are notable exceptions in speciality segments of the domestic horse industry that keep mature stallions together in direct contact. For example, the meat producers successfully keep large groups of maturing colts and stallions together at pasture. The pregnant mare urine ranchers in North America often keep their pasture breeding stallions together at pasture in the non-breeding season when the mares are housed together in barns. These pastured all-male herds appear to interact as a large bachelor band.

Modification of sexual behaviour for work and performance

Other than those bred for meat and farm work, most domestic breeding horses must first become distinguished as valuable breeding stock by excelling at some sort of athletic or show career. During training for these first careers, sexual response is generally considered undesirable. Normal response is actively discouraged, mostly using aversive conditioning and/or counter-conditioning techniques. Stallion behaviour has been shown to be sensitive to aversive experience, and so can usually be effectively suppressed (McDonnell *et al.*, 1985).

Domestic breeding systems and practices

Current practices for breeding domestic horses vary considerably between disciplines, regionally within and between areas of the world, and with economic value and purpose of the breeding stock. While large numbers of agricultural horses still breed under conditions similar to those of natural populations, most horse-breeding farms practice some level of controlled breeding that significantly affects sexual behaviour. In general, the more valuable the breeding stock, the more closely the animals are confined and managed, and the greater the degree of human intervention. Intervention includes selecting the pair to be bred as well as the time of breeding. Domestic horse breeding is done under a variety of farm conditions, ranging from specialized breeding farms or 'studs' with one or more stallions that are there primarily or exclusively for breeding, to farms on which both training and breeding occurs simultaneously with the same animals. For systems using artificial insemination, with the transportation of chilled or frozen semen, mares and stallions are not usually kept on the same farms.

Another important domestic practice which results from the domestic racing and show economics, is to try to have foals born as early in the yearly cycle as possible so that they have the advantage of age for show and racing events. As a result many segments of the equine industry start breeding before the natural springtime breeding season, using endocrine manipulation by artificial photoperiod or exogenous treatments to achieve earlier pregnancies and foalings.

Semi-feral breeding

Although a fairly rare breeding practice, semi-feral mixed-sex breeding herds can be found throughout the world. These are assemblages of horses that are kept together for most of the year or are kept together indefinitely. Typically these herds are confined within large enclosures, but have minimal management other than supplemental forage and selective culling as needed. These herds vary in sex ratios, but assembling and culling is often done to limit the number of mature stallions. The number of mares bred to a stallion varies with the number of stallions present, in much the same way that it does among natural breeding populations. This practice is most commonly used with ponies or stock horses. These horses are generally considered suitable for semi-feral management because they are physically hardy and of little individual economic value. For example, ranch or 'stock' horses, which are used for cattle ranch work in North America and Australia, are bred in semi-feral herds that live and breed with little human intervention.

Pasture breeding

The system known as pasture breeding involves one to several mares assembled for all or part of a breeding season within an enclosure, usually with a single mature stallion. The behaviour of the stallion and mares appears much like that of a harem under natural conditions. The stallion may settle to a more relaxed level, as the threat of bachelors is not present. New mares added to an established pasture breeding group may not be accepted by the existing mares and/or the stallion, so successful group pasture breeding programmes generally assemble all the mares to be bred at the outset and then add the stallion. Mares may have suckling foals, with no apparent effect on behaviour or fertility.

Stallions that are removed from a pasture group of mares will generally accept new groups of mares almost immediately. In some systems, for example the breeding of Icelandic horses in Iceland, it is customary for each breeding stallion to be grouped with each of two to three batches of mares successively for four to six weeks each over the course of each breeding season, from May through to July (Steinbjornsson, 2002).

Pasture breeding systems are known for generally high fertility, with foaling rates above 75%, even

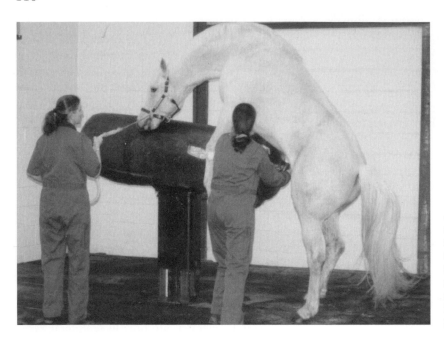

Figure 7.3. Semen collection, using a dummy mount and artificial vagina, from a stallion that had pasture-bred for years, covered by natural service in hand, and complies with semen collection by several techniques. (Photo by Elkanah Grogan, Equine Behavior Laboratory, University of Pennsylvania.)

with as many as 25–35 mares per stallion for two to three cycles' exposure. (Bristol, 1982; Hugason *et al.*, 1985).

Paddock breeding

Temporary pairings of stallions and mares for relatively free interaction within an enclosure is known as paddock breeding. This can be done for all or part of the oestrus period. Paddock breeding is often done for as little as a few minutes at a time with close observation to ensure that copulation occurs. In most cases the mare is teased with the stallion to be paddock-bred or with another stallion across a safe barrier before the two to be bred are placed together in an enclosure. For this and the other supervised 'natural cover' breeding methods it is common for the mares to either reside permanently on the farm where the stallion resides for all or part of the breeding season, or to be brought to the stallion farm for scheduled breeding visits of a few days to a few minutes. A less common practice is for a stallion to travel from farm to farm for breeding, either daily from the farm of residence or for longer visits.

In-hand breeding

Another practice of individual mating is commonly referred to as in-hand 'natural cover' breeding. In this system, typically a stallion and a mare are both restrained in-hand with a halter or breeding bridle and lead and are brought together briefly for copulation. There is considerable variation within the horse-breeding industry on the amount and type of restraint used with stallions and mares, ranging from many pieces of restraining equipment and many assistants to simply one person handling the stallion with the mare either loose in an enclosure or tethered along a rail.

Assisted reproductive technology

Only one major horse breed association, the Thoroughbred, still requires natural breeding for registration eligibility. A rapidly growing percentage of horse breeding is now done by insemination of mares with transported fresh, chilled or frozen semen. These practices eliminate the need for contact or copulation of individual mares and stallions. Most semen is still collected *in copula*, using either a mount mare, a dummy mount (Figure 7.3), or manual stimulation with the stallion standing on the ground (McDonnell, 1999a; McDonnell & Love, 1990). Although methods are available for pharmacologically inducing *ex copula* ejaculation (McDonnell, 2001), application is currently primarily with disabled stallions or on farms with limited facilities for *in copula* collection of semen. For mares bred by artificial insemination, behaviour in response to a teaser stallion may still be used alone or in combination with other methods for deciding when to inseminate. However, in some operations, no stallions are present on the farm and decisions on when to inseminate a mare are based solely on ovarian activity (palpated manually per rectum and/or imaged ultrasonographically).

Stallions appear readily to learn breeding routines and seem generally to benefit from consistency in breeding procedures, at least initially. However, experience in clinical and research settings suggests that with respectful and confident handling most horses can be quite flexible with changes in routine. Just as with training for work or performance tasks (described in Chapters 12, 13 and 14), horses can readily switch back and forth among widely varying general methods of breeding as well as specific procedures. So collection of semen does not adversely affect a stallion's ability to breed naturally, and vice versa. Ground semen collection does not appear to significantly retard mounting when the opportunity is offered to mount a live mare or dummy, and similarly mounting does not significantly retard learning of ground semen collection (McDonnell & Love, 1990).

Under domestic breeding conditions, most young females mature and are fertile at about one year of age, depending upon their season of birth. (If born late in the year, they may not have their first ovulation until the spring of their second year.) Young mares are usually not bred until three or four years of age or older. Custom also varies with how many foals are expected over the lifetime of a brood mare. Well-managed broodmares are often expected to foal and efficiently conceive (within one to three ovarian cycles) annually and if otherwise healthy, remain similarly fertile into their late teens or early twenties. Although colts mature sexually and are fertile as early as eight months to one year, their sexual behaviour in terms of stamina and their sperm production capacity generally continues to improve through five to six years of age. Most are not started in serious breeding programmes until about four to eight years of age, depending mostly on convention of various athletic, show or work purposes of the breeding programme.

It is quite remarkable how quickly domestic horses, regardless of their domestic lineage or their individual developmental experience, can return to normal natural social organization and herd behaviour when the opportunity is provided. They quickly (within hours of turn-out) organize socially into stable herds of harem and bachelor bands (S. McDonnell pers. obs.). As described in Chapter 4, domestic horses almost always thrive and reproduce prolifically in widely diverse environments without human intervention. Reproductive efficiency and productivity over time or between populations seem most affected by nutrition.

Sexual behaviour problems of domestic horses

While sexual dysfunction appears to be rare in horses living under natural conditions, it appears to be fairly common in domestic horses. This section describes the common types of sexual behaviour problems of domestically managed horses, and the types of management strategies and veterinary interventions that are currently recommended. In this brief outline, problems will be organized into the broad categories of: (1) normal sexual behaviour that is incompatible with domestic management or work; (2) normal equid behaviour that is misunderstood as aberrant behaviour in domestic horses; (3) inadequate reproductive behaviour under domestic management and breeding conditions; and (4) truly abnormal or aberrant behaviour for the species. More detailed reviews of clinical evaluation and treatment protocols are available (McDonnell, 1992, 1999a,b).

Problems with normal sexual behaviour in management and training of horses
Strong stallion-like behaviour in stallions While under natural sociosexual conditions it is normal for stallions readily to take on the role of a harem stallion, for example showing concern for mares or fighting off intruding males, these behaviours can be problematic under domestic conditions. As mentioned earlier, sexual behaviour in stallions during work or performance can usually be effectively inhibited or discouraged using aversive or counter-conditioning. Domestic breeding programmes to some degree probably select for temperament favouring less vigilant stallion-like behaviour, and the tendency to be easily distracted when interacting and working for humans. None the less, a common complaint of handlers is unwanted sexual behaviour in stallions. When sexual behaviour is problematic, it often involves a variety of factors, including strong motivation of the individual stallion as well as management conditions and handling

skills or attitudes. While judicious training is usually the most effective therapy, endocrine treatments that suppress androgen production may be a useful adjunct. Although not usually recommended because of adverse effects on physical co-ordination as well as mentation and possibly learning, general tranquilization is also used to quiet stallion-like behaviour for training or performance.

Residual stallion-like behaviour in geldings A considerable percentage of geldings retain stallion-like behaviour at levels that are problematic for management under usual domestic conditions or performance. In the author's experience of obtaining geldings from auction for use in behaviour research requiring minimal stallion-like sexual behaviour in standardized tests, it was typically found that of those horses that are confirmed as castrates (using endocrine challenge tests) approximately half will show a sexual response towards an oestrous mare within the range for normal intact stallions. This includes immediate precopulatory sexual interest and arousal, erection, mounting and ejaculation when given unrestrained access to a solicitous oestrous mare.

In a follow-up survey of 140 geldings that had been castrated three to nine years earlier, 20 to 30% of geldings were reported by their owners to exhibit stallion-like aggressive or sexual behaviour towards other horses and 5% exhibited stallion-like aggressive behaviour with handlers. Results were similar for geldings castrated before two years of age and those castrated after three years of age (Line *et al.*, 1985). In that study, for cases in which castration had been done specifically to eliminate objectionable or unmanageable stallion behaviour, owners reported that castration had effectively reduced stallion-like aggressive behaviour towards people in 60–70% of cases and towards horses in 40% of cases.

Clinical impression and research observations suggest that sexual behaviour is more easily suppressed in geldings than in intact stallions. Training is the recommended therapy.

Training and performance problems of mares related to oestrus and/or ovarian activity Trainers and riders often report variable performance or trainability of mares related to the oestrus cycle. In addition to undesirable oestrus behaviour in performance situations, complaints often concern periods of deterioration,

either mild or marked, of performance or temperament associated with a particular stage of the ovarian cycle. Detailed clinical evaluation of the behaviour and the ovarian cycle can be useful in identifying or ruling out associations. In some cases, the conclusion is that the performance or training problem is not associated with the ovarian cycle, but rather is a physical or medical problem, or a training issue. In many cases, evaluation bears out the complaints of association of temperament and performance changes with the ovarian cycle. Some mares are found to be simply generally hyperexcitable and difficult to handle at certain stages of the cycle. Some mares appear particularly sensitive to weight or manipulation that might affect the area of the ovaries during the periovulatory period of the cycle. In some cases, the problem behaviour is found to be associated with the dioestrus phase of the cycle, but in most it is associated with oestrus or ovulation. In many such cases, riders and trainers report that the mare experiences abnormally frequent and/or long oestrus periods. In a number of cases such complaints have been found to be associated with shorter than normal cycles. Therapy for such confirmed oestrus cycle-related problems is generally aimed at manipulating or suppressing the ovarian cycle using endocrine treatments to match training schedules or events.

Normal sexual behaviour misunderstood as abnormal

One area of education of horse keepers that should lead to better welfare and reduce the frequency of behaviour problems concerns normal equid behaviours that are misunderstood as aberrant or misbehaviour in domestic horses. Examples involving reproductive behaviour include masturbation, faecal and urine marking, inter-male sexual behaviour and play sexual behaviour.

'Masturbation' In most mammals studied periodic penile erections and movements are understood as normal behaviour. Spontaneous erections and penile movements are quite conspicuous in the horse. Within the horse industry, this behaviour has traditionally been known as *masturbation*. Descriptive field studies as well as extensive study of domestic horses has established that episodes of penile erection and movements occur at about 90-minute intervals in all of the equids, all ages, with little effect of environment

(McDonnell *et al.*, 1991). Attempting to inhibit or punish spontaneous erection and penile movements, which is still a common practice of horse managers regionally around the world, often leads to increased masturbation and disturbances of normal breeding behaviour (S. McDonnell and A. L. Hinze, unpub. data).

Faecal and urine marking behaviour A normal aspect of horse behaviour that typically remains intact under domestic conditions is elimination-marking behaviour of males. In the domestic situation, stallions and some geldings defecate in the same place in the stall, forming small 'stud piles'. They also perform faecal marking near gates or along fence lines in turn-out paddocks, resulting in stud piles. They typically investigate existing piles and their own freshly voided faeces, especially in paddocks that are sequentially shared. These behaviours are sometimes seen as peculiar by owners especially in the case of geldings, or as they emerge in young colts.

Inter-male sexual behaviour A normal aspect of sparring among young males and bachelors is intermale sexual behaviour. Partners play both the male and female role with approach, solicitation, teasing, standing for mounting and mounting, with and without erection (McDonnell & Haviland, 1995). In domestic horses, this behaviour is often seen as misbehaviour when it occurs, usually in young intact males grouped together. This stallion-like behaviour is often retained in geldings, and so often occurs also in grouped geldings. Common misinterpretations are that this bachelor-type sparring is abnormal sexual behaviour or frank aggressive behaviour. In the confines of domestic paddocks with square corners, poor footing and/or man-made obstacles that interfere with the normal submissive retreat, incessant sparring of grouped males that would cause no injury in wide-open spaces is often the cause of injuries.

Play sexual behavior Play sexual behaviour in young animals is a normal and frequent equine activity that when seen in domestic foals is commonly misunderstood as aberrant precocious sexual behaviour. In foals living in natural social groups under semiferal conditions, all major elements of both male and female behaviour were observed in all fillies and colts from a few days of age (McDonnell *et al.*, 1998; Figure 7.4). Play sexual behaviour directed toward the dam (Figure 7.4), including precopulatory investigation, elimination marking behaviour, and mounting with and without an erection, usually peak during the dam's post-parturient oestrus. Play sexual behaviour with foal cohorts continues for fillies until puberty at about one year of age; with colts it is more frequent and continues into maturity as a bachelor stallion. Simply understanding that these interactions are normal and possibly important to development should reduce managers' concerns about the behaviour and the practices aimed at eliminating them. As with sparring of grouped males and other incessant play behaviour in foals, play sexual behaviour can lead to injuries under domestic conditions, for example slippery flooring or poor footing in paddocks, where foals may slip and fall while playing.

Inadequate reproductive behaviour under domestic breeding conditions

Inadequate libido in stallions Failure to show normal sexual interest in mares or to respond adequately under various domestic breeding conditions is a common problem in domestic stallions. Common examples include slow starting novice breeders, experienced breeders that become uninterested or slow at breeding, and stallions with particular preferences or aversions regarding mare, location, handler or procedures. While factors such as developmental deprivation, breeding outside the natural season, bachelor sociosexual conditions, sub-optimal or severely restrained or no stimulus females, or deliberate suppression of sexual behaviour in other circumstances, are usually obvious to the ethologist, they are often not considered as potential contributing factors by modern managers who have come to expect all stallions to comply with all procedures. Evaluation and therapy involves comprehensive physical and behavioural evaluation, and best results are often obtained by initially reverting to attempting to breed the stallion in as natural conditions as possible. A variety of endocrine and other medications have been used as aids in retraining low libido stallions in a crisis situation (see Martin & McDonnell, 2003 for current review).

Intense libido in stallions Stallion behaviour can also be too vigorous for breeding under certain domestic conditions. The usual recommendation is skillful training (McDonnell *et al.*, 1994). Under those

Figure 7.4. Play sexual behaviour sequence in young foals (in this case two females), and a week-old foal (male) mounting dam during post-parturient oestrus.

circumstances, where the horse still needs to remain fertile and able to breed, hormonal manipulation that effectively suppresses stallion behaviour may also reduce fertility. Using tranquillizers may be unsafe. Tranquillizing intact stallions with certain common phenothiazine compounds have the potential to cause a paralyzed penis. Though this side-effect is rare, it usually ends the breeding career of a stallion and so is not recommended.

It has long been established that in equid breeding systems some of the mature stallions gain access to a harem or breeding territory, and some remain non-breeding 'bachelors' (Klingel, 1975; Keiper, 1985). Important behavioural and physiological differences resulting from harem or bachelor social status are at last being understood and appreciated. It has been found that harem status imparts an upgrading of reproductive endocrinology and function including increased androgen levels, sexual and aggressive behaviour, accessory sex gland size and character, testicular size and character, and semen quality; bachelor status imparts suppression of those characteristics (McDonnell & Murray, 1995). Traditional group housing of domestic stallions inadvertently imposes social bachelor status on breeding stallions. For example, stabling of stallions in barns with other stallions led to suppressed reproductive function, including lower sexual interest and arousal, compared to stabling as the only stallion with mares (McDonnell, 1995). Both in normal research animals and in clinical cases of sexual behaviour dysfunction, simply by manipulating the amount of ongoing exposure that a stallion has to mares and other stallions can markedly affect libido, testosterone levels, testicular volume and sperm production efficiency (McDonnell, 1995, 2000). On most farms, sociosexual conditions can usually be manipulated to effect physiological and behavioural conditions corresponding to those of bachelor and harem status.

Ejaculatory failure Slow or absent ejaculation, with or without normal libido, is another common stallion breeding problem. Ejaculation disturbance of course can be caused by physical or neurological problems that interfere with libido, musculoskeletal coordination necessary for copulation, or with the pelvic ejaculatory apparatus itself. However, ejaculatory disturbance also can be behavioural or 'psychogenic', in association with poor handling techniques or with past aversive experience or inadvertent training to interrupt copulation. A common example of handling induced ejaculation failure results from the handler rushing the stallion to dismount after ejaculation by jerking on the breeding bit or halter. This is done for a variety of reasons, for example, to quickly separate the stallion and mare, or to obtain a semen sample to confirm ejaculation. Some stallions become anxious in anticipation of such treatment, or the consequence of falling during a rushed dismount. They eventually appear unable or reluctant to ejaculate. Untreated physical pain, for example, hind limb pain, can also lead to a similar pattern of ejaculatory dysfunction. The challenge in evaluation and therapy is to ascertain and alleviate the various contributing factors.

Failure to show oestrus One of the most common problems described by breeders of domestic horses is difficulty in detecting oestrus and ovulation using domestic oestrus detection systems. This is one area where better education of mare managers would yield fewer problems. In particular, an understanding of the natural variation in the display of oestrus within and among mares, as well as the natural pattern of varying proceptivity and receptivity across the oestrus period of mares, described earlier, would improve efficiency of breeding and welfare for both mares and hand-bred natural cover stallions.

Another common problem of brood mares kept under domestic social conditions and bred in-hand involves failure to tolerate mounting and to facilitate copulation with the covering stallion, even when showing full receptivity when exposed to a 'teaser' stallion or under teasing conditions. This is commonly called 'failure to stand'. Careful evaluation often suggests that this failure to stand is related to the increased entrapment of the restraint applied to the mare to protect the valuable stallion during natural cover. Often the facility itself is designed to 'corner' the mare, which inadvertently provokes an explosively dangerous escape behaviour as the stallion approaches. Simple procedures such as directing the excited breeding stallion to the mare from the rear for immediate mount, rather than allowing the natural nose-to-nose approach and systematic procopulatory investigation, in many cases seems to contribute

to this phenomenon of a mare standing in the less restrained teasing situation, but moments later not for the covering stallion. Clinical experience indicates that failure to show oestrus and failure to stand are particularly common among young and/or maiden mares. This is consistent with observations of natural herds, and probably related to inexperience and lack of confidence of young animals, even under natural herd conditions.

Truly aberrant sexual behaviour

A final class of sexual behaviour problems in domestic horses includes a variety of behaviours that are currently understood as truly aberrant. These are usually not seen in horses breeding under natural herd conditions. Examples include: sexual arousal and response to other species, for example other farm animals or to people; hyperactivity or 'frenzied stallion syndrome', seen mostly in confined stallions, or stallions that are within sight of mares that they cannot contact; and self-mutilation of the type that looks like self-directed inter-male aggression. Stallion-like behaviour in mares is a truly aberrant behaviour resulting from exposure to androgens, either administered by managers or from an ovarian tumor. Adult mares can be quickly androgenized to show the full range of stallion-like harem formation and maintenance behaviours, as well as precopulatory and copulatory behaviour (McDonnell *et al.*, 1988). In the case of ovarian tumors, although the author is not aware of observations of stallion-like behaviour or ovarian tumors identified among feral or semi-feral herd conditions, the condition of stallion-like behaviour with an ovarian tumor has been identified in mares breeding under pasture breeding or semi-feral conditions.

Summary comments

A comparison of sociosexual environment and breeding behaviour of horses under domestic management with that of free-ranging horses reveals striking differences that may account for some of the more common reproductive behaviour problems. Standard stud farm practice includes varying degrees of social isolation of stallions from mares other than for the few minutes of breeding contact per service, and not at all outside the breeding season. In contrast, under natural social systems, a harem stallion interacts continuously year round with his harem mares. Similarly, direct contact of stallions is rarely permitted on modern breeding farms. If a stallion is the only stallion on a farm, he may be isolated from direct or fence line contact with any other horses. Where there are multiple stallions on a farm, they are typically stabled in barns together in individual stalls or paddocks within close proximity with other mature stallions and away from any mares. In contrast, under natural conditions, harem stallions interact regularly in rituals and battles, apparently aimed at establishing and maintaining an order of dominance. But at the same time it is normal for stallions under natural sociosexual conditions to be able to retreat and remain generally at a distance from other stallions most of the time. Problems ranging from inadequate libido to overly aggressive behaviour and self-directed inter-male aggression (self-mutilation) to specific apparent preferences and aversions of a significant percentage of breeding stallions can be logically attributed to inadequate or inappropriate social exposure (McDonnell *et al.*, 1986; Bedford & McDonnell, 2000). Simple modification of husbandry to model better the natural social environment of horses, while still considering the safety of animals and handlers, is often immediately effective in overcoming, as well as in preventing, behaviour problems.

In the breeding procedure itself, there are some key domestic practices and stallion-handling difficulties that appear to inadvertently lead to sexual behaviour dysfunction and infertility. For example, on stud farms, standard breeding shed protocols include varying degrees of mare restraint. The restraint interferes with normal interactive behaviour, receptive postures of the mare, the ability of the mare to adequately support the stallion and to normally facilitate and accommodate intromission and thrusting. Skilled and informed handling of breeding animals, while critical to preventing and rehabilitating sexual behaviour dysfunction in mares and stallions, appears to be a 'dying art' (McDonnell *et al.*, 1994).

Modern stud farm practices vary considerably around the world, but in general impose considerable differences in sociosexual environment and breeding behaviour from natural conditions for both mares and stallions. Some of the modern horse husbandry practices are for safety, hygeine or practicality, but many are more or less tradition, or for considerations other than behaviour. For example, with the globalization of semen transport and the growing concern for disease control, stallion stations are required to have disinfectable semen collection facilities. These must be indoors. The flooring chosen, and even recommneded by regulatroy agencies, for ease of disinfection, is often slippery. This can be off-putting and dangerous to animals and personnel, inadvertently creating behaviour and health problems. Similarly, the wall and ceiling materials used to meet disinfection standards, such as stainless steel and tile or sealed masonry, are typically noisy. This can also be intimidating to stallions.

While most domestic horses breed successfully under intense domestic management and sub-optimal conditions, there is a considerable amount of sexual behaviour dysfunction and/or sexual behaviour-related management, and performance problems and inefficiency in horses managed under these conditions. Applied ethology and veterinary clinical research addressing behavioural considerations with aggressive educational programmes for horse keepers should lead to better efficiency and welfare concerning sexual behaviour of domestic stallions and mares.

References

Asa, C. S. (1986). Sexual behavior of mares. *Veterinary Clinics of North America Equine Practice*, **2**(3): 519–34.

Asa, C. S., Goldfoot, D. A. & Ginther, O. J. (1979). Sociosexual behavior and the ovulatory cycle of ponies (*Equus caballus*) in harem groups. *Hormones and Behavior*, **13**: 49–65.

Bedford, S. J. & McDonnell, S. M. (2000). Theriogenology question of the month (diagnosis: specific aversion to handling for semen collection and to personnel approaching the genital area, probably associated with a previous negative experience). *Journal of the American Veterinary Medical Association*, **216**(4): 491–3.

Bristol, F. (1982). Breeding behavior of a stallion at pasture with 20 mares in synchronized oestrus. *Journal of Reproduction and Fertility*, **32**(Suppl.): 71–7.

Feist, J. D. (1971). Behavior of feral horses in the Pryor Mountain Wild Horse Range. M.S. Thesis, University of Michigan, Ann Arbor.

Hugason, K., Árnason, T. & Jónmundsson, J. V. (1985). A note on the fertility and some demographical parameters of Icelandic töelter horses. *Livestock Production Science*, **12**: 161–7.

Keiper, R. (1985). *The Assateague Ponies*. Centreville, MD: Tidewater Publishers.

Keiper, R. & Houpt, K. (1984). Reproduction in feral horses: an eight-year study. *American Journal of Veterinary Research*, **45**: 991–5.

Klingel, H. (1975). Social organization of reproduction in equids. *Journal of Reproduction and Fertility*, **23**(Suppl.): 7–11.

Line, S. W., Hart, B. L. & Sanders, L. (1985). Effect of prepubertal versus postpubertal castration on sexual and aggressive behavior in male horses. *Journal of the American Veterinary Medical Association*, **186**(3): 249–51.

Martin, B. B. & McDonnell, S. M. (2003). Lameness in breeding stallions and broodmares. In *Equine Lameness*, ed. M. Ross. Philadelphia: Saunders, pp. 1077–84.

McCort, W. D. (1984). Behavior of feral horses and ponies. *Journal of Animal Science*, **58**(2): 493–9.

McDonnell, S. M. (1992). Sexual behavior dysfunction of mares. In *Current Therapy in Equine Medicine*, vol. 3, ed. N. O. Robinson. Philadelphia: W. B. Saunders, pp. 633–7.

(1995). Stallion behavior and endocrinology – what do we really know? In *Proceedings of the Annual Meeting of American Association of Equine Practitioners, Lexington, KY, December 2–6*, pp. 18–19.

(1999a). Stallion sexual behavior. In *Equine Breeding Management and Artificial Insemination*, ed. J. Samper. Philadelphia: W. B. Saunders, pp. 53–66.

(1999b). Libido, erection, and ejaculatory dysfunction in stallions. *Compendium*, **21**(3): 263–6.

(2000). Reproductive behavior of stallions and mares: comparison of free-running and domestic in-hand breeding. *Animal Reproduction Science*, **60–61**: 211–19.

(2001). Oral imipramine and intravenous xylazine for pharmacologically induced *ex copula* ejaculation in stallions. *Animal Reproduction Science*, **68**: 153–9.

(2003). *The Equid Ethogram: A Practical Field Guide to Horse Behavior*. Lexington, KT: Eclipse Publications.

McDonnell, S. M. & Haviland, J. C. S. (1995). Agonistic ethogram of the equid bachelor band. *Applied Animal Behaviour Science*, **43**: 147–88.

McDonnell, S. M. & Love, C. C. (1990). Manual stimulation collection of semen from stallions; training time, sexual behavior and semen. *Theriogenology*, **33**(6): 1201–10.

McDonnell, S. M. & Murray, S. C. (1995). Bachelor and harem stallion behavior and endocrinology. *Equine Reproduction VI, Biology of Reproduction Monograph Series*, **1**: 135–48.

McDonnell, S. M., Diehl, N. K. & Oristaglio Turner, R. M. (1994). Modification of unruly breeding behavior in stallions. *Compendium*, **17**(3): 411–17.

McDonnell, S., Henry, M. & Bristol, F. (1991). Spontaneous erection and masturbation in equids. V International Equine Reproduction Symposium. *Journal of Reproduction and Fertility Supplement*, **44**: 664–5.

McDonnell, S. M., Hinrichs, K., Cooper, W. L. & Kenney, R. M. (1988). Use of an androgenized mare as an aid in detection of estrus in mares. *Theriogenology*, **30**: 547–53.

McDonnell, S. M., Kenney, R. M., Meckley, P. E. & Garcia M. C. (1985). Conditioned suppression of sexual behavior in stallions and reversal with diazepam. *Physiology and Behavior*, **34**(6): 951–6.

McDonnell, S. M., Kenney, R. M., Meckley, P. E. & Garcia M. C. (1986). Novel environment suppression of sexual behavior in stallions and effects of diazepam. *Physiology and Behavior*, **37**(3): 503–5.

McDonnell, S. M., Lutz, M. M., Ewaskiewicz, E. H. & Ruducha, A. (1998). Ontogeny of sexual behavior of foals in an established semi-feral population. *Proceedings VII International Equine Reproduction Symposium, Pretoria, South Africa*.

Miller, R. (1981). Male aggression, dominance, and breeding behavior in Red Desert feral horses. *Zeitschrift Tierpsychologie*, **57**: 340–51.

Salter, R. E. & Hudson, R. J. (1982). Social organization of feral horses in Western Canada. *Applied Animal Behavior Science*, **8**: 207–23.

Stahlbaum, C. C. & Houpt, K. A. (1989). The role of flehmen response in the behavioural repertoire of the stallion. *Physiology and Behavior*, **45**: 1207–14.

Steinbjornsson, B. (2002). Sexual behavior in herds of Icelandic horses. *Proceedings, Horse Behavior and Welfare, A Dorothy Russell Havemeyer Foundation Workshop, Holar, Iceland June 13–16, 2002*, www2.vet.upenn.edu/ labs/equinebehavior/hvnwkshp/ hv02/hv02tpcs.htm

Stevens, E. F. (1987). Ecological and demographic influences on social behavior, harem stability, and male reproductive success in feral horses (*Equus caballus*). Ph.D. Thesis, University of North Carolina, Chapel Hill.

(1990). Instability of harems of feral horses in relation to season and presence of subordinate stallions. *Behaviour*, **112**: 149–61.

Turner, J. W., Perkins, A. & Kirkpatrick, J. F. (1981). Elimination marking behaviour in feral horses. *Canadian Journal of Zoology*, **59**: 1561–6.

Tyler, S. J. (1972). The behavior and social organization of the New Forest ponies. *Animal Behaviour Monograph*, **5**: 85–196.

8

Maternal behaviour and mare–foal interaction

Sharon L. Crowell-Davis and Jennifer W. Weeks

Introduction

In a prey species, such as the horse, maternal care is important not only for the provision of nutrients to the young, but also for protection from predators. Overall patterns of maternal behaviour in this species are consistent with the need to maximize the chance that young survive to adulthood and minimize the chance that the young become victims of a predator.

Gestation and parturition

Pregnant mares generally remain active throughout pregnancy and engage in all maintenance behaviours typical for adult mares, including lying down and rolling over from side to side to groom themselves, despite a popular misconception that pregnant mares cannot do this. Normal gestation length ranges from 315 to 365 days, with the average being about 340 days (Rossdale, 1967; Jeffcott, 1972; Campitelli *et al.*, 1982). There is genetic variation in length of pregnancy that is independent of environmental factors (Rollins & Howell, 1951). 'Well-fed' mares,

mares carrying fillies and mares who have conceived toward the end of the year tend to have shorter pregnancies than do mares on maintenance rations, mares carrying colts and mares who conceive earlier in the year (Howell & Rollins, 1951; Rollins & Howell, 1951; Jeffcott, 1972; Campetelli *et al.*, 1982). Mares are monotocous, typically delivering one offspring at a time.

Mares usually give birth at night, even if they are kept in a barn with artificial lighting (Rossdale & Short, 1967; Jeffcott, 1972; Bain & Howey, 1975). While births can occur throughout the year, there is a strong peak in the spring, i.e. April, May and June in the northern hemisphere (Tyler, 1972; Feist & McCullough, 1975). Early conception, and thus early parturition, is possible with the use of artificial lighting, as mares have seasonal ovulation that is responsive to changes in photoperiod (Oxender *et al.*, 1977).

Parturition occurs in three phases. First-stage labour begins with uterine contractions during which the foetus is positioned for delivery. It ends when the allantochorion ruptures. Typically, the mare is restless

Figure 8.1. A mare in second-stage labor. All limbs are contracted as the mares' abdominal muscles contract. The hooves of the foal's forelimbs, covered by the amniotic membrane, are visible where they have just passed between the labia.

during this period of time. She will pace, paw, sweat, repeatedly look at her flanks and repeatedly lie down, then rise to pace some more. If she is in a pasture with other mares, there are several possible scenerios that have been observed. Firstly, she may actively move away from the herd. Secondly, she may be left behind when the others move on. Thirdly, she may remain with other mares who typically appear interested and gather around the foaling mare (S. McDonnell, pers. comm.). Isolation from other horses may facilitate appropriate imprinting of the foal to its own mother, rather than to another mare. In natural herd conditions, the band stallion becomes highly vigilant and may attempt to move the band away from other bands if they are nearby (McDonnell, 2003). Vigorous movements of the foetus during this period of time may result in visible, irregular movement of the abdominal wall. In domestic mares, observable first-stage labour may last from minutes to hours (Rossdale, 1967; Jeffcott, 1972). Some mares may appear to enter first-stage labour, for example by visible restlessness, then cease and resume hours or days later (Rossdale, 1968). In a feral situation, prolonged first-stage labor would potentially draw the attention of a predator.

Second-stage labour is typically very short in the mare, lasting only about 10–20 minutes, with multiparous mares having slightly shorter labours than primiparous mares (Rossdale, 1967; Jeffcott, 1972). Among the wild ancestors of the horse, mares that could complete second-stage labour rapidly would have had an adaptive advantage over mares that had a prolonged second-stage labour, since this is a period of time during which the mare would be highly vulnerable to predation. It begins with the rupture of the allantochorion and ends with the delivery of the foal. When the allantochorion ruptures, the mare typically lies down and initiates vigorous contractions of the abdominal muscles. She may periodically rise, only to lie down again and resume contractions (Figure 8.1). Usually the feet of the foals' forelimbs are delivered first, then the head, torso and hips. The mare is typically recumbent when the foal is delivered. Once the hips are delivered, the efforts to expel the foal will typically cease, and the feet of the hindlimbs may remain in the mares' vagina until she begins moving around. The umbilical cord will rupture when the mare first stands after delivery of the foal, or as a consequence of movements of the neonatal foal.

During third-stage labour, mares often show obvious and significant signs of exhaustion and abdominal discomfort. They may remain lying down for several minutes, and be covered in sweat. Third-stage labour

Figure 8.2. A mare in third-stage labour. The mare and foal are both standing and the mare is nuzzling her foal. The placenta has not been expelled. (Comment: Since the foal's hips have been delivered, the mare has made the transition from second to third stage. Mares are never 'between' stages of labour.)

Figure 8.3. A mare and her foal rest immediately after delivery. Note the darkness of sweat on the mare's neck and shoulder, and her nose resting on the ground.

begins when the foal is delivered and ends with the expulsion of the placenta (Figures 8.2 & 8.3). The majority of mares subsequently investigate the placenta. In one survey of 509 mares that had the opportunity to investigate and consume the placenta, 69% exhibited post-partum investigation of the placenta (Virga & Houpt, 2001). Placentophagia, or consumption of part or all of the placenta is uncommon but does occur. Five of the 509 mares in Virga and Houpt's (2001) study consumed placenta.

Figure 8.4. A mare nuzzling her newborn foal.

Immediate post-natal period

In addition to investigating the placenta, the mare will nuzzle and lick her foal, particularly around the head and the base of the tail (Figure 8.4). Maternal imprinting is the process by which a female identifies a given neonate as her own – i.e. identifies it as the individual or individuals that she will defend and nurse. Maternal labelling occurs when, in the process of licking her own or another neonate, a female deposits chemicals contained in her saliva, which are later used to aid in the identification and discrimination of her own offspring as compared to non-offspring (Klopfer *et al.*, 1964). After a mare imprints on a foal, she will typically reject attempts at suckling by any other foal.

No controlled studies have been published on maternal imprinting and maternal labelling in the horse, but subjective observations suggest that mares may imprint on their foal within the first 30 minutes or so post-partum, and certainly do so within the first few hours. Maternal activities directed towards the foal tend to decrease in frequency over the first 30 minutes of foal life (Houpt, 2002).

Among domestic mares, if a foal is born dead and there is no live replacement that the mare's maternal attention can be directed toward, she may spend several hours near the dead foal. Gradually, she will resume grazing, steadily grazing farther and farther away from the body. While some mares will exhibit intense responses to the immediate removal of their dead foal, whinnying and attempting to follow wherever the body is being taken, waiting until the mare has apparently lost interest on her own to quietly remove the body can avoid such incidents. Mares that have undergone the intense arousal that occurs subsequent to immediate removal of their dead foal appear to exhibit decreased conception rates (Barty, 1974).

In the first few hours after parturition, mares may exhibit increased aggressiveness towards other horses and towards humans (Tyler, 1972). However, they do not rise or fall in the dominance rank order. While a subordinate mare may threaten a higher ranking mare that approaches her offspring, she may be forced to retreat if the higher ranking mare is determined to

Figure 8.5. A mare nuzzling her foal while it suckles. The foal is nursing in the typical reverse parallel position.

be near the foal. New mothers will actively disengage themselves socially from the rest of the herd, and will be less likely to be near others members of the herd post-partum (Estep *et al.*, 1993).

Nursing

When a neonatal foal first attempts to suckle, the most useful thing that a mare can do is stand still while the foal investigates her body and learns, by trial and error, where it can productively suckle. Most experienced broodmares do this, either with all four limbs fully extended, or when the foal is successful in locating a teat, with the hind limb flexed opposite the foal. Sometimes mares gently push their foal with their head as the foal attempts to find the teat. This may be an attempt to assist the foal, but in some cases it is clearly more hindrance than help. Mares will often move their nose close to the foal's perineal area during nursing. This is probably for olfactory

verification that the suckling foal is, indeed, her own (Figure 8.5).

During the first week of foal life, mares terminate about half of the nursing bouts, mainly by walking away (Tyler, 1972; Duncan *et al.*, 1984; Crowell-Davis, 1985). In spite of this, this is the time period during which the nursing bouts are greatest in both frequency and duration (Crowell-Davis, 1985). Terminating nursing bouts by briskly walking away may be one mechanism by which the mare facilitates the foal's behaviour of following her. When the mare walks while the foal is still attempting to suckle, it follows her closely, repeatedly attempting to resume suckling. If she walks briskly enough to make this impossible, the foal simply follows closely by her hindquarters for a while.

At all ages of the foal, the mare may graze-walk while suckling occurs, instead of standing. Specifically, she lowers her head and grazes while walking

Figure 8.6. A foal nursing by reaching between its mother's hind limbs.

slowly. Younger foals may have to repeatedly reposition themselves in order to continue suckling, while older foals learn to walk backwards. Alternatively, foals confronted with mothers who repeatedly walk as they are trying to suckle may learn to suckle by reaching between their mothers' hind limbs (Figure 8.6).

A certain degree of maternal aggression directed towards the foal is normal. Most foal-directed aggression occurs in the context of nursing. Mares may lay back their ears, swish their tail violently against the foal, squeal, threaten to bite, bite, bunt, threaten to kick, kick at or smack at their foal during nursing. The smack is an audible threat in which the mare, while laying her ears back against her head, briskly opens her mouth, making a loud smacking noise (Crowell-Davis, 1985). These aggressions mainly occur during the initiation of nursing by the foal, when it is nuzzling the udder with its muzzle, and are probably in response to painful stimulus of the udder by the foal (Barber & Crowell-Davis, 1994). This normal maternal aggression is uncommon when the foal is very young, and peaks when the foal is about four to six months of age (Barber & Crowell-Davis, 1994).

Defence

Protection of the foal by the mare can be important to prevent harm by other members of the herd, as well as by predators (Joubert, 1972; Duncan, 1982). The mare may do this by simply positioning herself between the foal and another horse, or by actively aggressing against another horse. This is particularly likely during the first few hours of foal life (Figure 8.7). However, while the mare may defend the foal, she does not always do so, even if the foal approaches another horse in the herd which aggresses against it in response (Estep *et al.*, 1993). Domestic mares may be aggressive to stallions that the foal is near, even if the stallion is not being aggressive (Crowell-Davis & Houpt, 1986).

Recumbency response

Ungulates are classified into two groups depending upon whether the young follow and stay with the mother, or whether the young hide. These are referred to as followers and hiders, respectively (Lent, 1974). Horses are followers and, within a herd, mares and their foals spend more time close to each other than

Figure 8.7. A mare exhibiting a head threat to a horse that has approached a young foal, but is on the opposite side of the fence. In this case, the foal is an orphan that was fostered on to the mare when her foal was born dead less than a day later.

any other two individuals (Wells & Goldschmidt-Rothschild, 1979). During the first week of life, foals and their mothers are within 5 m of each other more than 90% of the time (Tyler, 1972; Crowell-Davis, 1986). When the foal is upright, this proximity is primarily due to the foal's behaviour as it, literally, follows (Crowell-Davis, 1986). This proximity steadily decreases over time. However, if humans do not interfere with herd structure, a mare's one- and two-year-old offsprings remain as her closest associates even with the coming of subsequent foals (Tyler, 1972; Wells & Goldschmidt- Rothschild, 1979).

When the foal lies down, however, the mother's behaviour changes as she is responsible for maintaining proximity to her sleeping, and defenceless, foal. This change is called the 'recumbency response' (Crowell-Davis, 1986). One of two major patterns of behaviour is typically adopted. First, if the foal lies down in a grassy area, the mare may continue grazing. However, instead of moving in the wavy-line pattern typical of horses moving across a pasture, she grazes in a small circle around the foal (Figure 8.8). If the foal sleeps for a long time, and the mare continues to graze, she gradually depletes all the grass in a roughly circular area around the foal. When the foal gets up and they leave, the area is easily visually discernible as a circle of shortened and flattened grass.

The second major pattern of behaviour exhibited during the recumbency response is rest while standing upright beside the foal (Figure 8.9). The anatomy of the horse is such that a mare can doze and even enter slow wave sleep while standing beside her foal. Rapid eye movement (REM) sleep is not possible in this position. In the position of standing resting upright, the mare can be observed to alternately doze with her eyes half closed and her lips flaccid, and looking around. In the domestic condition, the recumbency response notably wanes as the foal matures, so that mares may entirely leave their three- to four-month-old foals, grazing 100 m or more away (Crowell-Davis & Houpt, 1986).

The rate at which a mare and foal become independent of each other is unrelated to foal gender, but is specific to each mare (Crowell-Davis, 1986). The influence of the mare's maternal style can readily be seen in this pattern. In this study (Crowell-Davis, 1986), one mare, whose two colts and one filly, born over a three-year period, all became independent relatively rapidly, frequently graze-walked while her offspring were attempting to suckle, and rarely whinnied to them. Another mare, whose two fillies and one colt were about as likely to be close to her at 20 weeks of age as they had been in their first week of life almost invariably stood still when they were suckling and

Figure 8.8. A mare exhibiting the recumbency response by grazing beside her recumbent mule foal even though the grass in the area has become very short.

often whinnied to them when they strayed very far away (Crowell-Davis, 1986).

Whinnying appears to be used by mares and foals primarily to signal to each other where they are and to initiate reunification. Domestic mares appear to be able to discriminate between their own and other foals by vocal cues alone. They respond more to audio recordings of their own foals than to audio recordings of other foals (Wolski *et al.*, 1980). While coat colour seems to be an important means of foal identification at a distance (Leblanc & Bouissou, 1981), olfactory, visual and auditory cues (nickers) are used at close proximity when the mare and foal are reuniting after a separation (Wolski *et al.*, 1980).

Weaning While foals raised in modern, western society are often weaned when they are four to six months of age, natural weaning does not occur until a mare delivers her next foal, when a foal is at least 12 months old. At this time, the mare allows suckling by and defends the neonate, but may aggressively reject attempts by the older foal to suckle. Aggression to the older foal only occurs if they are very close or attempting to suckle, and is only of sufficient intensity to drive them off. The older foal is not typically injured by the mare's rejection. In some cases, the

mare tolerates continued nursing by the older foal even with a neonate at her side. If a mare does not have a subsequent foal for a longer time than 12 months, she often continues to allow the last foal born to suckle for several years. During this time suckling becomes less and less frequent, and is ultimately probably merely a comfort behaviour rather than a behaviour that results in the obtaining of any real nourishment. Feral mares have been observed to 'nurse' offspring as old as five years (R. Keiper, pers. comm.). Weaning methods for domestic horses are discussed in Ladewig, Chapter 9.

Maternal investment

Maternal investment is a sociobiological theory that involves any investment by the mother in her current offspring that will increase the offspring's chance of surviving. If the offspring survives to reproduce, it will increase the mother's reproductive success. This investment can be in any form of maternal care such as the provisioning of food through nursing, hunting, foraging, etc., and/or providing protection from predators or the aggression of conspecifics. This investment on the part of the mother is often to her detriment, as she tends to lose condition, or to the detriment of her future offspring. The more resources

Figure 8.9. A mare exhibiting the recumbency response by standing resting over her sleeping foal.

and energy she invests in her current offspring, the less resources she will have to invest in her future offspring, due to having lost some degree of body condition by investment in current offspring.

Trivers and Willard (1973) suggested that mothers should bias their investment into the offspring that will provide them with the greatest reproductive success. They suggested that mothers would bias their investment based on their own condition. Mothers that are in good condition, or have more resources to invest, should produce more sons than daughters, as, in most polygynous species, a male in good condition will have a better chance of gaining breeding capabilities than those in poorer condition. Mothers that are in poor condition would gain more reproductive success by having female offspring, as females in polygynous species will almost always reproduce regardless of size, condition or dominance rank. The theory of

Table 8.1. *Parameters of maternal investment that have been assessed in various studies*

Measure of maternal investment	Studies
Percentage of suckling episodes ended by the mother	Cameron & Linklater, 2000
Percentage of suck attempts that were unsuccessful	Cameron & Linklater, 2000
Maintenance of proximity between mother and foal	Cameron *et al.*, 2000; Weeks, 2000
Weaning dates	Cameron & Linklater, 2000
Foal dispersal	Cameron & Linklater, 2000
Suckle bout duration and frequency	Crowell-Davis, 1985; Barber & Crowell-Davis, 1994; Cameron *et al.*, 1999a, 2000; Weeks, 2000
Spatial relationships	Cameron & Linklater, 2000; Cameron *et al.*, 2000; Weeks, 2000
Birth sex ratios	Cameron *et al.*, 1999b
Recumbency response	Weeks, 2000

Trivers and Willard (1973) relies on several assumptions: (1) there is some physiological mechanism for females facilitating conception of and/or maintaining pregnancy of a particular gender of offspring; (2) in any given population of animals, body condition of females will range from poor to good and that females in good condition will be more capable of investing in their offspring than those in poor condition; (3) this difference in maternal body condition will result in offspring with body conditions that range from poor to good; and (4) the differences in body condition will be maintained into adulthood and will affect the reproductive success of the offspring.

Maternal investment in horses has been investigated in several studies done in 1985 and 2000. Several different measures of investment have been hypothesized and tested, as shown in Table 8.1.

Suckling has been the focus of several studies of maternal investment, as lactation is the most obvious form of investing in mammalian offspring. In studies of Welsh ponies and Belgians, Crowell-Davis (1985) and Barber and Crowell-Davis (1994), respectively, did not find any difference in nursing rate by colts and fillies. Duncan *et al.* (1984) reported that colts spent a greater percentage of time nursing during early lactation than did fillies. However, this result was obtained only after the exclusion of two female foals that had high percentages of suckling, which the authors considered to be statistical outliers. Weeks (2000) looked at nursing rates and nursing bout duration in relation to gender of the offspring and mare dominance rank. It was found that nursing rates, nursing bout duration and percentage of time spent nursing did not differ between genders. Additionally, when

split by foal gender and compared to mare dominance rank there was no significant correlation. That is, it was not the case, as had been predicted, that higher ranking mares invested more in male offspring and lower ranking mares invested more in female offspring.

However, Cameron *et al.* (1999a) found that suckle bout duration, suckle bout frequency and proportion of scan samples during which the foal was nursing were not significantly related to the amount of milk energy intake the foal was receiving. In their study, female foals sucked more than male foals but did not actually receive any more milk energy. This suggests that these nursing measures are not accurate measures of maternal investment for the horse. In a subsequent study, Cameron *et al.* (1999b) used percentage of sucking episodes ended by the mother and the percentage of sucking attempts that were unsuccessful to measure the mother–offspring conflict over this resource. Using such measures they found no significant differences between the genders of offspring at any age.

Cameron and Linklater (2000) looked at many other measures of maternal investment relating to mare body condition and found that change in the condition of the mare both pre- and post-natal did not differ with the gender of the offspring. They also found no significant relation between mare body condition and foal dispersal, and weaning ages.

Spatial relationships between mares and their offspring have been studied in relation to both age (Cameron & Linklater, 2000) and dominance rank (Cameron *et al.*, 1999b). During the first 20 days of a foal's life, an older mare remains closer to and approaches her foal more often, regardless of the foals' gender. After the first 20 days this vigilance decreases (Cameron & Linklater, 2000). In relation to a mare's dominance rank in a herd, proximity data does not differ significantly (Weeks, 2000). Mares in the herd studied by Weeks did not differ in their effort to maintain proximity with their offspring, regardless of foal gender or mare dominance rank. Coinciding with this, mares did not differ in their recumbency response to their foals either as a function of foal gender or dominance rank of the mare.

One last measure of maternal investment that has been investigated in horses is that of birth sex ratios. Cameron *et al.* (1999b) studied birth sex ratios in relation to mare condition and parity. They found that mare condition *at conception* was a significant predictor of offspring gender but mare condition at mid-gestation was not a significant predictor of offspring gender. There was no significant variation in sex ratios between primiparous and multiparous mares.

The study of maternal investment in the horse is too new to make any definitive statements as to any potential bias that may exist relating to offspring gender, mare parity, mare dominance rank or mare condition. The studies conducted to date all use different measures of investment and, as Cameron and Linklater (2000) demonstrated, these commonly used measures might not be valid measures of maternal investment. Also, the published studies were conducted with both captive and feral herds of horses. Captive herds may not be comparable to feral herds in that most captive herds are provisioned with supplemental feed. Though there are noticeable differences in body condition among members of captive herds, these differences may not be great enough to adequately represent the variation that occurs in feral herds. In addition, the selective advantage associated with differential maternal investment may have been lost as a result of domestication where competition for resources is less. At this point in time, there is no clear and consistent evidence that mares vary in their investment in sons or daughters.

Fostering foals on to a mare

Foals with their mothers suckle multiple times every hour, especially during the first few weeks of life. The need for frequent suckling makes it difficult to raise foals by hand because of the intense schedule required for bottle-feeding. In addition to the difficulty of bottle-feeding, hand-rearing presents the problem of possible *excessive bonding to humans*, which can lead to later behaviour problems. Thus, when a foal is orphaned, fostering it onto a mare is the ideal solution. However, as discussed above, mares typically imprint onto a foal in the immediate post-partum period and most mares do not readily care for foals that are presented to them after this period. It is generally easiest to foster a foal onto a mare during the first 30 minutes after the mare has given birth, although it is not too difficult to foster a foal for several hours post-partum so long as the mare has not had the opportunity to imprint onto another foal. The odours coming from the amniotic fluid on the foal are probably critical to this process, and smearing amniotic fluid on a foal facilitates acceptance of a foster foal (Tyler, 1972; Barty, 1974). It may be necessary to restrain the mare on a lead rope during initial interactions in case she becomes aggressive. Interactions between the foster mare and foal for the first several hours need to be closely supervised to ensure that appropriate bonding occurs. A small number of mares will readily accept foals other than their own, and may even allow suckling from multiple foals. Such mares, once identified, are often kept at horse-breeding farms as 'nurse' mares.

Foal rejection

There are two major types of foal rejection: avoidance of the foal and active aggression against the foal (Houpt, 1984; Houpt & Olm, 1984). The first type, foal avoidance, is the most common, and occurs primarily in primiparous mares. The primiparous mare may appear to be frightened of her foal, and actively avoids all of its approaches. If her udder is turgid and painful, attempts by the foal to suckle can result in squealing and avoidance, necessitating competent professional handling to ensure survival of the foal. Once the foal suckles a few times, the mare usually becomes more tolerant of further suckling. This may be because the mare learns that suckling by the foal results in decreased pain. This process can be facilitated by feeding the mare grain while she is standing, thereby associating the foal and suckling with the pleasant experience of eating palatable food. Likewise, restraining her while she habituates to the simple presence of the foal will result in her becoming more tolerant of its presence.

If the mare is so aroused that competent handling is not sufficient, she can be sedated or further physically restrained, for example by a bar positioned at the level of her barrel that allows the foal access to the udder (Crowell-Davis & Houpt, 1986). However, sedatives and anxiolytics should only be used if essential, as they enter the milk.

Sometimes, mares reject their foals more intensely, and may engage in significant aggression against the foal, biting and even kicking it. Bites are usually directed to the neck and withers (Houpt, 1984). Even mares that reject their foals aggressively are not usually aggressive to it when it is recumbent, i.e. the recumbency response persists even in these mares. Aggression occurs when the foal is standing. Aggressive maternal rejection can occur in any breed but there is evidence of genetic effects, with certain families *and breeds* exhibiting the problem more frequently than others (Juarbe-Diaz *et al.*, 1998).

Maternal behaviour directed to non-offspring

The great majority of mares reject all attempts by foals other than their own to suckle. During the periparturient period, mares are ready to identify and accept a foal as their own and various circumstances during this period can lead to their adopting a non-own foal. In one case that occurred at the University of Georgia, a low ranking mare gave birth hours before a high ranking and very assertive mare in the herd gave birth. During the intervening period between the two births, the higher ranking mare stole the low ranking mare's foal, stood for it to suckle, and repeatedly drove the low ranking mare away as she attempted to get the foal back. When the higher ranking mare went into labour, she ignored the stolen foal and subsequently imprinted on and bonded to her own foal, successfully raising it. As soon as the higher ranking mare was ignoring the first foal, the low ranking mare successfully retrieved it and also successfully raised it (S. L. Crowel-Davis pers. obs.). In rare cases, a mare who has a foal will be tolerant of suckling by foals other than her own. Some will even allow young of other species to suckle, for example a mule mare who had delivered a horse foal after embryo transfer adopted a calf after the horse foal was removed (Shaw & Houpt, 1985). While such behaviour is evolutionarily maladaptive in this species, such mares are very useful as nurse mares for orphaned foals.

When adult mares that are mother and daughter are in the same herd, they may engage in co-operative offspring care, allowing each other's offspring to suckle (Cameron *et al.*, 1999c). In this case, the behaviour is probably evolutionarily adaptive, though rare.

Summary and conclusion

Mares are monotocous and appear to imprint on and label their own offspring shortly after birth. Most mares subsequently reject solicitation of care by any other foal. This pattern of behaviour makes fostering of orphaned young difficult in this species. Care of the foal that the mare imprints on is intense, regardless of the foal's gender, the mare's dominance rank or the mare's body condition, with frequent nursing and active defence of the young occurring. The mare also modifies her behaviour in order to 'stand guard' over her foal when it is sleeping. She will continue to care for a given foal until a subsequent foal is born. Altogether, these behaviours maximize the likelihood of the foal surviving to adulthood.

To be most likely to be beneficial to domestic horses, future research on maternal behaviour should focus on the genetic and experiential factors that affect foal rejection, maternal aggression towards foals, and willingness to care for a non-own foal. We must be careful not to selectively breed poor mothers just because they are conformation or performance champions if their poor maternal skills are genetically influenced. Conversely, information that helps us decrease the incidence of foal rejection and better identify or raise mares that can be used as nurse mares will help improve the survival and rearing conditions of domestic foals.

References

Bain, A. M. & Howey, W. P. (1975). Observations on the time of foaling in Thoroughbred mares in Australia. *Journal of Reproduction and Fertility*, **23** (Suppl.): 545–6.

Barber, J. A. & Crowell-Davis, S. L. (1994). Maternal behavior of Belgian (*Equus caballus*) mares. *Applied Animal Behaviour Science*, **41**: 161–89.

Barty, K. J. (1974). Observations and procedures at foaling on a Thoroughbred stud farm. *Australian Veterinary Journal*, **50**: 553–7.

Cameron, E. Z. & Linklater, W. L. (2000). Individual mares bias investment in sons and daughters in relation to condition. *Animal Behaviour*, **60**: 359–67.

Cameron, E. Z., Linklater, W. L., Stafford, K. J. & Veltman, C. J. (1999b). Birth sex ratios relate to mare condition at conception in Kaimanawa horses. *Behavioral Ecology*, **10** (5): 472–5

Cameron, E. Z., Linklater, W. L., Stafford, K. J. & Minot, E. O. (1999c). A case of co-operative nursing and offspring care by mother and daughter feral horses. *Journal of Zoology*, 249: 486–9.

(2000). Aging and improving reproductive success in horses: declining residual reproductive value or just older and wiser? *Behavioral Ecology and Sociobiology*, 47: 243–9.

Cameron, E. Z., Stafford, K. J., Linklater, W. L. & Veltman, C. J. (1999a). Suckling behavior does not measure milk intake in horses, *Equus caballus*. *Animal Behaviour*, 57: 673–8.

Campitelli, S., Carenzi, C. & Verga, M. (1982). Factors which influence parturition in the mare and development of the foal. *Applied Animal Ethology*, 9: 7–14.

Crowell-Davis, S. L. (1985). Nursing behaviour and maternal aggression among Welsh ponies (*Equus caballus*). *Applied Animal Behaviour Science*, 14: 11–25.

(1986). Spatial relations between mares and foals of the Welsh pony (*Equus caballus*). *Animal Behaviour*, 34: 1007–15.

Crowell-Davis, S. L. & Houpt, K. A. (1986). Maternal Behavior. In *The Veterinary Clinics of North America: Equine Practice, Behavior*, vol. 2, ed. S. L. Crowell-Davis & K. A. Houpt. Philadelphia: W. B. Saunders. pp. 557–72.

Duncan, P. (1982). Foal killing by stallions. *Applied Animal Ethology*, 8: 567–70.

Duncan, P., Harvey, P. H. & Wells, S. M. (1984). On lactation and associated behaviour in a natural herd of horses. *Animal Behaviour*, 32: 255–63.

Estep, E. Q., Crowell-Davis, S. L., Earl-Costello, S. A. & Beatey, S. A. (1993). Changes in the social behavior of drafthorse (*Equus caballus*) mares coincident with foaling. *Applied Animal Behaviour Science*, 35: 199–213.

Feist, J. D. & McCullough, D. R. (1975). Reproduction in feral horses. *Journal of Reproduction and Fertility* 23(Suppl.): 13–18.

Houpt, K. A. (1984). Foal rejection and other behavioral problems in the postpartum period. *Compendium on Continuing Education for the Practicing Veterinarian*, 6: S144.

(2002). Formation and dissolution of the mare-foal bond. *Applied Animal Behaviour Science* 78: 319–28.

Houpt, K. A. & Olm, D. (1984). Foal rejection: a review of 23 cases. *Equine Practice*, 6: 38.

Howell, C. D. & Rollins, W. C. (1951). Environmental sources of variation in the gestation length of the horse. *Journal of Animal Science*, 10: 789–96.

Jeffcott, L. B. (1972). Observations on parturition in crossbred pony mares. *Equine Veterinary Journal*, 4: 209–12.

Joubert, E. (1972). The social organization and associated behaviour in the Hartmann zebra *Equus zebra hartmannae*. *Madoqua*, Series I(6), 17–56.

Juarbe-Diaz, S. V., Houpt, K. A. & Kusunose, R. (1998). Prevalence and characteristics of foal rejection in Arabian mares. *Equine Veterinary Journal*, 30: 424–8.

Klopfer, P. H., Adams, D. K. & Adams, M. S. (1964). Maternal imprinting in goats. *Proceedings of the National Academy of Sciences of the United States of America*, 52: 911–14.

Leblanc, M. A. & Bouissou, M.-F. (1981). Mise au point d'une epreuve distinée a l'étude de la reconnaissance de jeune par la mère chez le cheval. *Biology of Behaviour*, 6: 283–90.

Lent, P. C. (1974). Mother–infant relationships in ungulates. In *The Behaviour of Ungulates and Its Relation to Management* (New Series), No. 24., ed. V. Geist & W. Walther. Morges, Switzerland: International Union for conservation of Nature and Natural Resources, pp. 14–55.

McDonnell, S. (2003). *The Equid Ethogram: A Practical Field Guide to Horse Behavior*. Lexington, KT: Eclipse Press.

Oxender, W. D., Nodan, P. A. & Hafs, H. D. (1977). Estrus, ovulation, and serum progesterone, estradiol, and LH concentration in mares after an increased photoperiod during winter. *American Journal of Veterinary Research*, 38: 203–7.

Rollins W. C. & Howell, C. E. (1951). Genetic sources of variation in the gestation length of the horse. *Journal of Animal Science*, 10: 203–7.

Rossdale, P. D. (1967). Clinical studies on the newborn Thoroughbred foal. I. Perinatal behaviour. *British Veterinary Journal*, 123: 470–81.

(1968) Perinatal behavior in the Thoroughbred horse. In *Abnormal Behavior in Animals*, ed. M. W. Fox. Philadelphia: W. B. Saunders, pp. 227–37.

Rossdale, P. D. & Short, R. V. (1967). The time of foaling of Thoroughbred mares. *Journal of Reproduction and Fertility*, 13: 341–3.

Shaw, E. & Houpt, K. A. (1985). Pre- and post-partum behavior in mares impregnated by embryo transfer. In *Equine Embryo Transfer*, ed. W. R. Allen & D. F. Antczak. *Equine Veterinary Journal*, 73(Suppl. 3).

Trivers, R. L. Willard, D. E. (1973). Natural selection of parental ability to vary the sex ratio of offspring. *Science*, **179**: 90–1.

Tyler, S. J. (1972). The behaviour and social organization of the New Forest ponies. *Animal Behaviour Monographs*, 5: 85–196.

Virga, V. & Houpt, K. A. (2001). Prevalence of placentophagia in horses. *Equine Veterinary Journal*, 33: 208–10.

Wells, S. M. & Goldschmidt-Rothschild, B. V. (1979). Social behaviour and relationships in a herd of Camargue horses. *Zeitschrift fur Tierpsychologie*, **49**: 363–80.

Weeks, J. W. (2000). Maternal Investment in *Equus caballus*. PhD Thesis, The University of Georgia.

Wolski, T. R., Houpt, K. A. & Aronson, R. (1980). The role of the senses in mare-foal recognition. *Applied Animal Ethology*, 6: 121–38.

9

Ontogeny: preparing the young horse for its adult life

Jan Ladewig, Eva Søndergaard and Janne W. Christensen

From the time of birth till adulthood the young horse goes through a number of phases during which its species-specific behaviour pattern develops. The behavioural development is a result of hereditary behaviour programmes and learned behaviour patterns. Some aspects of the development occur relatively independent of the environment, but others depend upon a certain stimulation from the environment. Thus, specific key factors must be present (or absent) in the environment during specific periods in order for a foal to develop normally. In contrast, if these specific key factors are absent (or present), the development does not proceed normally or the development of an abnormal behaviour pattern may result. In some cases such a course may be permanent and result in behaviour problems that last throughout life.

In addition to the development of natural behaviour patterns, domestic horses must learn an additional behaviour repertoire. Much of this learning occurs unnoticed, but part of it is brought about through training. Learning about the domestic environment undoubtedly follows the pattern of the development of natural behaviour. If so, this learning as well as the training of various skills may occur more effectively during specific periods.

In order to raise horses that are optimally prepared for the various tasks for which horses are used, it is important to understand the different phases of behaviour development and what factors are important for this development. It is important that the young horse learns the social communication of horses, and since most domestic horses live in some kind of social relationship with humans, it is equally important that the young horse learns social communication with humans. In addition, since normal development in many cases requires specific features of the environment, the way young horses are housed is important. Considering the importance of these different aspects of horse husbandry, it is somewhat surprising how relatively few studies deal with how factors in the domestic environment affect the behavioural development of a foal.

The Domestic Horse: The Origins, Development, and Management of its Behaviour, ed. D. S. Mills & S. M. McDonnell.
Cambridge University Press. © Cambridge University Press 2005.

Ontogeny of behaviour patterns

Behavioural development is the result of the continuous interaction of genetic and environmental factors. The genetic code is fixed, but the individual will develop appropriately only with the right cues from the environment. Some behaviours are innate, such as standing, nursing, running, neighing and snapping (mouth clapping). These behaviours are nearly complete the first time they are expressed although they are still fine-tuned and modified by learning (Mills & Nankervis, 1999). Innate behaviours tend to be triggered by fairly general events or impressions (sign stimuli), i.e. they are less open to modification by the environment. Learned behaviours take much longer to develop and are obviously much more affected by the environment. The advantage of learning is that it increases an animal's flexibility and the ability to adapt within one lifetime. Thus, adaptation of behaviour occurs over generations by selection (innate behaviour), and is modified within the lifetime of the individual (learned behaviour). The way in which behaviour develops does not support a clear division, and the definitions of learned and innate behaviours are relative.

The time period from the embryonic state through birth till adulthood can be divided into various stages. It is important to realize, however, that behavioural development is a continuous process and that the transition from one stage to the next is far from clear-cut and that a large degree of individual variation can be seen as far as the age at which transition occurs. The age at which the onset of various behaviour patterns has been reported in the scientific literature is shown in Table 9.1.

During the neonatal period, which lasts only about two hours, the foal begins to stand, walk and search for the udder. Although the search for the udder is inherited to a large extent, its perfection appears to be learned. Thus, the initial search for the udder is based primarily on the tactile and olfactory senses and later on vision. The following two weeks represent the time of greatest sensory development. In other species, too little sensory exposure in this period has been shown to result in reduced sensory capacity (Blakemore & Cooper, 1970). During the first four weeks of life the foal is maximally dependent on its mother for sustenance. It remains close to her and learns behaviour strategies from her responses to the environment as well as from its own experiences. The close association between mare and foal gradually decreases as the foal matures. For instance, a foal's mother was found to be its nearest neighbour 92% of the time during the first month, whereas by the second month of life, the mother was the nearest neighbour only 66% of the time. This change results from the foal's increasing contact with other herd members when it proceeds to develop relationships with other foals or yearlings (Tyler, 1972; Crowell-Davis, 1986a,b).

Although social interactions with other foals may be observed as early as the first week in foals that have young herdmates (McDonnell & Poulin, 2002), there tends to be an increase in social behaviour during the second and third month of life when the foal enters the socialization phase (Crowell-Davis, 1986a). Social play behaviour and mutual grooming are frequently shown during this phase, and an increase in snapping behaviour (mouth clapping) may also be observed as the foal begins to make frequent ventures from the dam to approach and contact other horses (Crowell-Davis et al., 1985a; Crowell-Davis, 1986a). Snapping behaviour is commonly viewed as a submissive behaviour (McDonnell & Haviland, 1995), but has also been suggested to have a calming effect on the submissive individual (Boyd, 1988). Crowell-Davis et al. (1985a) suggested that snapping behaviour may be a displacement activity developed from nursing.

Grazing behaviour in foals is infrequent during the first week, but time spent grazing increases gradually over the next few months (Tyler, 1972; Boy & Duncan, 1979). Before foals begin to ingest grass or drink water, they appear to go through a brief developmental stage during which they seem to mimic the grazing or drinking behaviour of adults in a playful investigative manner, without ingesting (Blakeslee, 1974, quoted by McDonnell & Poulin, 2002). The playful phase of drinking and eating is very short, and once a foal begins ingesting grass or drinking water, the playful, investigative form is not seen again (McDonnell & Poulin, 2002). Up to the age of three to four months, coprophagy (eating faeces) can be observed in foals. The behaviour, which is considered to be a normal behaviour in foals, is particularly common during the first two months, after which it gradually decreases (Crowell-Davis & Houpt, 1985a). Foals generally show a strong preference for the fresh faeces of their mother in both field

Table 9.1. *Ages at which onset of behaviours are described in the scientific literature*

Age	Behaviour	Reference
0.5–2 hours	Stand	(Rossdale, 1967; Waring, 1982)
0.5–3 hours	Nurse	(Rossdale, 1967; Jeffcott, 1972; Waring, 1982; Houpt, 2002)
0.5–1.5 hours	Vocalise (nicker, whinny)	(Waring, 1982)
Day 1	Graze using nibbling motions	(Tyler, 1972; Crowell-Davis *et al.*, 1985b; Boyd, 1988)
	Graze and drink	(McDonnell & Poulin, 2002)
	Groom self by rolling	(Boyd, 1988)
	Groom self by rubbing	(Boyd, 1988; Zharkikh, 2000)
	Neigh and nicker	(Boyd, 1988)
	Paw	(Boyd, 1988)
	Play by self – running	(Crowell-Davis *et al.*, 1987)
	Shake	(Boyd, 1988)
	Snap (mouth clap)	(Boyd, 1988)
	Stand-rest	(Boyd, 1988; Crowell-Davis, 1994)
Day 2	Flehmen	(Crowell-Davis & Houpt, 1985b)
	Groom self by chewing	(Boyd, 1988)
	Groom self by scratching with hindleg	(Boyd, 1988)
	Play by self – running	(Boyd, 1988)
Day 3	Graze using tearing motions	(Boyd, 1988)
	Play by self – bucking	(Boyd, 1988)
Day 5	Coprophagy	(Crowell-Davis & Houpt, 1985a)
Day 6	Coprophagy	(Boyd, 1988)
	Eat hay and grain	(Boyd, 1988)
	Kick defensively	(Boyd, 1988)
Week 1 (day not given)	Flehmen	(Weeks *et al.*, 2002)
	Graze using nibbling motions	(Tyler, 1972; Fraser, 1980; Carson & Wood-Gush, 1983)
	Mutual groom	(McDonnell & Poulin, 2002)
	Social play	(McDonnell & Poulin, 2002; Tyler, 1972)
Week 2	Flehmen	(Boyd, 1988)
	Group running and bucking	(Crowell-Davis *et al.*, 1987)
	Mutual groom with dam	(Boyd, 1988; Crowell-Davis *et al.*, 1986)
	Social play	(Crowell-Davis *et al.*, 1987)
Week 3	Drink	(Crowell-Davis *et al.*, 1985b; Boyd, 1988)
	Graze and walk simultanously	(Carson & Wood-Gush, 1983)
	Play by self – mount dam	(Boyd, 1988)
	Snort	(Boyd, 1988)
	Social play	(Boyd, 1988)
Week 4	Social play	(Francis-Smith, 1979)
Week 6	Grazing behaviour (adult-like)	(Marinier & Alexander, 1995)
Month 2	Mutual groom with other horses	(Boyd, 1988)
Month 10	Masturbation by colts	(Boyd, 1988; McDonnell, 1989; McDonnell, 2003)
Month 11	Marking of elimination by colts	(Boyd, 1988)

studies (Tyler, 1972; Francis-Smith & Wood-Gush, 1977; Crowell-Davis & Houpt, 1985a) and in a test situation (Crowell-Davis & Caudle, 1989). Crowell-Davis and Houpt (1985a) suggested that foals may consume faeces in response to a maternal pheromone which signals the presence of deoxycholic acid or other acids in which the foal may be deficient and which it may require for gut immuno-competence and myelination of the nervous system. In addition, coprophagy may provide nutrients and introduce normal bacterial flora to the gut. Grazing behaviour and food selection was studied by Marinier and Alexander (1995) and they noted that adult-like grazing behaviour and selection of plants develop between the ages of four and six weeks. They also found that coprophagy of maternal faeces decreased during the same period and suggested that coprophagy may also function to teach the foal food selection.

The development of behaviour patterns is gradual and subject to much individual variation. However, behavioural development during the first few days is characterized by less individual variation, compared to later development. For instance, some foals engage in social play already during their first week of life (Tyler, 1972; McDonnell & Poulin, 2002), whereas Crowell-Davis *et al.* (1987) first observed this type of behaviour during the second week. Francis-Smith (1979) first observed social play at four weeks of age in Thoroughbred foals. The observations may reflect differences in schedules and methods of observation, and/or they may suggest that there are some differences in the ontogeny of play behaviour dependent on the social environment in which a foal develops. Since play frequently appears to be stimulated by novel conditions or stimuli (McDonnell & Poulin, 2002), it is important to note that foals of different studies do not get similar environmental cues, and it is possible that less individual difference occurs within one environment. Thus, the process of behavioural development is not simply a function of age but is affected by external factors.

From the fourth month onward, foals gradually become more independent from their mothers and progress towards adult patterns of spatial relationships, social interactions and maintenance behaviours (Crowell-Davis, 1986a). Play sexual behaviours are frequent in foals of both sexes and a juvenile form of elimination marking behaviour may be observed even during the first week of life. Marking behaviour becomes more frequent once physical puberty is reached and was first observed in the adult form in Przewalski's colts at the age of 11 months (Boyd, 1988). These signs of development of the hormone secreting glands, which control sexual behaviour, mark the end of the foalhood. Sexual maturation is also subject to much individual variation, and probably also depends on physical conditions.

Behavioural development under husbandry conditions

Rearing methods for foals under domestic conditions differ considerably from natural rearing conditions, e.g. in that many foals are stabled for part of the day or grow up with the dam only. In addition, most foals are prematurely weaned or may be fostered by humans or mechanical means.

The effect of maternal deprivation

The attachment (or bonding) of the mare to the foal begins during and immediately after birth when she is licking the birth membranes and the foal (Waring, 2003), and it may continue to strengthen up to two weeks after birth (Houpt, 2002). The attachment of the foal to the mare begins slightly later when the foal stands and begins suckling. During the first days the attachment of the foal to the mare seems less strong than that of the mare to the foal (Waring, 2003). Again, the attachment seems to grow stronger until the foal is about two weeks old (Houpt, 2002), at which point it slowly weakens with no definite end point (Waring, 2003). This process is discussed further by Crowell-Davis and Weeks in Chapter 8

Depriving the foal of the presence of the dam is not a usual part of management but happens occasionally when a dam dies or rejects her foal. The mother can be replaced by another mare, the foal can be raised by a foster mother from another species, including humans, or the foal can be raised mechanically. It is important to be aware that there are two aspects of maternal deprivation. One is the feed and the way of feeding, the other is the upbringing and the behavioural development without the dam. Naturally, most emphasis has been put on the first aspect since feeding is essential for the survival of the foal. Getting a mare to accept a foal that is not her own can be difficult because it is not natural for her to foster an offspring which she did not deliver. In feral horses cross-fostering between relatives was observed in a mare and her daughter together bringing up the foal of the daughter (Cameron *et al.*, 1999). Because the foal–mare attachment develops more slowly, the foal usually has no trouble accepting another mare. If the only purpose of a foster mother is to provide feed it may be safer to feed the foal in another way although it may be difficult to provide the number of feedings that is required. Elze *et al.* (1996) recommend 15–16 feedings per day decreasing to 10 feedings per day by the end of the second week. In that case an automatic feeding machine might be a good solution, as described by Glendinning (1974). When the foals were healthy at birth, no physical problems with artificial rearing by man or machine have been reported.

The other aspect of maternal deprivation is the social environment of the foal. If the foal has absolutely no contact with other horses it may not even recognize horses when subjected to them later in life (Grzimek, 1949) or it may prefer human to equine company (Williams, 1974). Even when foals are raised together with other foals, but without a mare, it may affect their behavioural development. Williams (1974) noted that artificially reared foals snap towards humans whereas for naturally raised foals, snapping was only observed in one very timid foal. Houpt and Hintz (1983) found no difference in activity between orphaned and mothered foals when kept in a group on pasture, but orphan foals scratched more than mothered foals. The scratching could be related to the removal of insects, as mothered foals stood next to the mare and benefited from her tail. Orphaned foals spent more time with other foals than did mothered foals but when taking into account the time that mothered foals spent with their dam or another mare there was no difference in the time the foals spent with other horses. These observations indicate that horses are social by nature and that they do not necessarily need to learn social interactions from their mother. On the other hand, social behaviour improves with practice. When placed in a novel environment mothered foals reacted more to being alone than orphan foals indicating that the bond between a mare and her foal is stronger than the bond between orphaned foals that have been grouped together all their life. The learning ability of orphaned foals was no different than for mothered foals as assessed in a maze (Houpt *et al.*, 1982), but orphaned foals spent more time in the maze when first exposed to it.

The effect of weaning method

Weaning of the domestic foal occurs when the foal is permanently separated from the mare. Apart from the separation weaning implies other changes for the foal, such as changes in feeding and in management. In feral and wild horses weaning is gradual and starts when foals are about eight months old but the process is often not terminated until the foal or the young horse is more than two years old. Under husbandry conditions weaning can occur as early as immediately after birth but weaning is more common at an age of four to seven months. An argument for early weaning is the condition of the mare whereas arguments for later weaning concern feeding and behavioural aspects for the foal. In any case weaning is considered to be stressful for the foal (Apter & Householder,

1996) and various experiments have investigated the possibility to minimize weaning stress.

Habituating foal and mare to separation by separating them for short periods prior to weaning was found to have no effect on either mare or foal behaviour or cortisol response at weaning (Moons & Zanella, 2001). Short-term separation affected first-parity mares more than mares of later parity and younger foals reacted less than older foals (Søndergaard, 1998). It is thus important to correct for these factors when investigating effects of short-term separation prior to weaning.

Partial weaning in which mare and foal had visual, auditory and olfactory contact seems to be less stressful than abrupt weaning (McCall *et al.*, 1985), but unfortunately it was not investigated how these foals later reacted to total separation from their mothers. In this study foals were weaned in pairs or triplets which in other experiments has been shown to be more stressful than weaning singly (Malinowski *et al.*, 1990; Hoffmann *et al.*, 1995). However, Houpt *et al.* (1984) found that foals weaned alone vocalised more than foals weaned in pairs indicating that they were more stressed although there was no difference between treatments in other behaviours or plasma cortisol. Heleski *et al.* (2002) compared foals weaned singly in stalls to foals weaned in groups on pasture and concluded that the latter had a better welfare although they did not find evidence that the former had a poor welfare. Perhaps the least stressful method of weaning is the gradual weaning described by Holland *et al.* (1996). They compared abrupt weaning where all mares were removed from a group of foals and mares, to gradual weaning where mares were removed one or two at a time every two days. All foals reacted to weaning behaviourally but abruptly weaned foals reacted more than gradually weaned foals. The authors concluded that foals adapted to weaning better when left on pasture than when weaned into stalls, a method they had tried in a previous experiment (Hoffmann *et al.*, 1995). In another part of the experiment they found indications that diet around weaning may affect the behaviour of the foal; fat and fibre possibly having an advantage over starch and sugar in enabling the foals to cope with weaning.

From the few experiments on the effect of weaning it seems that weaning singly may be better than or at least not worse than weaning in pairs or groups but the question could be: Why wean singly if young horses are better off growing up with other horses? To answer this question it is necessary to analyse not only the effects of the social environment in relation to housing (Glade, 1984; Christensen *et al.*, 2002; Søndergaard & Schougaard, 2000), but also the effect of confinement on locomotion behaviour (Barneveld *et al.*, 1999) and sleeping behaviour (Glade, 1984).

Effects of housing

Although horses are social by nature their social skills still have to be refined and practiced. Søndergaard and Halekoh (2003) housed young horses either singly or in groups of three from weaning until two and a half years of age. During this period several aspects of their social behaviour and of their reactions towards humans were investigated. Singly housed horses showed more interest in having contact with humans and were more easily approached by humans in their home environment (Søndergaard & Halekoh, 2003). This result corresponds to observations by Hughes *et al.* (2002) who found that foals kept without contact with other foals interacted more readily and spent more time with an observer than foals kept on pasture with other foals. In the experiment by Søndergaard and Ladewig (2004) singly housed horses also interacted more with a trainer during weekly training sessions than group-housed horses. The interaction consisted mainly of non-aggressive biting indicating that singly housed horses were motivated for physical interaction. This result was confirmed by observations during summer when horses from both treatment groups were kept on pasture in groups. Singly housed horses spent more time performing social grooming than group-housed horses (Christensen *et al.*, 2002). The social behaviour of group-housed horses seemed better developed as shown by the lower frequency of aggressive behaviour and higher frequency of subtle agonistic interactions such as displacement and submissive behaviour. These differences imply that by housing young horses singly, and thereby not giving them the opportunity to practice their social skills, they may be more prone to the risk of injuries when they are later in contact with other horses.

Considering that many adult horses are housed singly, an argument could be made for depriving horses of social contact throughout life. There are, however, several reasons that speak against this practice. First, we know that lack of social contact probably is one of the main reasons for horses to develop stereotypic behaviour (See Mills, Chapter 15). Apart from the welfare aspect, stress reactions cost energy, which could limit the performance of the horse. Second, one of the reasons why horses were domesticated was because of their social skills. These social skills include communication and acceptance of a social hierarchy, characteristics that are important for handling and training the horse. For example, in an experiment by Rivera *et al.* (2002) it was found that singly housed horses needed more time for initial training than group-housed horses. Since the singly housed horses were kept in stalls, however, whereas the group housed horses were kept on pasture, environmental enrichment and stimulation could be part of the difference in their results. In the experiment by Søndergaard and Ladewig (2004), however, similar results were found. Singly housed horses needed more training sessions to complete a task and they were biting more during training sessions than group housed horses.

There are other arguments for rearing young horses in groups rather than singly. Group-housed horses exercise more when in a paddock compared to singly-housed horses (Søndergaard & Schougaard, 2000) and the lack of exercise for young foals has been shown to alter their locomotive behaviour (Barneveld *et al.*, 1999). Besides, the lack of exercise retards the development of the musculo-skeletal system. Because the exercise in group-housed horses not only consists of forward locomotion but also, for example, rearing, bucking and play fighting, it is very likely that group-housed horses develop a better co-ordination of movements than singly housed horses.

Finally, Glade (1984) studied the social sleeping behaviour in young horses and found that when space allowance was not a limiting factor, group-housed young horses had fewer but longer periods of recumbency than young horses housed singly. Young horses housed singly in stalls spent about one-third of the day in a recumbent position. This may be a reflection of the quieter environment in the stalls or of the fact that the restriction of the environment limited the possibility to perform behaviour patterns compared to young horses in outdoor pens which had the opportunity of interacting with neighbours in the adjoining pens.

Development of abnormal behaviours

Abnormal behaviours are non-existent in feral or wild horses indicating that the environment is the main reason for the development of these behaviours. Waters *et al.* (2002) found that horses initiated weaving at a median age of 60 weeks, box walking at 64 weeks, wood-chewing at 30 weeks and cribbing at only 20 weeks indicating that the rearing period is very important in the development of abnormal behaviours. The mechanism behind the development of abnormal behaviours is not established but learning may be an important feature (Mills, 1999) as well as physical problems like gastric ulceration and mucosal inflammation (Nicol *et al.*, 2001). These matters are discussed further by Mills in Chapter 15.

The degree to which a foal will learn abnormal behaviour from the dam has not been investigated, although this type of learning was ruled out in an investigation by Vecchiotti and Galanti (1986) and horses have not been shown to learn effectively by observation alone (See Nicol, Chapter 12).

The effect of handling and habituation on behaviour development

Optimal behaviour development means encouraging the development of the correct skills at the appropriate time. In the wild this often happens as a result of interaction with certain stimuli. In terms of the survival of the individual, the stimulation from conspecifics constitutes the most important type of stimulation. In the domestic situation most horses are deprived of some of this opportunity to interact and learn from others. Specific and goal directed training can help to ensure the development of a confident and well-mannered horse with the fundamental skills it will require later in life.

As far as the development of natural behaviour is concerned, specific periods may exist during which learning of certain skills is achieved. Such periods have been termed critical or sensitive. The postnatal period during which bonding between a mare and her foal is established, may be called critical because if it does not occur within a relatively short time after birth, it may not occur at all. In contrast, habituation towards people may occur faster in a young foal, i.e. at a time when its natural tendency to react fearfully is less, as opposed to later when it is more inclined to react fearfully. Consequently, the earlier period could be termed

sensitive. Thus, critical periods can be defined as periods during which a certain exposure must occur in order for behaviour change to happen, whereas sensitive periods can be defined as periods during which exposure exerts a larger effect than at other times (Bateson, 1979).

In terms of handling and training of the young horse, it is important to realize that there can be different reasons why training may be more effective during particular periods. One reason is related to neuronal development, particularly the establishment of synaptic connections between neurones in the central nervous system. This development has been described in other species (e.g. cats: Blakemore & Cooper, 1970; and chickens: Gunnarsson *et al.*, 2000) but not in horses. A second reason is related to the fact that in terms of habituation to potentially threatening stimuli, desensitization occurs faster in naïve animals than in animals that have already developed fear of the stimuli. In other words, it is easier to habituate a horse to something before it has become afraid of it, i.e. at a young age. But apart from these neurological and psychological reasons, we should not underestimate the fact that handling a smaller and less strong horse is easier and less risky for the trainer than handling a larger and stronger horse. The less risky the handling, the bigger is the chance that it will be done in a calm and controlled way, something that in itself will exert a calming effect on the horse and improve learning (see Waran and Casey, Chapter 13).

'Early training'

Exposing a newborn foal to handling during the first 14 days of life, the first session starting immediately after birth, has been claimed to have a permanent effect on the foal's acceptance to handling later in life. The practice has been termed imprint training (Miller, 1989; 2001), the term 'imprint' suggesting that a critical period for handling exists in the horse. Considering that horses obviously can learn to accept handling also during later stages of life, the term imprint training is somewhat misleading (see Nicol, Chapter 12, for further discussion of this misnomer). But whether foals are more sensitive to handling early in life, i.e. that a certain handling procedure has a greater or longer lasting effect when performed early in life, has been analysed in several studies. Whereas no major differences between early and later handling were found in two studies (Mal *et al.*, 1994; Williams *et al.*,

2002), a greater effect of early handling was found in two other studies (Mal & McCall, 1996; Larose & Hausberger, 1998). A further study, which compared the effect of handling during the postnatal period and the post-weaning period, concluded that handling after weaning had a greater effect than postnatal handling (Lansade *et al.*, 2004), a difference that has also been seen in cattle (Boivin *et al.*, 1991). The most interesting result of the study was that the effect of both types of treatment decreased over a 10-month period. In other words, handling must be repeated in order to have a permanent effect. This conclusion is supported by a study by Heird *et al.* (1986) who concluded that horses handled for 18 months reacted least emotionally to novel stimuli, as compared to horses handled for one to three weeks. A similar result was found by Visser *et al.* (2002) who handled one group of horses from the time of weaning till adulthood whereas another group received no handling. At the age of 9, 10, 21 and 22 months both groups were exposed to different novel situations. Handled horses showed less emotional reaction as determined by heart rate response and heart rate variability compared to non-handled horses.

Habituation and training of the young horse

The result of the study by Lansade *et al.* (2004) is important because it implies that it is not so much the specific time at which training takes place but rather that it is repeated with certain intervals over an extended period in order to have a permanent effect. Unfortunately, no systematic studies have been conducted to evaluate how often training must be repeated, how long the intervals between training sessions can be, and whether the extended period should last from weaning till adulthood or whether a shorter period is sufficient.

Apart from concerns about when handling should be done and how often it should be repeated, an equally important question is what the young horse should learn in order to function satisfactorily as an adult. Obviously, socialization to humans, i.e. learning the ability to obey commands to stand still, to lift a hoof, to be led, etc., is only part of what is required of a horse. Other skills such as remaining calm in traffic or when separated from its conspecifics, are equally important. Unfortunately, also in this aspect our knowledge is based primarily on practical experience rather than systematic investigation. There is

little doubt, however, that in order to produce horses that fulfil the demands of human trainers and handlers, we need a much better understanding of what type of training a young horse should receive before its education as a riding horse starts.

Conclusion

The behavioural development of a horse from birth till adulthood is a continuous process that needs the presence of certain environmental features, as well as the absence of others, to occur optimally. For some behaviour patterns (e.g. maternal bonding) exposure must occur during specific periods in order to exert an effect, but for other behaviour patterns (e.g. socialization on conspecifics or on humans) repeated exposure over an extended period appears to be necessary in order for learning to be permanent. Since these ontogenetic mechanisms seem to operate similarly under natural as well as husbandry conditions, optimal housing and management of the young developing horse is necessary to obtain an adult horse that functions the way we expect.

The ontogeny of natural behaviour patterns has been investigated in a number of studies, with the result that our present understanding of the process is primarily based on systematic analysis. But as far as the behavioural development of those behaviour patterns that are important for an adult horse to function harmoniously in a domestic environment, our understanding is primarily based on practical experience rather than systematic analysis. Questions such as which stimuli are essential for normal development, which stimuli must be avoided to prevent abnormal development, when exposure should occur, how often it should be repeated and similar questions have not yet been subjected to scientific investigation. Considering that the answers to these questions affect not only the welfare of the domestic horse but also the welfare of their riders and trainers, such research is urgently needed.

References

Apter, R. C. & Householder, D. D. (1996). Weaning and weaning management of foals: a review and some recommendations. *Journal of Equine Veterinary Science*, 16: 428–35.
Barneveld, A., van Weeren, R. & Knaap, J. (1999). Influence of early exercise on the locomotion system. Paper presented at the *50th Annual Meeting of the European Association for Animal Production, Zurich, August 1999*.

Bateson, P. (1979). How do sensitive periods arise and what are they for? *Animal Behaviour*, 27: 470–86.

Blakemore, C. & Cooper, G. F. (1970). Development of the brain depends on the visual environment. *Nature*, 228: 477–8.

Boivin, X., LeNeindre, P. & Chupin, J. M. (1991). Establishment of cattle-human relationships. *Applied Animal Behaviour Science*, 32: 325–35.

Boy, V. & Duncan, P. (1979). Time-budgets of Camargue horses: I. Developmental changes in the time-budgets of foals. *Behaviour* 71, 187–202.

Boyd, L. E. (1988). Ontogeny of behavior in Przewalski horses. *Applied Animal Behaviour Science*, 21: 41–69.

Cameron, E. Z., Linklater, W. L., Stafford, K. J. & Minot, E. O. (1999). A case of cooperative nursing and offspring care by mother and daughter feral horses. *Journal of Zoology*, 249: 486–9.

Carson, K. & Wood-Gush, D. G. M. (1983). Equine behaviour: I. A review of the literature on social and dam foal behaviour. *Applied Animal Ethology*, 10: 165–78.

Christensen, J. W., Ladewig, J., Søndergaard, E. & Malmkvist, J. (2002). Effects of individual versus group stabling on social behaviour in domestic stallions. *Applied Animal Behaviour Science*, 75: 233–48.

Crowell-Davis, S. L. (1986a). Developmental behavior. *Veterinary Clinics of North America. Equine Practice*, 2: 573–90.

(1986b). Spatial relations between mares and foals of the Welsh pony (*Equus caballus*). *Animal Behaviour*, 34, 1007–17.

(1994). Daytime rest behavior of the Welsh pony (*Equus caballus*) mare and foal. *Applied Animal Behaviour Science*, 40: 197–210.

Crowell-Davis, S. L. & Caudle, A. B. (1989). Coprophagy by foals: recognition of maternal feces. *Applied Animal Behaviour Science*, 24: 267–72.

Crowell-Davis, S. L. & Houpt, K. A. (1985a). Coprophagy by foals: Effect of age and possible functions. *Equine veterinary Journal*, 17: 17–19.

(1985b). The ontogeny of flehmen in horses. *Animal Behaviour*, 33: 739–45.

Crowell-Davis, S. L., Houpt, K. A. & Burnham, J. S. (1985a). Snapping by foals of *Equus caballus*. *Zeitschrift für Tierpsychologie*, 69: 42–54.

Crowell-Davis, S. L., Houpt, K. A. & Carini, C. M. (1986). Mutual-grooming and nearest neighbour relationships among foals of *Equus caballus*. *Applied Animal Behaviour Science*, 15: 113–23.

Crowell-Davis, S. L., Houpt, K. A. & Carnevale, J. (1985b). Feeding and drinking behaviour of mares and foals with free access to pasture and water. *Journal of Animal Science*, 60: 883–9.

Crowell-Davis, S. L., Houpt, K. A. & Kane, L. C. (1987). Play development in Welsh pony (*Equus caballus*) foals. *Applied Animal Behaviour Science*, 17: 119–31.

Elze, K., Schmidt, E., Reinisch, F. & Erices, J. (1996). Analyse des Leipziger Modells der mutterlosen Fohlenaufzucht am Patientengut der Ambulatorischen und Geburtshilflichen Tierklinik von 1961 bis 1995. *Pferdeheilkunde*, 12: 227–30.

Francis-Smith, K. (1979). Studies on the feeding and social behaviour of domestic horses. Ph.D. Thesis, University of Edinburgh.

Francis-Smith, K. & Wood-Gush, D. G. M. (1977). Coprophagia as seen in thoroughbred foals. *Equine Veterinary Journal*, 9: 155–7.

Fraser, A. F. (1980). Ontogeny of behavior in the foal. *Applied Animal Ethology*, 6: 303.

Glade, M. J. (1984). Social sleeping behavior in young horses. *Equine Practice*, 6: 10–14.

Glendinning, S. A. (1974). A system of rearing foals on an automatic calf-feeding machine. *Equine Veterinary Journal*, 6: 12–16.

Grzimek, B. (1949). Ein fohlen, das kein Pferd kannte. *Zeitschrift für Tierpsychologie*, 6: 391–405.

Gunnarsson, S., Yngvesson, J., Keeling, L. J. & Forkman, B. (2000). Rearing with early access to perches enhances spatial awareness in the domestic hen. *Applied Animal Behaviour Science*, 67: 217–28.

Heird, J. C., Whitaker, D. D., Bell, R. W., Ramsey, C. B. & Lokey, C. F. (1986). The effects of handling at different ages on the subsequent learning ability of 2-year-old horses. *Applied Animal Behaviour Science*, 15: 15–25.

Heleski C. R., Shelle, A. C., Nielsen, B. D. & Zanella, A. J. (2002). Influence of housing on weanling horse behavior and subsequent welfare. *Applied Animal Behaviour Science*, 78: 291–302.

Hoffmann, R. M., Kronfeld, D. S., Holland, J. L. & Greiwe-Crandell, K. M. (1995). Preweaning diet and stall weaning method influences on stress response in foals. *Journal of Animal Science*, 73: 2922–30.

Holland, J. L., Kronfeld, D. S., Hoffman, R. M., Greiwe-Crandell, K. M., Boyd, T. L., Cooper, W. L. & Harris, P. A. (1996). Weaning stress is affected by nutrition and weaning methods. *Pferdeheilkunde*, 12: 257–60.

Houpt, K. A. (2002). Formation and dissolution of the mare-foal bond. *Applied Animal Behaviour Science*, 78: 319–28.

Houpt, K. A. & Hintz, H. F. (1983). Some effects of maternal deprivation on maintenance behavior, spatial relationships and responses to environmental novelty in foals. *Applied Animal Ethology*, 9: 221–30.

Houpt, K. A., Hintz, H. F. & Butler, W. R. (1984). A preliminary study of two methods of weaning foals. *Applied Animal Behaviour Science*, 12: 177–81.

Houpt, K. A., Parsons, M. S. & Hintz, H. F. (1982). Learning ability of orphan foals, of normal foals and of their mothers. *Journal of Animal Science*, 55: 1027–32.

Hughes, C. F., Goodwin, D., Harris, P. A., & Davidson, H. P. B. (2002). The effect of social environment on the development of object play in domestic horse foals.

Proceedings of the Havemeyer Workshop on Horse Behavior and Welfare, Hólar, Iceland, June 13–15 2002, pp. 21–2.

Jeffcott, L. B. (1972). Observations on parturition in crossbred pony mares. *Equine Veterinary Journal*, 4: 209–12.

Lansade, L., Bertrand, M., Boivin, X. & Bouissou, M.-F. (2004). Effect of handling at weaning on manageability and reactivity of foals. *Applied Animal Behaviour Science*, 87(1–2): 131–49.

Larose, C. & Hausberger, M. (1998). Effects of handling on the behaviour of foals. *Proceedings of the 32nd International Congress of the ISAE, July 21–25, Clermont-Ferrand, France*, p. 120.

Mal, M. E., McCall, C. A. (1996). The influence of handling during different ages on a halter training test in foals. *Applied Animal Behaviour Science*, 50: 115–20.

Mal, M. E., McCall, C. A., Cummins, K. A., Newland, M. C. (1994). Influence of preweaning handling methods on post-weaning learning ability and manageability of foals. *Applied Animal Behaviour Science*, 40: 187–95.

Malinowski, K., Hallquist, N. A., Helyar, L., Sherman, A. R. & Scanes, C. G. (1990). Effect of different separation protocols between mares and foals on plasma cortisol and cell-mediated immune response. *Journal of Equine Veterinary Science*, 10: 363–8.

Marinier, S. L. & Alexander, A. J. (1995). Coprophagy as an avenue for foals of the domestic horse to learn food preferences from their dams. *Journal of Theoretical Biology*, 173: 121–4.

McCall, C. A., Potter, G. D. & Kreider, J. L. (1985). Locomotor, vocal and other behavioral responses to varying methods of weaning foals. *Applied Animal Behaviour Science*, 14: 27–35.

McDonnell, S. M. (1989). Spontaneous erection and masturbation in equids. *Proceedings, American Association of Equine Practitioners, Boston, Massachusetts, December 3 and 4*, pp. 567–80.

(2003). *The Equid Ethogram: Practical Field Guide to Horse Behavior*. Lexington, KY: Eclipse Press.

McDonnell, S. M. & Haviland, J. C. S. (1995). Agonistic ethogram of the equid bachelor band. *Applied Animal Behaviour Science*, 43: 147–88.

McDonnell, S. M. & Poulin, A. (2002). Equid play ethogram. *Applied Animal Behaviour Science*, 78: 263–95.

Miller, R. (1989). Imprint training the newborn foal. *Large Animal Veterinarian*, 44: 18–22.

(2001). Fallacious studies of foal imprint training. *Journal of Equine Veterinary Science*, 21: 102–5.

Mills, D. S. (1999). The origin and development of behavioural probelems in the horse. In *Proceedings of the Specialists Days on Behaviour and Nutrirtion*, ed. P. A. Harris, G.Gomersall, H. P. B. Davidson & R. Green. Newmarket: Equine Veterinary Journal, pp. 17–21.

Mills, D. & Nankervis, K. (1999). *Equine Behaviour: Principles and Practice*. London: Blackwell Science Ltd.

Moons, C. & Zanella, A. J. (2001). Effect of short-term separations on weaning stress in foals. *Proceedings of the 35th International Congress of the International Society of Applied Ethology Davis, California, August 4–9, 2001*, p. 39.

Nicol, C. J., Wilson, A. D., Waters, A. J., Harris, P. A. & Davidson, H. P. B. (2001). Crib-biting in foals is associated with gastric ulceration and mucosal inflammation. *Proceedings 35th International Congress of the International Society of Applied Ethology, Davis, California, August 4–9, 2001*, 40.

Rivera, E., Benjamin, S., Nielsen, B., Shelle, J. & Zanella, A. J. (2002). Behavioral and physiological responses of horses to initial training: the comparison between pastured versus stalled horses. *Applied Animal Behaviour Science*, 78: 235–52.

Rossdale, P. D. (1967). Clinical studies on the newborn Thoroughbred foal I. Perinatal behavior. *British Veterinary Journal*, 123: 470–81.

Søndergaard, E. (1998). Reaction to separation on mares and foals. Proc. of the *32nd International Congress of the International Society of Applied Ecology Clermont-Ferrand, France, July 21–25*, 120.

Søndergaard, E. & Halekoh, U. (2003). Young horses' reactions to humans in relation to handling and the social environment. *Applied Animal Behaviour Science*, 84: 265–80.

Søndergaard, E. & Ladewig, J. (2004). Group housing exerts a positive effect on the behaviour of young horses during training. *Applied Animal Behaviour Science*, 87(1–2): 105–18.

Søndergaard, E. & Schougaard, H. (2000). The effect of social environment on feed intake, growth and health in young Danish Warmblood horses. Paper presented at the *51st Annual Meeting of the European Association for Animal Production, The Hague, August 2000*.

Tyler, S. J. (1972). The behaviour and social organization of the New Forest ponies. *Animal Behaviour Monograph*, 5–6: 85–96.

Vecchiotti, G. G. & Galanti, R. (1986). Evidence of heredity of cribbing, weaving and stall-walking in Thoroughbred horses. *Livestock Production Science*, 14: 91–5.

Visser, E. K., van Reenen, C. G., van der Werf, J. T. N., Schilder, M. B. H., Knaap, J. H. & Blokhuis, H. J. (2002). Heart rate and heart rate variability during a novel object test and a handling test in young horses. *Physiology and Behavior*, 76: 289–96.

Waring, G. H. (1982). Onset of behaviour patterns in the newborn foal. *Equine Practice*, 4: 28–34.

Waring, G. H. (2003). *Horse Behavior*, 2nd edn. Park Ridge: Noyes Publications.

Waters, A. J., Nicol, C. J. & French, N. P. (2002). Factors influencing the development of stereotypic and redirected behaviours in young horses: findings of a four year prospective epidemiological study. *Equine Veterinary Journal*, 34: 572–9.

Weeks, J. W., Crowell-Davis, S. L. & Heusner, G. (2002). Preliminary study of the development of the flehmen response in *Equus caballus*. *Applied Animal Behaviour Science*, 78: 329–35.

Williams, J. L., Friend, T. H., Toscano, M. J., Collins, M. N., Sisto-Burt, A. & Nevill, C. H. (2002). The effects of early training sessions on the reactions of foals at 1, 2, and 3 months of age. *Applied Animal Behaviour Science*, 77: 105–14.

Williams, M. (1974). The effect of artificial rearing on the social behaviour of foals. *Equine Veterinary Journal*, 6: 17–18.

Zharkikh, T. L. (2000). Comfort behaviour of Przewalski horses (*E. przewalskii*) at the Biosphere Reserve "Askania' Nova". In *News Biosphere Reserve "Askania-Nova": Protection and Preservation of Rare Species*, ed. V. S. Havrylenko, Ukraine: Askania Nova, 40–9.

10

Equine play behaviour

Deborah Goodwin and Carys F. Hughes

Many authors consider play more difficult to define than other categories of behaviour (Lorenz, 1956; Wilson, 1975). Some consider play impossible to define, though acknowledge that in practice this is inescapable (e.g. Hinde, 1976). The diversity and complexity of play can confound attempts to produce a meaningful definition of play that acknowledges its origin and unity (Bekoff & Byers, 1998). The purposive play categorization suggested by Purton (1978) is probably most salient in recognizing fundamental unity in forms of play.

Play is widely recognized as a primarily juvenile activity, though it is also seen in adults of many species (Muller-Schwarze *et al.*, 1982). In a comprehensive review of animal play, Fagen (1981) considers play important in accelerating development, providing a foundation for cognitive and motor skills, developing adaptive behavioural flexibility, facilitation of social interaction and also acting as an interface between genes and culture. Therefore, there may be selective advantages in the performance and

persistence of play in the behavioural repertoire of horses.

Play has been described in a range of equid species in varying environmental contexts, for example Przewalski's horses in reserves (Zharkikh, 1999, 2003) and in the domestic environment (Christensen *et al.*, 2002), plains-living zebra (*Equus burchelli*, Klingel, 1974), free-ranging New Forest ponies (Tyler, 1972), semi-feral ponies (McDonnell & Poulin, 2002) and domestic Thoroughbred horses on stud farms (Fagen, 1981).

In common with many aspects of equine social behaviour, play behaviour, is routinely restricted in many domestic settings, due to concerns over risk of injury (Goodwin, 2002). This restriction often extends to limitation of social and even solitary play opportunities, due to owners' perceptions of associated injury risk to horses and humans (Hughes, 2002). Owners' concerns over injury risks are most prevalent where unstable social groups exist, for example in boarding stables (USA) and livery yards (UK) where the equine population is commonly subject to continual change.

Play is also mentioned in chapters by Boyd and Keiper on behavioural ecology (Chapter 4), Feh's on relationships (Chapter 5), Crowell-Davis and Weeks on mare–foal interactions (Chapter 8), and in Ladewig *et al.* on development (Chapter 9). In this chapter the discussion of play will consider evolutionary origins of play, its function as an ethosystem and describe selected research on play in non-domestic, domestic and feral horses.

Evolutionary origins of play

Studies of play seeking to identify function of play within extant species may reveal little about the conditions that facilitate the evolution of play, or the origin of play behaviour. Few studies of the evolution of play address the origins of playfulness (Burghardt, 1998). As little can be learned about play behaviour from the fossil record, comparative studies of extant species could provide phylogenctic analysis of the evolution of play.

Play as an ethosystem

A number of important studies of play have been published that support its role as a factor in the evolutionary and adaptive behaviour of some species, including the horse (Fraser, 1992). Play begins to develop in foals during the first weeks of life. It is initially evident as solitary locomotor play, play directed towards inanimate objects and play directed towards the mother. As foals develop, the performance of play in peer groups fulfils an important role in the development of social and physical motor skills. Up to 75% of the kinetic activity of foals is devoted to play (Fraser, 1992). An element of learning is evident in play as individuals refine the activities according to feedback from playmates and the environment. This is the basis for interpreting play as an ethosystem in equid species that are active, reactive and social.

Play is frequently characterized through the performance of serious or adult behaviour patterns in a non-serious context. As such, the play performed may take the form of incomplete patterns of behaviour, repetition or exaggeration of particular components of play sequences. The non-serious nature of play means that real injury is rare as escalation of rough play bouts frequently results in their termination. Figure 10.1 illustrates examples of play and serious fighting. In social

(a)

(b)

(c)

Figure 10.1. (a) Playful sparring in yearling colts; and (b) similar play sparring in mature bachelor stallions in contrast to (c) frank fighting in mature stallions. (Photos by Elkanah Grogan, Equine Behavior Lab, University of Pennsylvania.)

play the concepts of games and the rules are introduced. Fraser (1992) recognizes four types of foal
games known as nip and shove, chase and charge,
pair of pals and foal capers. He then goes on to define
eight features of horse play in general. These are based
on the characteristic play features proposed by Fagen
(1981) and include:

• Seeking opportunities for play.
• Inhibition to avoid injury.
• Use of objects for play.
• Signals of playful intent preceding episodes of play.
• Play sequences that are exaggerated, unordered or
  repetitious.
• Frequent return to a play eliciting stimulus.
• Play is terminated by stronger stimuli or unpleasant
  events.
• Units of play are not bound into functional
  sequences.

Signals that are used to establish and maintain play
and avoid confusion between other similar action patterns occur in many social species (Bekoff & Allen
1992, 1998; Bekoff & Byers, 1998). These signals
establish play in context and tend to be clearly identi-
fiable due to the consequences associated with misunderstanding. Signals that occur in social and
solitary play have been identified in observations
of chimpanzees (Spijkerman *et al.*, 1996) and lions
(Manning & Dawkins, 1998) and also in domestic
dogs (Bekoff, 1995) and cats (Bradshaw, 1993). Identification of an equine play signal has received less
attention and is worthy of further investigation.

Many play bouts are initiated by nudging or nipping a conspecific (McDonnell & Poulin, 2002).
However, we consider that an equine play signal may
exist as a composite signal observable as a high state
of arousal, for example elevated postural outlines
of head, neck and tail, plus rapid expulsion of air
through the nostrils, often referred to as high blowing. The postural component of the composite play
signal may prove similar to prance as defined by
McDonnell and Poulin 2002, though they state this
often occurs following play bouts. Humans mimic the
high blowing signal during inter-specific interactions
in the Arabian show ring, in order to effect a high
level of equine arousal. As such this may resemble the
use of the canine play bow by humans in dog–human
play interactions (Rooney *et al.*, 2001). However, the
existence of an equine play signal warrants further

investigation in horse–horse interactions before
extrapolation to horse–human interactions can be
reliably supported.

Play in non-domestic equids

There have been relatively few studies of play in
non-domestic equids and where recorded this has
generally comprised descriptive accounts accompanying other aspects of social behaviour and ecology. Moehlman (1974) observed play in free-ranging
African wild asses *Equus asinus* in Death Valley, California. Gardner (1983) reported play in Grévy's zebra
Equus grevyi in Kenya. Schilder *et al.* (1984) recorded
social interactions in plains zebra in a semi-reserve
and classified facial play expressions separately from
aggressive signals. Penzhorne (1984) reported play in
Cape Mountain zebra *Equus zebra zebra* in South
Africa. Boyd *et al.* (1988) included play behaviour
in their study of time budgets in Przewalski's horses
in a semi-reserve.

Social and solitary locomotor play has been
recorded in Hartmann's zebra foals (Joubert, 1972)
and also in captive juvenile Przewalski's horses
(Zharkikh, 1999). Young plains zebra stallions are
reported to leave their natal band earlier if they have
no playmates and then find playmates in another
band (Klingel, 1974). McDonnell and Poulin (2002)
provided further details of play in non-domestic
equids. These authors included information drawn
from organized studies of play in non-domestic equids
in their equid play ethogram.

Studies of Przewalski's horses at Askania Nova
Reserve in the Ukraine, 1993–2002 by Zharkikh
(2003) concentrated on play behaviour and its development. These studies resulted in a play ethogram,
which described 26 play behaviour patterns, for this
population. The play behaviour patterns, are grouped
into two main categories (social and solitary), each
with two subdivisions:

• solitary:
  manipulative play (1 behaviour pattern);
  locomotor play (5 behaviour patterns);
• social:
  non-contact play (10 behaviour patterns)
  contact play (10 behaviour patterns).

In this study the definition of manipulative play was
restricted to familiar objects to eliminate investigative
behaviour and required the object to be picked up

Figure 10.2. Three-week-old colt taking off quickly from standing to a canter during a session of solitary locomotor play common at this age, involving repeated quick starts, stops and turns, as if 'practicing' (McDonnell & Poulin 2002; McDonnell, 2003). (Photo by Sue McDonnell, Equine Behavior Lab, University of Pennsylvania.)

in the mouth or pawed. No objects were introduced to the site and few were objects available on the steppe. These authors also differed from McDonnell and Poulin (2002) by categorizing nipping and pulling the mane of a conspecific as social contact play rather than object play, as the mane constitutes part of the body of a conspecific, and so object play was seldom recorded. In this population, age and sex determined frequency and duration of play activity. Colts were reported to play for longer and more vigorously than fillies, which agrees with behaviour of domestic foals (Tyler, 1972). Adverse weather conditions were also found to reduce play behaviour in Przewalskis on this reserve, which agrees with the findings in the UK of Capps (2001) in free-ranging Exmoor ponies and Farrelly (1998) in free-ranging Dartmoor ponies.

Play in caballine horses

Several authors have published ethograms of the behaviour of the domestic horse that include play behaviour. Aspects of adult play are detailed in the inter-male agonistic ethogram of McDonnell and Haviland (1995), juvenile play is described in McDonnell and Poulin (2002), and play has also been included in McDonnell's (2003) practical field guide to horse behaviour. These are all based primarily on observations of a semi-feral group of ponies of Shetland type (Figure 10.2). Waring (2002) also includes a discussion of play behaviour.

Play in free-ranging and feral horses

Social, solitary locomotor and object play have been recorded in free-ranging and feral horse populations in the UK and USA (Tyler, 1972; Berger, 1986; Waring, 2002). In these groups, foals exhibited the majority of play behaviour. Social play behaviour patterns included play fighting, neck wrestling and chasing. Solitary locomotor play included gambolling, high-speed turns and sudden stops. Object play was defined as the manipulation of inanimate objects and occurred in solitary and social contexts. Free-ranging New Forest pony foals have been reported to play with sticks and pieces of paper (Tyler, 1972). In general, play is reported to be similar in colts and fillies for the first month of life (Waring, 2002). After this time, colts are reported to play more frequently than fillies (Tyler, 1972).

Play in domestic horses

Social and object play have also been recorded in juvenile and adult domestic horses, although less frequently than in foals (Fraser, 1992; Mills & Nankervis, 1999; Hughes, 2002; McDonnell & Poulin, 2002). Social play, comfort and maintenance behaviour, for example mutual grooming, appear to be necessary for bonding and cohesion in equid social groups (Fraser, 1992). Juvenile and adult domestic horses have been reported to play with a variety of objects including sticks, footballs, clothing and feed sacks (Hughes, 2002). As illustrated in Figure 10.3 novel objects typically elicit exploration and play, especially by younger animals. The functional role of object play in the behavioural repertoire of equids has not been greatly explored in the current literature. However, it would appear to function as a means of acquiring information about the environment and feed handling skills (Hughes, 2002).

McDonnell and Poulin (2002) used over 100 hours of direct observation plus photographic records gathered over several years from a semi-feral herd of ponies maintained at the University of Pennsylvania. In their play ethogram they described 38 play behaviour patterns and grouped these into four main categories, locomotor play (14 behaviour patterns), object play (14 behaviour patterns), play fighting (7 behaviour patterns) and play sexual behaviour (3 behaviour patterns). Though play events appear more frequent and diverse in this domestic group than

Figure 10.3. Object play commonly elicited by novel objects. Novel object play often proceeds in a fixed order of playful investigatory responses including approach, sniff at distance, sniff touching, snort or nuzzle to move object, manipulate with lips, mouth, lick, pick-up or pull, drop or toss. The series is often repeated several times with a single object as if each time it is dropped it is again a novel object (McDonnell and Poulin, 2002, McDonnell, 2003). (Photos by Sue McDonnell, Equine Behavior Lab, University of Pennsylvania.)

in the Przewalskis at Askania Nova (Zharkikh, 2003) a number of factors other than species differences may contribute to the variation. There are a number of differences in definitions between the two studies: habitat at the Pennsylvania site appears less barren than that described on the steppe and also contained more objects that could be manipulated.

The effects of domestication on play

As in other domestic species, the effects of domestication on the behaviour of horses may have resulted in retention of juvenile characteristics plus an increased propensity to play; this requires further investigation. Such work is likely to prove more challenging than for the dog, as the ancestral species of the horses are generally considered extinct (Clutton-Brock, 1992; Jansen *et al.*, 2002). Comparative work of this type would, therefore, be reliant on assessment between domestic horses and other extant equid species or, feral and free-ranging domestic horses (Goodwin, 2002). It could be argued that the persistence of play behaviour in domestic animals may be attributable to the lack of selection against play behaviour, however, artificial selection for appealing behavioural traits is likely to have played a positive role in the high level of play recorded in many domestic species. Domestication has been associated with progressive retention of juvenile behavioural and morphological characteristics in many species, e.g. the dog (Goodwin *et al.*, 1997) and the cat (Hall, 1998); often with a corresponding increase in play behaviour in comparison with the ancestral species.

Studies of play behaviour in free-ranging and pastured breeding groups of Dartmoor (Farrelly, 1998) and Exmoor ponies (Capps, 2001) have recorded higher incidences of social and object play in pastured ponies than those that are free-ranging in their native environment. These short-term studies suggest that confinement to pastures elevates propensity to play and this is an area worthy of further investigation. The differences between the two populations may relate to the absence of a range of demands in the pasture that are present on the moor, allowing more time and energy for play. However, it may also be explained as a response to a restricted and unvarying environment. If domestication has resulted in increased playfulness in horses, confinement to paddocks and stables may, therefore, result in a corresponding increase in motivation to play while at the same time thwarting its expression.

Play in horse–human interactions

Fraser (1992) sought to explain most equestrian disciplines in terms of games theory (Maynard-Smith, 1982) with horses learning the rules of a game of play. Other authors (e.g. Rees, 1984) have pointed out that most equestrian activities must appear relatively pointless to the equid participants, e.g. travelling in repeated circles or jumping over easily avoidable obstacles. However, as a social prey species that places great survival value on co-operative behaviour, and one that is highly motivated by play, the horse appears a relatively willing participant in most horse–human interactions. It is such a pity, therefore, that much of the history of horse–human interaction has been interpreted by the human participants using a dominance/submission paradigm (Goodwin, 1999).

As the majority of the horse population in developed countries now assumes a role as a companion or leisure horse, greater appreciation of the playful nature of horses by horse owners and trainers could provide a liberating and rewarding route away from the currently prevalent pre-occupation with human dominance in many equestrian training programmes.

Adopting an approach to horse–human interactions based on mutual learning and appreciation of the rules of the games of play could present a route towards more rewarding interactions and relationships between the species, with potential welfare benefits to both. However, playing horse games between humans and horses must be approached with great caution and is not generally advisable. Horses, even small ponies, are large, powerful and sometimes unpredictable creatures. Over-handled foals that do not learn the difference between horse–horse play and horse–human play can grow into unintentionally dangerous companions as adults. Leisure horse owners are frequently tempted to breed a foal from a favoured mare. Single foals born to mares without other foals to play with have been recognized for many years as potentially problematic. Horse games, for example nip and shove, chase and charge, (Fraser, 1992) are, therefore, most appropriate to intra-specific equine interactions. Human–horse play interactions based on puzzle solving and mutual enjoyment of locomotion, novelty and exploration can be safe, stimulating and surprisingly intellectually challenging even for the human participants.

Restriction of play opportunities in current management practices

Horses are generally maintained in conditions convenient to humans, which often restrict the highly motivated adaptive behaviour of horses. The rise of the leisure horse and single horse ownership amongst people who do not own their own land, has lead to increased provision of rented horse accommodation. Such accommodation usually consists of a single use loosebox and limited access to pasture. The population of horses kept in barns (USA) or at livery (UK) is often transient and, therefore, social groups are rarely stable. The resultant rate of injury may lead to concerned owners limiting social interactions, including play, to avoid the associated veterinary costs.

Owners generally acknowledge the importance of play to horses and some provide object play opportunities in their stables through horse toys. However, there are no published studies on the efficacy of these, with the exception of a foraging device (Henderson & Waran, 2001). Other owners deliberately limit object play opportunities for their solitary housed horses, through a well-meaning but erroneous understanding of the factors associated with the development of stereotypical behaviour patterns (Hughes, 2002).

One of the risks of restricting play opportunities is that it may not extinguish the motivation to play, but merely frustrate its performance. Removal of these restrictions can result in explosive play bouts, for example when dray horses are turned out to pasture for their annual holiday their play bouts can be dangerously energetic (Fraser, 1992). If these play bouts occur in unstable groups, especially if they include horses that have not had the opportunity to learn the rules of normal social interaction, injury is a risk. Such injuries are likely to reinforce the beliefs of horse owners that play bouts should be restricted or prevented. However, from the ethogram of the horse it is clear that horses also perform solitary play bouts, although it remains uncertain whether object play may act as a substitute for play partners. If it does, this may present a practical solution for owners of single horses. The real solution lies in changes in the social management of horses that reflect more closely their natural adaptive behaviour.

Exploring environmental enrichment using play

Given the rise in solitary horse keeping and the behavioural importance of play to the domestic horse, one approach to enriching the restricted environment might be to enhance object play opportunities (Hughes *et al.*, 2000). The aim of providing stabled horses with play objects is to reduce the monotony of the stable environment by increasing its diversity. As interaction with play objects rapidly decreases with time in other domestic species (e.g. the cat, Hall, 1995) it is possible that stabled horses will require several toys that can be alternated in order to maintain an enrichment effect. Hall (1995) reported a post-inhibitory rebound in cats when a single sensory stimulus was changed in the play object. This is yet to be investigated in horses, but is open to exploration by scientists and most horse keepers, in keeping with the great tradition of informed lay contributions to the understanding of horse behaviour.

References

Bekoff, M. (1995). Play signals as punctuation – the structure of social play in canids. *Behaviour*, **132** (Part 5–6): 419–429.

Bekoff, M. & Allen, C. (1992). Intentional icons – towards an evolutionary cognitive *Ethology*, **91**(1): 1–16.

(1998). Intentional communication and social play: how and why animals negotiate and agree to play. In *Animal Play: Evolutionary, Comparative and Ecological Perspectives*, ed. M. N. Bekoff & J. A. Byers. Cambridge: Cambridge University Press, pp. 97–114.

Bekoff M. N. & Byers, J. A. (1998). *Animal Play: Evolutionary, Comparative and Ecological Perspectives*. Cambridge: Cambridge University Press.

Berger, J. (1986). *Wild Horses of the Great Basin*. Chicago: University of Chicago Press.

Boyd, L. E., Carbonaro, D. A. & Houpt, K. A. (1988). The 24-hour time budget of Przewalski horses. *Applied Animal Behaviour Science*, **21**(1–2): 5–17

Bradshaw, J. W. S. (1993). *The True Nature of the Cat*. London: Boxtree.

Burghardt, G. M. (1998). The evolutionary origins of play revisited. In *Animal Play: Evolutionary, Comparative and Ecological Perspectives*. ed. M. N. Bekoff & J. A. Byers. Cambridge: Cambridge University Press, pp. 1–26.

Capps, F. (2001). Social and object play in free-ranging and pastured Exmoor ponies. Undergraduate thesis, University of Southampton.

Christensen, J. W., Zarikkh, T. L., Ladweig, J. & Yasinetskaya, N. I. (2002). Social behaviour in stallion groups (*Equus prezwalskii* and *Equus caballus*) kept under natural and domestic conditions. *Applied Animal Behaviour Science*, **76**: 11–20.

Clutton-Brock, J. (1992). *Horse Power. A History of the Horse and the Donkey in Human Societies*. London: Natural History Museum.

Fagen, R. M. (1981). *Animal Play Behavior*. Oxford: Oxford University Press.

Farrelly, T. (1998). An investigation into the effect of human management and a controlled environment on play duration in horses. Undergraduate thesis, University of Southampton.

Fraser, A. F. (1992). *The Behaviour of the Horse*. Wallingford: CAB International.

Gardner, C. D. (1983). Grevy's zebra of Samburu Kenya: mothers and foals project in wildlife ecology. M.S. Thesis, Yale University, Newhaven.

Goodwin, D. (1999). The importance of ethology in understanding the behaviour of the horse. *Equine Veteterinary Journal*, **28** (Suppl.): 15–19.

(2002). Horse behaviour: evolution, domestication and feralisation. In *The Welfare of Horses*, ed. N. Waran. Kluwer, pp. 1–18.

Goodwin, D., Bradshaw, J. W. S. & Wickens, S. M. (1997). Paedomorphosis affects visual signals of domestic dogs. *Animal Behaviour*, **53**: 297–304.

Hall, S. L. (1995). Object play in the domestic cat. Ph.D. Thesis. University of Southampton.

(1998). Object play by adult animals. In *Animal Play: Evolutionary, Comparative, and Ecological Perspectives*, ed. M. Bekoff & J. A. Byers. Cambridge: Cambridge University Press, pp. 45–60.

Henderson, J. V., Waran, N. K. (2001). Reducing equine stereotypies using the Equiball^{TM}. *Animal Welfare*, **10**: 73–80.

Hinde, R. A. (1976). Multiple review of Wilson's *Sociobiology. Animal Behaviour*, **24**: 706–7.

Hughes, C. F. (2002). Object play in the domestic horse. PhD Thesis, University of Southampton.

Hughes, C. F., Goodwin, D., Harris, P. A. & Davidson, H. P. B. (2000). The role of play in enriching the environment of the domestic horse. Poster Presentation. *Companion Animal Behaviour Therapy Study Group. Birmingham, UK, April 2000*.

Jansen, T., Forter, P., Levine, M., Oelke, H., Hurles, M., Renfew, C., Weber, J. & Olek, K. (2002). Mitochondrial DNA and the origins of the domestic horse. *Proceedings National Academy of Science USA*, **99**(16): 10905–10.

Joubert, E. (1972). Activity patterns shown by mountain zebra *Equus zebra hartmannae* in South West Africa with reference to climatic factors. *Zoologica Africana*, **7**(1): 309–31.

Klingel, H. (1974). A comparison of the social behaviour of the Equidae. In *The Behaviour of Ungulates and its Relation to Management*. ed. V. Geist & F. R. Walther. Calgary, Alberta: IUCN Publication, pp. 124–32.

Lorenz, K. J. (1956). Play and vacuum activities. In *L'instincte dans le comportement des animaux et de l'homme*, ed. M. Autoria. Paris: Masson et Cie, pp. 633–45.

Manning, A. & Dawkins, M. S. (1998). *An Introduction to Animal Behaviour*, 5th edn. Cambridge: Cambridge University Press.

Maynard-Smith, J. (1982). *Evolution and the Theory of Games*. Cambridge: Cambridge University Press.

McDonnell, S. M. (2003). *The Equid Ethogram: A Practical Field Guide to Horse Behavior*. Lexington: Eclipse Press.

McDonnell, S. M. & Haviland, J. C. S. (1995). Agonistic ethogram of the equid bachelor band. *Applied Animal Behaviour Science*, **43**: 147–88.

McDonnell, S. M. & Poulin, A. (2002). Equid play ethogram. *Applied Animal Behaviour Science*, **78**: 263–95.

Miller, M. N. & Byers, J. A. (1998). Sparring as play in young pronghorn males. In *Animal Play: Evolutionary, Comparative and Ecological Perspectives*, ed. M. N. Bekoff, & J. A. Byers. Cambridge: Cambridge University Press, pp. 141–60.

Mills, D. S. & Nankervis, K. J. (1999). *Equine Behaviour: Principles and Practice*. Oxford: Blackwell Science.

Moehlman, P. D. (1974). Behavior and ecology of feral asses (*Equus asinus*). Ph.D. Thesis. University of Wisconsin, Madison.

Muller-Schwarze, D., Stagge, B. & Muller-Schwarze, C. (1982). Play behaviour: persistence, decrease and energetic compensation during food shortage in deer fawns. *Science*, **215**: 85–7

Penzhorne, B. L. (1984). A long-term study of social organisation and behaviour of Cape Mountain zebra *Equus zebra zebra*. *Zeitschrift fur Tierpsychologie*, **64**(2): 97–146.

Purton, A. C. (1978). ethological categories of behaviour and some consequences of their conflation. *Animal Behaviour*, **26**: 653–70.

Rees, L. (1984). *The Horse's Mind*. London: Stanley Paul.

Rooney, N. J., Bradshaw, J. W. S. & Robinson, I. H. (2001). Do dogs respond to play signals given by humans? *Animal Behaviour*, **61**(4): 715–22.

Schilder, M. B. H., Vanhooff, J., Vangeerplesman, C. J. & Wensing, J. B. (1984). A quantitative analysis of facial expressions in the plains zebra. *Zeitschrift fur Tierpsychologie*, **66**(1): 11–32.

Spijkerman, R. P., Dienske, H., vanHooff, J. A. R. A. M. & Jens, W. (1996). Differences in variability, interactivity and skills in social play of young chimpanzees living in peer groups and in a large family zoo group. *Behaviour*, **133**(9–10): 717–39.

Tyler, S. (1972). The behaviour and social organisation of the New Forest ponies. *Animal Behaviour Monographs*, **5**(2): 86–194.

Waring, G. H. (2002). *Horse Behavior: The Behavioral Traits and Adaptations of Domestic and Wild Horses, Including Ponies*. Park Ridge, NJ: Noyes Publishing.

Wilson, E. O. (1975). *Sociobiology*. Cambridge, MA: The Bleknap Press.

Zharkikh, T. L. (1999). Development of the Przewalski horse, *Equus przewalski* (Perissodactyla) play behavior at the 'Askania-Nova' Reserve. *Zoologicheskii Zhurnal*, **78**(7): 878–84.

Zharkikh, T. L. (2003). Play ethogram and the onset of play behaviour in Przewalski horses at Askania Nova Reserve. *37th International Congress of ISAE, Abano Terme, Italy*.

III

The impact of the domestic environment on the horse

11

The rider–horse relationship

Daniel S. Mills and June McNicholas

Just as the role of the horse in society varies from one culture to another and from one time to another (see Clutton-Brock, 1992; Hall, Chapter 2) so have our relationships with horses. Lawrence (1986) has reviewed the relationship between horses and their keepers from an anthropological perspective in a range of societies including that of the Crow Indians, Gypsies, the rodeo and the police. However, several authors have commented on the lack of sound psychobiological studies into this relationship (Jones, 1983; Crowell-Davis, 1992; Robinson, 1999). This lack of information is perhaps surprising given the shift in the role of the horse in Western society towards one of service in recreation and its overall economic importance. For example, it has been estimated by the Barents Group that in the USA there are 6.9 million horses and 7.1 million individuals involved in the 'horse industry', which employs around 1.4 million individuals. This is worth around US $112.1 billion with $1.9 billion being gained annually by government from related tax. The industry contributes more to gross domestic product in the USA than either the

motion picture or tobacco industries. Racing, showing and recreation account for the majority of this business (Table 11.1). In the UK it has been estimated that the industry accounts for 125 000 full-time equivalent employees, making a direct economic contribution in excess of $1.5 billion and an indirect contribution of several times this (Graham-Suggett, 1999).

The human–horse relationship within the context of interest in companion animal bonds

It has been suggested (Lagoni *et al.*, 1994) that the human–horse relationship is marked by mutual trust, respect and affection, similar in many ways to the relationship between two people who spend several hours together each day doing something enjoyable. While this may seem a very reasonable assumption, there remains remarkably little published research to determine whether this is the case.

Jones (1983) reported that the majority of riders considered their horse to be part of the family and that most related to the horse as if it were a child

The Domestic Horse: The Origins, Development, and Management of its Behaviour, ed. D. S. Mills & S. M. McDonnell. Cambridge University Press. © Cambridge University Press 2005.

Table 11.1. *Number of horses and participants in each aspect of the horse industry*

Activity	US	
	No. of horses	No. of participants
Racing	725 000	941 400
Showing	1 974 000	3 607 900
Recreation	2 970 000	4 346 100
Other[a]	1 262 000	1 607 900
Total	6 931 000	7 607 900[b]

Figures from The American Horse Council Foundation web-site http://www.horsecouncil.org/statistics.htm (accessed 19/02/04)

[a] Includes, farm and ranch work, police work, rodeo and polo.
[b] The sum of participants by activity does not equal the total number of participants because individuals could be counted in more than one activity.

(although there was a suggestion that male riders may more frequently relate to the horse as an adult family member). However, despite this child-like projection, they would confide in it and speak to it as if it were a person of the same age. A similar relationship has been described by pet owners (Berryman *et al.*, 1985). Driscoll (1995) in a study of public attitudes to a range of species found that horses tended to be grouped with animals such as dogs in terms of their rating for usefulness, importance, smartness, responsiveness, lovableness and safety. This group was distinct from one that contained several other domestic species including cats, sheep and rabbits, who tended to be rated lower in all these dimensions except safety. This might reflect a greater ability for humans to identify with horses and dogs compared to these other species due to their long historical mutual dependence and suggest a greater emotional affinity with them, but this has not been investigated empirically.

The horse is increasingly described as a companion animal species in Western society but it seems reasonable to assume that our relationship with the horse is different to that shared with other companion animals. Horses tend not to be kept for the whole of their natural lives, but tend to be kept for more instrumental purposes, being sold on when they no longer fulfil this role, e.g. when a child outgrows its first pony, or the rider seeks a higher performance animal to reach their own potential. The

horse's size and nature also mean that it rarely shares our homes in such an intimate way and it requires considerably more effort for its maintenance. Indeed most scales developed for assessing the relationship between humans and companion animals, including the Companion Animal Bonding Scale (Poresky *et al.*, 1987), Pet Attitude Scale (Templer *et al.*, 1981) and Pet Relationship Scale (Lago *et al.*, 1988) cannot be applied to the human–horse relationship because of inappropriate items relating to these properties of the horse.

Horses not only demand far more from their keepers both economically and personally but also pose a greater risk to their keepers compared to other companion animal species. The rate of injury associated with riding is high, being greater than that associated with motor-cycling (Chitnavis *et al.*, 1996). It has been estimated that in the USA riding injuries account for 2300 hospital admissions each year, with an overall injury rate of 0.6 per 100 hours of riding (Christey *et al.*, 1994). In the UK it has been estimated that the rate of injury from riding is around 35.4/1000 occasions, with a fatality risk of 0.34 deaths per million days of participation (Anon, 1991). Keeling and colleagues (1999) concluded that the relationship that existed between a horse and its rider was also an important factor when determining the risk of injury while riding.

It has even been suggested (Jones, 1983) that risks such as these, together with the variable success of riding experiences and keeping a horse, contribute to an intermittent reinforcement schedule that may result in a stronger relationship between human and animal than one which has the predictability of the continuous reinforcement provided by an adoring pet.

It has been argued that our relationship with horses is also shaped by the position the horse occupies in society (Sahlins, 1976). Horses in Anglo-American culture, like dogs, increasingly play a significant role as subjects within our society, they are frequently given personal names and spoken to as if capable of engaging in personal conversation; however, because horses do not co-habit with us like dogs, it has been suggested they are less like family (Fifield, 2001) and more like befriended servants (Sahlins, 1976). By contrast, in France where horses are more readily eaten, it appears that the names used tend to be less personal and more like surnames, perhaps reflecting the different symbolic position of horses in this society (Sahlins, 1976). It is therefore dangerous to extrapolate data on human–horse relationships beyond the society examined.

The human–horse relationship in Anglo-American society

Serpell (1983) examined the key features of the owner–dog relationship by examining the structure of the match between ratings by owners for their ideal dog and their actual pet. Working on the hypothesis that those traits which are most closely matched in the two groups are those which are most important in the bond, he concluded that expressiveness, enjoyment of walks, loyalty/affection, welcoming behaviour and attentiveness are important features of the relationship. By contrast, it has been suggested (Jones, 1983) that horses are 'hard to get at emotionally', although they clearly play an important role as a passive listener and confidante as discussed above. S. J. Buckley and D. S. Mills (unpubl. data) have used a similar methodology to Serpell (1983) to explore the key traits underpinning the attraction of owners to leisure horses. Semi-structured interviews were conducted with leisure-horse owners in the first instance. During these, owners were asked a range of questions relating to their relationship with their horse, their reasons for owning it and what they found attractive in their horse's personality. From transcriptions of these interviews, 20 aspects of behaviour were identified, which were subsequently used to develop a questionnaire for rating the appealing aspects of a potential leisure horse. Each item, consisted of a dichotomous scale with contrasting expressions relating to each trait at the extremes and a 5-point bar scale between (semantic differential scale, Osgood *et al.*, 1957). These are listed below, with the low end of the scale stated first and the abbreviated term used in the description of the results in parentheses:

1. Responds positively to only one person versus anyone (selectivity).
2. Always gets excited and will not perform versus always appears to know how to behave in a given situation (appropriateness).
3. Always friendly to other horses versus never sociable towards other horses (sociability).
4. Always an individual in its behaviour versus never shows any individuality (individuality).

5. Never responds well to human contact versus always responds well to human contact (human contact).
6. Shows no reluctance when ridden out alone versus refuses to be ridden without other horses (group riding).
7. Never changes its behaviour with its owner's moods versus always changes it behaviour if owner's mood changes (sensitivity).
8. Always greets owner vocally versus never greets owner vocally (greeting).
9. Always tense in work or new situations versus always relaxed and takes everything in its stride (tension).
10. Always obedient versus never obedient (obedience).
11. Never playful with owner or other horses versus always making playful gestures towards owner or other horses (playfulness).
12. Always enthusiastic in work versus never keen to do anything (eagerness).
13. Does not do anything that amuses the owner versus always doing things that owner finds amusing (entertaining).
14. Easy to ride versus a challenge to ride (rideability).
15. Never seeks attention or company from the owner versus always seeking attention or company of the owner (attention seeking).
16. Always expresses its feelings through its body language versus never expresses itself in a visual way (expressiveness).
17. Behaviour is always unpredictable versus always predictable (predictability).
18. Always willing to work versus always reluctant to work (willingness).
19. Changeable character and mood swings versus always of the same temper (moodiness).
20. Willing to take the initiative versus never willing to take the initiative (initiative).

Fifty leisure horse-owners were then surveyed on two occasions (one to two weeks apart) with a similar questionnaire. In the first instance, owners were asked to score the 20 traits for their ideal leisure horse, and on the second occasion for their actual horse. Traits were then ranked and compared (Figure 11.1). Z-scores were then calculated to determine the significance of any difference between the actual and ideal

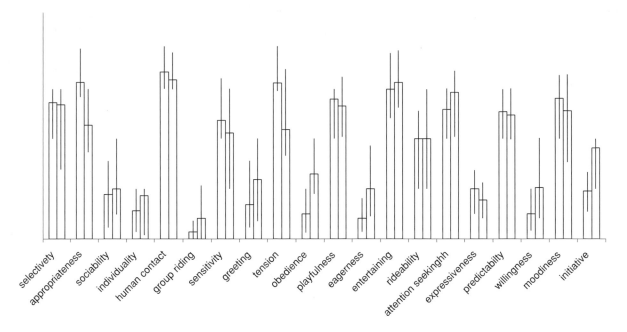

Figure 11.1. Median and inter-quartile ranges of ideal (left-hand column) and actual (right-hand column) scores for 20 traits described by leisure horse owners (*n* = 50). See text for a full explanation of the traits.

score for each trait (Figure 11.2). It has been suggested by Serpell (1983) that those traits which are most consistent between the ideal and actual animal – i.e. have small *z*-scores and relatively low variation in scores – reflect the most important aspects of the relationship. Thus it would seem that a high level of individuality and expressiveness in the horse with a moderate amount of attention seeking are perhaps the most consistent features of the leisure owner–horse relationship. By contrast, there are large differences in many of the traits relating to riding performance, notably enthusiasm, appropriateness, obedience and willingness. These results might suggest that the emotional content of the relationship between horse and rider is of more importance than the instrumental value of the horse. Thus it seems that in many respects the leisure horse may be fulfilling a psycho-social role, as has been reported for non-human animals in other human–pet relationships (Serpell, 1983).

Further evidence for this comes from a study by McNicholas and colleagues (2001). A questionnaire survey of 80 horse owners recruited from riding clubs, polo clubs and hunt associations evaluated the attitudes of these owners versus the frequency of riding, average length of ownership before selling on, time spent caring for the horse each week, expense incurred, membership of equestrian groups, and the use of the horse in competition. Attitudinal statements were rated on a 7-point Likert Scale and consisted of clusters of questions focusing on affection in the relationship (Cronbach's alpha = 0.74), social aspects of horse ownership (Cronbach's alpha = 0.67), involvement with horses as a healthy lifestyle (Cronbach's alpha = 0.80), the pleasure derived from ownership (Cronbach's alpha = 0.70) and resentment of chores (Cronbach's alpha = 0.81). The results suggested that the affection in the relationship was not associated with the tendency to compete, greater social activity, the level of resentment of chores, length of ownership, time spent with the horse or whether the horse was viewed as a pet or not. However, people who rode their horses reported significantly more affection in their relationship ($F_{1,77} = 4.301$, $P < 0.05$) as did people who spent more than £37 on their horse per week (£37 being the median amount spent by subjects

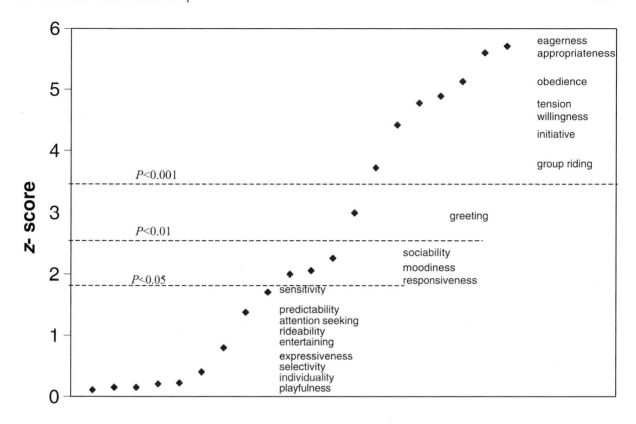

Figure 11.2. Values of Z for the difference between the actual and ideal scores of each trait (*n* = 50).

on their horses). There was no significant interaction between any of these factors which were predictive of the affectionate relationship. There were also no significant associations between any of the subject categories and how much importance they attached to social activities, although it might have been expected that those who competed would give greater weight to the social aspects of equestrianism. Competitors did however report deriving greater pleasure from horse ownership than non-competitors ($F_{1,76} = 4.86$, $P < 0.05$), although this was not influenced by the level of competition. By contrast, non-competitors reported deriving more prestige from horse ownership than those that competed with their horse ($F_{1,77} = 6.86$, $P = 0.01$). Interestingly the pleasure of ownership was inversely related to the average length of ownership, with more pleasure reported by owners who had or expected to have their horse for less than five years. This might be related to the finding that ownership and competition interacted with owners of

a newer horse deriving more pleasure ($F_{1,54} = 6.37$, $P < 0.05$). None of the other factors investigated were predictive of the pleasure derived from horse ownership. This included the average time spent with the horse each week, which ranged from 2–40 hours (median 15 hours). The belief that horse ownership contributed to a healthy lifestyle was only significantly associated with whether or not the horse was ridden ($F_{1,77} = 4.3$, $P < 0.05$) and the tendency to spend less than £37 a week on the horse ($F_{1,71} = 4.74$, $P < 0.05$). Only people who had owned their horse for less than five years and who did not take part in competition differed significantly from any other group in their resentment of chores, with these individuals reporting less resentment ($F_{1,55} = 4.42$, $P < 0.05$). This might reflect an initial difference in the emotional content of the relationship. One attitudinal statement examined the pleasure derived from the prestige of horse ownership and this was examined in isolation. Competition riders reported enjoying this prestige more than non-competitors ($F_{1,77} = 6.86$, $P = 0.01$). Thus, it seems that the relationship between an owner and their horse is characterized by high levels of affection and pleasure and relatively low levels of resentment of the amount of work involved, with a willingness to invest both time and money freely in the relationship. In general it seems that less importance is attached to success in competition than the relationship itself. This is reinforced by the finding that when asked to describe their relationship with their horses, 43% of respondents said they thought of their horses as pets and the remainder described them as friends. Even in high level competition work, the relationship has been best described as a partnership with a strong emotional involvement (Wipper, 2000), since success depends on close communication and mutual confidence between rider and horse.

In both of these studies (S. J. Buckley & D. S. Mills, unpub. data; McNicholas *et al.*, 2001 colleagues) over 90% of respondents were female and it might be that the results are limited to the feelings of this sex. Kidd and colleagues (1983) found that male horse-owners tended to be aggressive and dominant while female horse-owners were more easy going and non-aggressive. This report is supported by the behavioural observations of Brown (1984) who found female handlers tended to be more affectionate and interactive with their animals compared to males who were more domineering. However, the behavioural reactions of the sexes described in these studies may not relate directly to the feelings of those involved. The suggestion that the emotional involvement of male riders in the relationship can be just as great as that of females is supported by a study by G. Wilson and D. S. Mills, (unpub. data) which found no significant difference between the sexes among UK males. In a data set of 139 subjects of whom 40 were male, the effect of sex, current health status, age, growing up with horses and interactions between sex and health and sex and age were examined on scores derived from a modified version of the Pet Relationship Scale (Lago *et al.*, 1988). This scale consists of a series of statements relating to the behaviour and feelings of the respondent towards their pet. It is believed that these items belong to three subscales relating to affectionate companionship, equal family member status and mutual physical activity, although they have not been validated in the horse. In the modified scale used by G. Wilson and

D. S. Mills (unpub. data), some of the items that were inappropriate in the context of horses were removed. These were 'My pet and I watch TV together frequently', 'I take my pet along when I go jogging or walking', 'I enjoy having my pet ride in the car with me', 'My pet and I often take walks together'. One item relating to bathing was modified to relate to grooming. This left seven items relating to affectionate companionship and four relating to mutual physical activity with no change in the family status group. Responses to the items were scored on a five-point Likert Scale (strongly agree, agree, neutral, disagree and strongly disagree). Both leisure horse- and professional horse-owners were surveyed ($n = 104$ and 35 respectively), but all were instructed to interpret the term pet to refer to their horse. An initial factor analysis of the results suggested a three-factor solution as had been found in the original studies of Lago and colleagues (1988). These explained, 22.2%, 12.2% and 11.5% of the variance respectively (Table 11.2). The first factor appears to describe the emotional bond between the owner and horse, the second the equality in the relationship and the third the physical activity invested in the relationship and so appeared to reflect a similar structure to the relationship as that described by Lago and colleagues (1988) for other pet relationships. An analysis of variance revealed that only marital status and an interaction between sex and age were predictive of score on this modified scale ($P < 0.05$ for both), with no significant effect of sex, age, health status, a childhood with horses or the tendency to compete on the total score. Individuals in a stable, long-term relationship scored significantly higher than individuals who were currently divorced or separated. A similar analysis of variance of the scores for each of the three factors found a similar effect in each case, with the additional significance of growing up with horses on the emotional bond factor. Individuals who had not grown up with horses scored significantly higher on this factor ($P < 0.05$).

These studies are consistent with the suggestion by Jones (1983) that the human–horse relationship consists of a cycle of events, starting with a high level of romantic involvement based on the thrill of riding. This is then dampened by a period characterized by a greater awareness of the horse's limitations and the rider's desire to fulfil their potential, before a final equilibrium is reached; imposed on this is the cumulative emotional impact of several severed relationships

Table 11.2. *Varimax Factor Analysis of items in the modified Pet Relationship Scale. Bold values relate to the factor on which the item loads heaviest*

Item	Factor 1	Factor 2	Factor 3
There are times I'd be lonely except for my pet	**0.384**	0.061	0.293
I give gifts to my pet for birthdays and special occasions	0.377	**0.516**	0.233
My pet is a valuable possession	0.056	−0.188	**0.339**
I talk to my pet about things that bother me	**0.643**	0.136	0.255
I miss my pet when I am away	**0.479**	0.218	0.411
Making me laugh is part of my pet's job	**0.527**	0.109	0.111
My pet gives me a reason for getting up in the morning	**0.342**	0.133	0.133
My pet is an equal member of the family	**0.795**	0.127	0.231
I share my food with my pet	0.376	**0.671**	0.023
My pet knows when I'm upset and tries to comfort me	**0.640**	0.447	−0.021
My pet is constantly at my side	**0.624**	0.303	−0.026
My pet is an equal member in this family	**0.707**	0.279	0.103
I treat my pet to anything I happen to be eating if he/she seems interested	0.321	**0.796**	0.061
In many ways my pet is the best friend I have	**0.617**	0.386	0.093
My pet helps me to be more physically active	0.180	0.159	**0.363**
I spend a lot of time cleaning and grooming my pet	0.206	0.321	**0.792**
My pet is given regular checks by the veterinarian	0.003	−0.047	**0.243**
I groom my pet regularly	0.162	0.229	**0.813**

as horses are relinquished as they lose their instrumental value. It seems reasonable to suppose that the severing of such a relationship may alter future expectations and possible future emotional investments, but somewhat surprisingly, the impact of this process does not appear to have been investigated to date, although the grief associated with the loss of a horse through euthanasia is more widely recognized (Brackenridge & Shoemaker, 1996a,b,c; Endenberg *et al.*, 1999).

The horse's perspective on the relationship: an afterword

Human–animal interactions are believed to have derousing and stress relieving effects in people (Katcher, 1981; Friedman *et al.*, 1983) and recent neuro-physiological studies have suggested a complementary change may occur in the non-human animal subject (Odendaal & Meintjes, 2003). Horse heart rates have also been found to decline following various forms of affiliative contact with humans (Lynch *et al.*, 1974; Feh & de Mazzieres, 1993; Hama *et al.*, 1996) and human contact may also help reduce agitation (Crawley & Chamove, 1997). Whether this suggests that the relationship between an owner and his or her horse is a mutual one is debatable but worthy of further investigation. None the less, the rider–horse relationship is certainly characterized by an affectional bond similar to that found between the owners of smaller domestic pets and their charges. This appears to be the case even when the horse is kept primarily for competition purposes and perhaps it is time that our attitude to their management and training was re-examined in this light. It often seems that in equestrianism too much space is given to ideas built on the concept of domination of the horse, when really what both horse and owner seek may be the co-operation which comes from mutual affection.

References

Anon (1991). *Horse Riding*. Report to the Sports Council, November 1991, Section 9. London: Sports Council, pp. 23–5.

Berryman, J. C., Howells, K. & Lloyd-Evans M. (1985). Pet owner attitudes to pets and people: A psychological study. *Veterinary Record*, 117: 659–61.

Brackenridge, S. S. & Shoemaker, R. S. (1996a). The human/horse bond and client bereavement in equine practice. Part 1. *Equine Practice*, 18: 19–22.

(1996b). The human/horse bond and client bereavement in equine practice. Part 2. *Equine Practice*, 18: 23–5.

(1996c). The human/horse bond and client bereavement in equine practice. Part 3. *Equine Practice*, 18: 20–3.

Brown, D. (1984). Personality and gender influence on human relationships with horses and dogs. In *The Pet Connection – Its Influence on Our Health and Quality of*

Life, ed. R. K. Anderson, B. Hart & L. A. Hart, Minneapolis: University of Minnesota, pp. 216–23.

Chitnavis, J. P., Gibbons, C. L., Hirigoyen, M., Lloyd Parry, J. & Simpson, A. H., (1996) Accidents with horses: what has changed in 20 years? *Injury*, **27**: 103–105.

Christey, G. L., Nelson, D. E., Rivara, F. P., Smith, S. M. & Condie, C. (1994). Horseback riding injuries among children and young adults. *Journal of Family Practice*, **39**: 148–52.

Clutton-Brock, J. (1992). *Horse Power*. London: Natural History Museum Publications.

Crawley, J. & Chamove, A. S. (1997). Thoroughbred race horse behavioural responses to human contact. In *Looking Back and Moving Forward: Fifty Years of New Zealand Psychology. Proceedings of the 1997 Annual Conference of the New Zealand Psychological Society*. ed. G. M. Habermann, Palmerston North: Massey University Printery, pp. 16–21.

Crowell-Davis, S. L. (1992). The effect of the researcher on the behavior of horses. In *The Inevitable Bond: Examining Scientist–Animal Interactions*, ed. H. Davis & D. Balfour. New York: Cambridge University Press, pp. 316–33.

Driscoll, J. W. (1995). Attitudes toward animals: species ratings. *Society and Animals*, **3**: 139–50.

Endenberg, N., Kirpensteijn, J. & Sanders N. (1999). Equine euthanasia: the veterinarian's role in providing owner support. *Anthrozoos*, **12**: 138–41.

Feh, C. & de Mazzieres, J. (1993). Grooming at a preferred site reduces heart rate in horses. *Animal Behaviour*, **46**: 1191–4.

Fifield, S. (2001). What does 'member of the family' mean? Species and attitudinal differences in the attribution of family member status. In *Human Animal Conflict – Exploring the Relationships with Conflict between Humans and other Animals. Tenth Conference of the International Society of Anthrozoology, University of California Davis, August 2–4, 2001*. Available on line at http://www.vetmed.ucdavis.edu/CCAB/isaz2001.htm.

Friedman, E. S., Katcher, A. H., Thomas, S. A., Lynch, J. J. & Messent P. R. (1983). Social interaction and blood pressure influence of animal companions. *Journal of Nervous and Mental Disease*, **171**: 461–5.

Graham-Suggett, R. H. (1999). Horses and the rural economy in the United Kingdom. *Equine Veterinary Journal*, **28**(Suppl.): 31–7.

Hama, H., Yogo, M. & Matsuyama, Y. (1996). Effects of stroking horses on both humans' and horses' heart rate responses. *Japanese Psychological Research*, **38**: 66–73.

Jones, B. (1983). Just crazy about horses: the fact behind the fiction. In *New Perspectives on our Lives with Companion Animals*, A. H. Katcher & A. M. Beck, Philadelphia: University of Pennsylvania Press, pp. 87–111.

Katcher, A. H. (1981). Interrelations between people and their pets: form and function. In *Interrelations between People and Pets*, ed. B. Fogle. Springfield, IL: Thomas, pp. 46–67.

Keeling, L. J., Blomberg A. & Ladewig J. (1999). Horse-riding accident: when the human–animal relationship goes wrong! In *Proceedings of the 33rd International Congress of the International Society for Applied Ethology*, ed. K. E. Boe, M. Bakken & B. O. Braasted. Ås: Agricultural University of Norway, p. 86.

Kidd, A. H., Kelley, H. T. & Kidd, R. M. (1983). Personality characteristics of horse, turtle, snake and bird owners. *Psychological Reports*, **52**: 719–29.

Lago, D., Kafer, R., Delaney, M. & Connell, C. (1988). Assessment of favourable attitudes toward pets: development and preliminary validation of self-report pet relationship scales. *Anthrozoos*, **1**: 240–54.

Lagoni, L., Butler, C. & Hetts, S. (1994). *The Human–Animal Bond and Grief*. Philadelphia: W. B. Saunders.

Lawrence, E. A. (1986). *Hoofbeats and Society. Studies of Human–Horse Interactions*. Bloomington: Indiana University Press.

Lynch, J. L., Fregin, G. F, Mackie, J. B. & Monroe, R. R. (1974). Heart rate changes in the horse to human contact. *Psychophysiology*, **11**: 472–8.

McNicholas, J., Collis G. M. & Browne, G. (2001). The hassles and uplifts of horse ownership. Paper presented at the 9th *International Conference on Human–Animal Interactions. Rio de Janeiro, September 15th, 2001*.

Odendaal, J. S. J. & Meintjes, R. A. (2003). Neurophysiological correlates of affiliative behaviour between humans and dogs. *Veterinary Journal*, **165**: 195–200.

Osgood, C. E., Suci, G. J. & Tannenbaum P. H. (1957). *The Measurement of Meaning*. Urbana, IL: University of Illinois Press.

Poresky, R. H., Hendrix, C., Mosier, J. E. & Samuelson, M. L. (1987). The companion animal bonding scale: internal reliability and construct validity. *Psychological Reports*, **60**: 743–6.

Robinson, I. H. (1999). The human–horse relationship: how much do we know? *Equine Veterinary Journal*, **28**(Suppl.): 42–5.

Sahlins, M. (1976). *Culture and Practical Reason*. Chicago: Chicago University Press.

Serpell, J. (1983). The personality of the dog and its influence on the pet-owner bond. In *New Perspectives on our Lives with Companion Animals*, A. H. Katcher & A. M. Beck. Philadelphia: University of Pennsylvania Press, pp. 57–63.

Templer, D. I., Salter, C. A., Dickey, S., Baldwin, R. & Veleber, D. M. (1981). The construction of a pet attitude scale. *Psychological Record*, **31**: 343–8.

Wipper, A. (2000). The partnership: the horse–rider relationship in eventing. *Symbolic Interaction*, **23**: 47–70.

12

Learning abilities in the horse

Christine J. Nicol

Equine learning in an evolutionary context

There has not been a great deal of research on equine learning, and the studies that have been conducted cover only selected topics, leaving large gaps in our knowledge (Nicol, 2002). Some studies have taken careful account of the evolutionary origins of the horse and its ecological niche, looking at learning abilities that might be essential for survival in the wild. Other studies have treated the horse in the same way as any other laboratory animal, looking for more general principles in learning, or seeking to adapt general experimental paradigms. There is value in both approaches, although studies that combine the rigour of an experimental psychology experiment with a deep knowledge of horse ethology are likely to be the most successful and of most general value.

Any particular learning or cognitive ability should contribute to the inclusive fitness of an individual, increasing its chances of contributing genetically to the next generation. Thus, certain abilities are likely to be linked exclusively with certain lifestyles. Different aspects of intelligence will be of adaptive benefit to different species, making it difficult or logically impossible to say whether one species is 'more intelligent' than another, although it could be proposed that intelligence is a function of environmental and social complexity. Animals that are born into constant environments that change little between generations may survive perfectly well using instinctive responses. In contrast, when environments are varied and unpredictable a wide range of learning abilities will be essential. Studies of feral horses show that they tend to organize themselves into single-male harems, and bachelor bands that occupy undefended home ranges. Horses may face many problems including food of variable quality or unpredictable distribution, predators whose identities, locations and habits may change, and a complex social environment where the identities and roles of group members must be discovered (Boyd and Keiper, Chapter 4; Feh, Chapter 5). So there are good *a priori* reasons for believing that horses will be able to learn new associations and that learning may increase the flexibility of their behavioural strategies. Studies of spatial and social learning and cognition are relatively new, but research in this area is growing. There are also some good recent studies examining the effects of early handling and habituation on later horse behaviour, and the role of learning in diet choice and food selection.

Learning and development

The first thing a newly born foal must learn is the identity of its mother. The process of bond formation in horses involves mostly olfaction and taste, as the mare licks the foal and foetal membranes extensively. But the foal must also recognize its mother, as both are responsible for maintaining the bond once the herd is rejoined shortly after birth (Crowell-Davis and Weeks, Chapter 8). The following behaviour of the foal is refined through learning in the first week of life and is supported through smell when close to one another, but also through visual and vocal cues (Waran, 2001). This process of early bond formation is often termed imprinting and was originally thought to have many special features differentiating it from other forms of associative learning. The current view is that imprinting is a form of associative learning that is strongly biologically constrained (Bateson, 1990). Young birds and mammals are born with innate predispositions that lead them to attend to the relevant visual or olfactory stimuli required to recognize their mother. Thus, while it is possible to get a young animal to imprint on an object or a member of another species, this will occur only if the animal is not exposed to more relevant stimuli, and it results in a weaker bond that, under certain conditions, can be reversed (Bolhuis & Bateson, 1990).

Unfortunately, the term imprinting has been misused in the horse industry to describe a set of handling procedures conducted shortly after birth, with the aim of producing a horse that is easier to train when it is older (Miller, 1991). 'Foal imprint training' forces the foal to undergo an intensive and wide ranging programme of handling procedures within the first 48 hours after birth (Waran and Casey, Chapter 13). The effect of these procedures has been studied by Simpson (2002). It was found that foals handled as neonates were generally easy to handle and more responsive to people than controls at four months of age, although the effects appeared to be quite general. Specific procedures, such as tapping the

Figure 12.1. Foal being restrained during 'imprint' training. This is normally done in the presence of the mare and involves habituating the animal to contact over a wide range of parts of its body. (Picture courtesy of J. Williams.)

sole of the foot, did not seem to affect the reaction to related procedures such as tapping with a hammer when older. The results of a far larger trial have been reported by Williams *et al.* (2002). A comparison was made between the behaviour of 'imprint trained' foals and control foals at one, two and three months of age (Figure 12.1). Although there were some differences in behaviour at one and two months, with control foals scoring more highly on behavioural measures but trained foals exhibiting lower heart rates during testing, by three months of age there were no significant differences between the two groups. In a further study (Williams *et al.*, 2003), a comparison was made between control foals that received no additional handling, foals that received one session of 'imprint training' at either birth, 12, 24, 48 or 72 hours after birth, and foals that received four imprint training sessions at birth, 12, 24 and 48 hours after birth. After this early handling period, foals were left with their mares on pasture until weaning at six months. About one week after weaning, foals were subjected to a variety of tests. The time taken to catch and halter each foal was determined, and the foal's heart rate was measured at baseline, during exposure to stimuli used in the imprint training procedures, and during exposure to a novel object. It took longer for foals that were imprint trained for one session immediately after birth to stop resisting during an ear rub test than foals in most other groups, and control foals had a greater change in heart rate during a fly-spray test than some other groups. However, overall no consistent effects of early imprint training were found and the authors concluded that such early and intensive training was not effective (see also Ladewig *et al.*, Chapter 9).

Less intensive and more sustained forms of early handling may be beneficial. There is clear evidence that increased handling of foals can produce a prolonged reduction in fearfulness and animals that are easier to manage in later life. Mal and McCall (1996) handled foals for 10 minutes a day either between 1 and 42, or between 43 and 84 days. The early handled foals performed better in a test where they were led away from their dams. Similarly, Konik horses handled for 10 minutes five days a week as foals showed strongly significant reductions in heart rate, and were much easier to manage on a variety of tests given at 6, 12, 18 and 25 months, than non-handled foals (Jezierski *et al.*, 1999). Horses given additional training from the age of five months showed reductions in heart rate and heart rate variability when given novel object and handling tests between 9 and 22 months of age (Visser *et al.*, 2002).

Similar effects of early handling are noted in farm animals and a number of mechanisms have been proposed to explain the change. Repeated exposure to people where no negative consequences are experienced may produce habituation (decreased responsiveness after repeated application of a neutral stimulus), which may be responsible for some of the reduced fear. If this is the case, then the initial fear response to people might be expected to be reinstated if the animals do not have further exposure to people over a long period of time (Rushen *et al.*, 2001). This does not appear to be the case in horses. Mal and McCall (1996) found that foals handled between

1 and 42 days of age were easier to handle at 85 days of age than foals handled between 43 and 84 days, suggesting that the early handling had a more permanent effect on their behaviour. Classical conditioning provides another route by which horses' responses to people can be altered if people are associated either with rewards such as food, or punishments, such as shouting or hitting. Many farm animals can learn to approach or avoid individual people according to the type of treatment they receive (Rushen *et al.*, 2001) and, anecdotally, the differential response of horses to men and women, or to the veterinarian, suggests they can do the same. An interesting question is whether social contact with people can in itself be a primary reinforcer for a horse but this remains to be fully investigated. It has been found that stroking a horse, particularly at preferred allo-grooming sites and by people who have a positive attitude to horses, is accompanied by a reduction in equine heart rate (Lynch *et al.*, 1974; Feh & de Mazieres, 1993; Hama *et al.*, 1996). The full answer may depend on how the horse categorizes the human: as a threat, a neutral object, or as an affiliate.

As well as considering the mechanisms of learning that may modify the horse's response to people and to new situations, it is worth considering how early handling can affect the horse's own ability to learn. Although Mal *et al.* (1994) found no evidence that handling foals affected their performance on a one-trial learning test, and Marinier and Alexander (1994) reported no link between current handling and maze performance in adult horses, Heird *et al.* (1981, 1986b) found that foals handled frequently after weaning performed in a modified T-maze test better than non-handled foals. The mechanisms by which handling might affect equine learning potential are many and varied, but in rats, environmental enrichment can stimulate the development of additional neural connections in the brain and improve learning ability. It has been strongly argued that the provision of a cognitively interesting environment will enhance learning abilities in horses (Kiley-Worthington, 1997: 109).

Learning what to eat and what to avoid

Grassland areas provide diets that are heterogenous in nature, and varied in spatial and temporal distribution. In ruminants, social learning is important in food selection. Coprophagy (ingestion of faecal material) may be one route of acquiring information about the value of different plants in horses (Marinier & Alexander, 1995) but this needs empirical investigation. It is reasonable to propose that social learning plays a role in food selection in horses. Recent work on a variety of species, including lambs, rats and chickens, suggests that a function of social learning may be to induce preferences for novel feeds, rather than to induce aversion to toxic or harmful substances (Mirza & Provenza, 1994; Galef, 1996; Sherwin *et al.*, 2002). It is possible that horses might avoid toxins by avoiding eating anything that has not been safely demonstrated by their dam. However, horses do not appear to display such extreme levels of neophobia, suggesting that they use other mechanisms of learning to avoid toxic foods (see Houpt, Chapter 6).

Individual learning based on post-ingestive feedback from nutrients and toxins, is perhaps the most likely mechanism used. Individual learning requires that an association is formed between the flavour of the food and the consequences of ingestion, resulting in subsequent changes in intake of that food. In the wild this may be difficult because of the delay between feeding and the eventual consequences of digestion, and because horses rarely eat discrete meals of one plant (Houpt *et al.*, 1990). However, under experimental conditions it has been shown that horses can select between different types of concentrate pellet based, most probably, on the consequences of post-ingestive feedback. Cairns *et al.* (2002) showed that horses were able to select pellets with a higher caloric value over pellets with a low caloric value, when the pellets were differentially flavoured to provide a cue that could be associated with post-ingestive consequence. The emergence of the preference for the higher energy feed was not immediate, but emerged after the horses had consumed at least 10 meals. The horses were unable to make the same choice when the pellets were unflavoured (Cairns, 2002) suggesting that the choice was not made on the basis of some initial difference in taste or texture of the high and low energy pellets. Both these pieces of evidence suggest the horses were making choices based on post-ingestive feedback, rather than by the initial orosensory properties of the feeds.

Individual learning could also be a useful mechanism to facilitate avoidance of harmful or toxic feeds. In taste aversion learning, the expected outcome is that an animal will learn to avoid a novel feed if it is paired with a delayed experience of sickness (Garcia & Koelling, 1966). However, in an experiment to test aversion learning, Houpt *et al.* (1990) showed that ponies learned to avoid a feed only if it was followed by an immediate injection of apomorphine, and only if it was not a highly palatable feed. The ponies did not acquire taste aversions when the injection was given 30 minutes after ingestion of a novel feed, suggesting that the ponies were reacting to some immediate property of the injection, rather than any post-ingestive experience of sickness. Pfister *et al.* (2002) questioned the effectiveness of apomorphine as a conditioning agent, as it has a short duration of action and it works centrally in the central nervous system, rather than causing a long lasting gastrointestinal nausea. Instead, they administered lithium chloride immediately after consumption of a novel feed and were able to condition taste aversions to palatable feed and to locoweed *Oxytropis sericea*. Locoweed is a serious poison for horses grazing on infested rangelands in the USA. It is interesting to speculate why horses do not develop taste aversion conditioning to the plant itself, given their apparent sensitivity to the plants' alkaloids, which cause a variety of clinical symptoms. Possibly the delay between feeding and the post-ingestive consequences of eating a poisonous plant is too long to support associative learning in the field. The importance of the delay period could be investigated by increasing the interval between consumption of a novel feed and administration of lithium chloride in an experimental situation.

Mechanisms of learning
Association

Horses are well able to form associations where the relationship between two events guides behavioural change. In classical or Pavlovian conditioning, an environmental event or stimulus is followed predictably by some other events. Stabled horses become experts at detecting the subtlest signs that predict a feed or a period of exercise. The detection of a conditioned stimulus such as hearing the owner opening the door of the feed room, can set off a chain of anticipatory, and sometimes frustrated, responses such as pawing. Classical conditioning allows horses to predict events but gives them little control or influence over the actual timing of food delivery. A second form of associative learning is called instrumental or operant conditioning, where the first event is a response made by the animal and the second event is the associated reinforcing consequence. In this case, the horse gains not only the ability to predict what might happen next, but also the ability to control the timing of reward. An example would be the use of a 'foodball', an environmental enrichment device that delivers small amounts of pelleted feed when it is rolled around on the floor (Winskill *et al.*, 1996). Stabled horses learn to manipulate the foodball by pushing it with their noses or forelegs, controlling for themselves the arrival of food. The ability to exert control over the environment is thought to be an important component of good animal welfare (Wiepkema & Koolhaas, 1993; Jones & Nicol, 1998) and more thought should be directed towards allowing stabled horses instrumental control over aspects of their environment. It is not difficult to think of ways in which horses could be granted control, for example, over their own lighting or sound environments.

Frequently, horses learn using a mixture of classical and instrumental conditioning. Horses can be trained to perform the behaviour of walking up a ramp and into a trailer to obtain a food reward (Ferguson & Rosales-Ruiz, 2001). For horses trained in this way, the subsequent arrival of the trailer on the yard may become a conditioned stimulus indicating imminent food reward, and increasing the horses' level of anticipation. The horse cannot control when the trailer arrives, but once the trailer is present and the ramp lowered, it will be well prepared to enter.

Reinforcement and conditioned reinforcement

The importance of reward as the guiding force in establishing learning leads directly to the question: What is reinforcing for a horse? Psychologists have generally concluded that there is nothing beyond the animal's own reaction that can be used to define a situation as good or bad, which leads to an animal-defined concept of reinforcement (Staddon, 1983). Primary reinforcers are stimuli or resources with innate hedonic value. If an animal performs a response in order to obtain something it desires, then the desired resource can be described as a positive reinforcer. Food is a good example of a primary reinforcer that seems to guide a wide variety of equine learning in experimental contexts. If an animal responds in order to avoid something it dislikes, then the disliked stimulus is a negative reinforcer. Physical pressure is a good

example of a primary negative reinforcer, much used in horse training. Other primary reinforcers such as social contact or neutral thermal conditions may be important in guiding learned behaviour in the wild, but they are difficult to use in experimental or training situations. Because of this, conditioned reinforcers, initially neutral but paired with a primary reinforcer, are claimed to be important tools in promoting horse training.

There are many studies that show that a neutral stimulus may serve as an instrumental reinforcer by virtue of being paired with a primary reinforcer. An experiment by Hyde (1976) linked the delivery of food with a tone conditioned stimulus (CS) for one group of rats, and presented a tone CS and food randomly in respect to each other for a second group of rats. When rats were given the opportunity to press a lever to hear the tone, rats from the first group pressed the lever many times over the first few sessions, even though no food was ever presented. This suggests that pairing the tone with food resulted in it becoming a conditioned reinforcer.

The use of conditioned, or secondary, reinforcers is of considerable interest to horse trainers (see Waran and Casey, Chapter 13; McGreevy and McClean, Chapter 14), as it avoids the inconvenience of repeatedly providing a primary reinforcer such as food. Interest in this area is exemplified by articles and books, for example on 'clicker training' in the popular press (e.g. Karrasch *et al.*, 2000). During training, secondary reinforcement can best be achieved by presenting the new stimulus just before presentation of the primary reinforcer (McCall, 1990; Cooper, 1998). Recently McCall and Burgin (2002) examined the use of a conditioned reinforcer in the acquisition of an instrumental response, and in its maintenance under extinction conditions. All horses acquired an initial instrumental task to obtain a food reward. For one group of horses the delivery of food was linked with a tone CS, for the other group of horses the tone was not presented. An extinction trial with no food delivery was then conducted, with the tone CS applied throughout for horses from the first group. No differences in the number of trials to extinction were noted between the two groups. In a second phase of the study the same horses were re-trained to perform the original instrumental response using the original reinforcer combinations, and then were required to learn a new instrumental response in the presence of the tone CS only. In this situation, the horses that had previously experienced the tone linked with food, made more novel instrumental responses than the controls. This provides some support for the increasingly popular use of conditioned reinforcers in training, particularly when used as a 'bridge' signalling that a primary reinforcer will follow. However, because conditioned reinforcement did not delay the extinction of an instrumental response in horses, contrary to the results with rats, the efficacy of using conditioned reinforcement to replace primary reinforcement in the horse during training is therefore less certain.

Williams (2002) conducted a study to examine the effects of various combinations of conditioned reinforcement (a click presented 5 seconds before a food reward) and continuous or variable primary reinforcement schedule on the acquisition and extinction of an instrumental response. Ten horses were assigned to each of six different treatment combinations. Williams found that neither the number of training trials required to reach criterion, nor the number of responses made under extinction conditions were influenced by the presence of the conditioned reinforcer. This casts doubts on the claims made in the more popular press that clicker training promotes faster learning and greater retention.

However, it should be remembered that testing under extinction conditions is difficult, and McCall and Burgin (2002) could not rule out differences in the underlying motivation of their subjects. Another factor that might influence behaviour under extinction conditions is whether the horse is known to perform any stereotypic behaviour. Persevering with a previously trained, but currently unrewarded, response is a characteristic that is associated with the performance of stereotypic behaviour in laboratory rodents (Garner & Mason, 2002). It is not known whether this is also the case for horses but, in a preliminary study, J. Eardley and D. S. Mills (pers. comm., 2001) found that stereotypic horses continued to perform an instrumental response under extinction conditions for longer than normal horses, suggesting that they may have developed a general inability to suppress non-functional behaviour. Work is continuing in this area.

The classic approach to training a horse to perform in hand, or while being ridden under saddle, is to use negative reinforcement whereby the horse is rewarded for moving away from the leg or the whip, or from

the pressure of the spur or bit, but all such methods rely in large part on the rider's skill and subtlety to avoid imposing pain or frightening the horse. Traditional training methods have been developed over centuries because providing positive reinforcement, such as food, during a sequence of complex manoeuvres appears to be so impractical. The lure of secondary reinforcement as a training technique is that potentially it could bridge the delay between the response and the administration of food reward. Further studies of secondary reinforcement are still needed – in particular the effectiveness of human touch and the human voice need to be examined for both their primary and secondary reinforcing characteristics. However, one implication of the seemingly limited effectiveness of clickers as secondary reinforcers in horse training, is that new methods need to be developed to reward the horse during handling or riding. Work in this area is being conducted at the University of Sydney by Dr Paul McGreevy who is evaluating the effectiveness of a simple remotely operated pump that delivers a small food reward into the horse's mouth, thus permitting direct positive reinforcement of desired behaviour as the horse is being ridden or driven.

Discrimination

The focus of much research since the early 1990s has been on the stimulus control of instrumental learning. Discrimination learning examines the extent to which horse behaviour is sensitive to control by environmental cues. This type of learning enables some stabled horses to influence the timing of arrival of their feed. Noticing the owner approaching the feed room door, the horse starts to kick and bang on its stable door in such an annoying manner that it is fed before the other horses on the yard. This is a good example of discriminative control as the horse gains little by banging on its stable door when the owner is not present, and soon learns not to do so. In most experiments on discrimination, horses have to learn the relevance of one stimulus over another in predicting reinforcement. The relevant stimulus then starts to control behaviour, such that a horse will perform an operant response to obtain a food reward in the presence of one stimulus but not the other (McCall, 1990). Horses appear to discriminate using spatial cues more easily than other stimulus features, as indicated both by the speed of initial task acquisition and by the extent to which acquired discriminations can be reversed (Nicol, 2002). However, they can use visual stimulus features to discriminate between doors (Mader & Price, 1982) or buckets (Clarke *et al.*, 1996). In many practical situations the discriminative stimulus used by the horse may not be immediately apparent. A horse that will readily enter a trailer during practice sessions where it simply gets fed in the trailer and then leaves again, may refuse to enter the same trailer on a day when it has to be transported. The owner will need to think hard about the discriminative stimuli available to the horse in this situation – the use of a tail bandage in one situation but not the other, the different position of the trailer in the yard, or simply the more anxious demeanour of the owner are all plausible candidate cues.

Discrimination is facilitated if there is a differential outcome in reward. Thus, in a task where the colour of a centre panel signalled that either a left- or a right-lever response was correct, horses performed best when different rewards (chopped carrot for one level, solid food pellets for the other) were linked with each lever, than when rewards were randomly assigned or identical (Miyashita *et al.*, 2000). This highlights the fact that horses are able to form complex representations that encode information about the relationships between stimuli and outcomes. This should be borne in mind when training horses to make difficult discriminations when under saddle. The use of the rider's leg to indicate one type of movement could be paired with a physical reward, for example stroking the neck, while a subtly different leg cue to indicate a different type of movement should perhaps be paired with a distinctly different outcome, such as verbal praise.

Discrimination studies can be extended to investigate the limits of auditory and visual perception as the relevant stimulus predicting reward is taken to the limits of the horse's perceptual range (Timney & Keil, 1992; Bubnalittitz *et al.*, 1995; Smith & Goldman, 1999). Discrimination learning tasks have also been used to investigate whether horses are able to transfer information between the two brain hemispheres via the corpus callosum, as there are anecdotal reports that this ability may be limited in horses (Hanggi, 1999a). Reportedly, a horse may habituate to an object observed with one eye, but show a startle response to the same object the first time it is viewed with the other eye. To test this, two horses

were trained to perform a discrimination between two visual stimuli using just one eye. When the trained eye was covered and the untrained eye was uncovered, the horses were able to perform the same discriminations without detriment, demonstrating that neural transfer of visual information does occur in horses (Hanggi, 1999a). Perhaps startle responses occur because the same object has a different appearance when approached from the opposite direction, and not because of any neural deficit in information processing.

It is also important to consider how horses revise and re-learn behaviour, perhaps as they move from one owner to another. One way of examining flexibility is to see how easily horses can reverse an acquired discrimination. Kratzer *et al.* (1977) used a simple maze test where selection of the correct route required each horse to turn either left or right at the start of the maze. When horses were performing well on this spatial discrimination, the positions of the correct and incorrect routes were reversed. Horses were able to re-learn a new route over a period of three days. Fiske and Potter (1979) revealed a more marked ability for discrimination reversal learning in horses trained to negotiate a Y-maze to obtain food. Each arm of the maze was consistently signalled with a coloured bucket. On day 1, horses had to learn to obtain food from one arm of the maze, on day 2 from the alternate arm, and this pattern of alternation was repeated for 21 days. Twelve trials were given each day, and all but one subject reached a tight learning criterion on more than one day, indicating a general ability to revise a previously acquired discrimination. However, Sappington *et al.* (1997) found it hard to achieve reversal learning when the only consistent cue was bucket colour.

Stimulus generalization

An adaptive feature of learning is that animals are able to generalize many of the properties of the stimuli involved in the expression of a classical or instrumental response. Rather than focusing on some small and unique feature of a given stimulus, animals are able to form associations with a broader range of stimulus features than were present during training. For the ridden horse this can be either beneficial or detrimental. A riding-school pony has to respond appropriately to many riders using leg, seat and hand cues in different ways. The adaptive response of the pony is to generalize and respond in the same way to a broad range of different cues. For the advanced dressage horse, such generalization would be disastrous and dressage riders aim for precision and consistency in their aids in order to foster discrimination and avoid generalization.

When stimuli extend along some continuous gradient, animals show a gradation in their responses, which depends on how similar the newly presented stimulus is to the original training stimulus. In discrimination trials an animal will give its maximum response to the exact stimulus that predicts a reward, but will respond to a lesser extent to other stimuli that share features of the original stimulus. If a second stimulus predicting a negative outcome is introduced, animals are likely to 'play safe' and a phenomenon known as 'peak shift' can be observed. Dougherty and Lewis (1991) showed that, as expected, when horses were exposed to successive presentations of circles of various sizes, they showed a peak in instrumental responding to circles of the same size as that used in their initial training stimulus. When a small circle, signalling a negative outcome, was introduced, horses adapted their original behaviour and their peak responding now occurred to circles that were larger than the initial training stimulus. This suggests that the need to avoid an aversive event may dominate and interfere with the acquisition of a food reward, which is something that has considerable implication for training methods, as discussed in the section on stress and learning.

Social learning

Social influences on food selection have already been discussed, but social interaction can also facilitate the acquisition of new instrumental responses in animals of many species (Nicol, 1995). Observational learning occurs when a naïve animal (the observer) watches a trained conspecific (the demonstrator) carrying out a task, causing the observer to learn the appropriate response more quickly or more effectively than controls not exposed to a demonstrator. There is a need to examine the social learning capacities of horses, as the unsubstantiated beliefs of owners can have profound effects on the way in which they manage their horses. In a survey of UK horse owners, 72% of more than 1000 respondents believed that horses could copy abnormal behaviour from conspecifics (McGreevy et al., 1995). This belief leads many owners to isolate

horses showing abnormal behaviours, reducing the welfare of these social animals. Although there is some evidence that observational learning can play a part in the development of abnormal behaviour in rodents (Cooper & Nicol, 1994) there are no good data to support this proposition in horses. Clusters of horses showing abnormal behaviour can usually be explained either by genetic relatedness and/or by the exposure of a group of horses to the same eliciting environmental conditions. A first step in deciding whether horses can copy abnormal behaviour would be to show that they can socially acquire any behaviour in a controlled experimental setting.

In contrast to results with other species it has been surprisingly difficult to show an effect of social learning in horses. Three experiments have investigated whether horses acquire a discrimination between two differently coloured or marked buckets more readily if they have previously observed a trained demonstrator perform the task correctly. Baer et al. (1983) and Baker and Crawford (1986) reported no significant enhancement of learning in observers compared with controls. Various methodological issues may have been partly responsible for the negative results, prompting Clarke et al. (1996) to modify the procedure used by Baer et al. (1983). Despite ensuring that colour and pattern cues were available, that food was available in both buckets when observers were tested, providing no opportunity for individual instrumental learning, and controlling for olfactory cues, Clarke et al. (1996) also reported no significant effect of prior observation on the discrimination accuracy of their subjects. However, in contrast to Baker and Crawford (1986), Clarke et al. (1996) reported that observers were faster to approach the goal area on the first trial, suggesting that they had learned something about the general location of food. Lindberg et al. (1999) examined whether observation facilitated the acquisition of an operant foot press response. A demonstrator was pre-trained to obtain food by pressing a foot panel, and observer horses were divided into three treatment groups: those that observed the demonstrator acquire food, those that observed the demonstrator when the operant apparatus was not present, and those that observed the operant apparatus in the absence of the demonstrator. Although many of the horses acquired the instrumental response, social observation did not significantly facilitate the acquisition of the new behaviour.

Equine social systems appear to offer plentiful opportunities for observing and possibly learning from conspecifics but clearly laboratory experiments to date have failed to replicate the situations in which this might be most likely. As social and familial relationships between horses are of importance perhaps more attention should be given to examining whether foals learn by observation of their dams, or whether factors such as social dominance, shown to influence social learning in other species (Nicol & Pope, 1999), are relevant. Potentially, allowing young horses to observe experienced horses partaking in a variety of handling and management procedures could shorten and facilitate training regimes.

Stress and learning

As prey animals horses are susceptible to stress, which is generally thought to inhibit learning ability. The fact that horses will overcompensate and move away from optimum responding in order to avoid an aversive event has already been discussed in the section on stimulus generalization. Severe stress can interfere with learning and memory (Mendl, 1999) and for this reason, as well as for ethical reasons, horses should be treated calmly and not subjected to pain or deliberately frightened during training. Punishment, which involves the use of an aversive stimulus to inhibit or stop a behaviour, can easily produce fear and stress in the horse and should therefore have a limited place in the trainer's repertoire (Mills, 1998).

However, the effects of stress are complex. Animals often perform best on learning and memory tasks when stress levels are at some intermediate level (Mendl, 1999). An animal in a low state of arousal may simply not pay attention to relevant stimuli, or may not be motivated to obtain a reward. However, high stress levels may result in the animal shifting attention away from the task at hand to search for the source of the threatening stimulus, and may result in hasty decisions reached with inadequate information.

The effects of stress may also depend on the age of the animal. In rats, periods of temporary separation from the mother affect learning ability in a differential manner depending on the age of the offspring (Lehmann et al., 1999). It would be highly relevant to examine any analogous effects in horses, where different management practices are likely to produce a large range in maternal stress levels, and the degree of contact between mare and foal. The practice of weaning is itself likely to cause profound stress. Weaning is strongly associated with the initial onset of stereotypic behaviour (Waters et al., 2002) and it seems likely that there could be long-lasting effects of weaning method and age at weaning on learning ability. Although artificially reared six- to eight-month-old foals performed just as well on a simple maze test as foals that had been reared by their mothers and weaned a month prior to testing (Houpt et al., 1982), it would be interesting to test the learning abilities of older horses that had been weaned naturally by the dam, and to examine performance on a wider range of tasks.

Learning and temperament

An important aspect to consider before training a horse is whether it is able to learn and benefit from a training programme. In general, younger horses learn faster and more consistently than older horses (Houpt et al., 1982; Mader & Price, 1982) and are more investigative in trial and error learning (Lindberg et al., 1999). Trends for females to outperform males have been noted in discrimination reversal (Sappington et al., 1997) and spatial detour tests (Wolff & Hausberger, 1996). In contrast, social dominance appears unrelated to learning ability in visual discrimination (Mader & Price, 1982), simple maze (Haag et al., 1980; Houpt et al., 1982) and avoidance learning tests (Haag et al., 1980). There are likely to be genetic effects on learning but these have rarely been systematically examined in the horse (Hausberger and Richard-Yris, Chapter 3). Houpt et al. (1982) found no correlation between maze learning of individual mares and their foals. Trends towards significance in breed effects are noted in many studies but small sample sizes frequently preclude further conclusions. Mader and Price (1982) found that Quarter Horses learned a visual discrimination significantly faster than Thoroughbreds and suggested this might be because Quarter Horses were on average less reactive. However, individual reactivity scores were not correlated with learning score. Lindberg et al. (1999) found that horses classified as 'non warmbloods' learned an operant task faster than 'warmbloods', either because they were less reactive or because they were more motivated to obtain a food reward. A lower motivation to obtain food was possibly a reason for the finding that noted that fat horses had higher error scores on a discrimination task than thinner horses (McCall, 1989).

Recent work has shown that it is possible to measure individual variation and consistency of behaviour variables related to temperament. Visser *et al.* (2001) gave repeated novel object and handling tests to 41 young Dutch Warmblood horses and measured a wide range of behaviours performed by the horses during the tests. The variables could be statistically grouped into underlying components such as 'patience' and 'willingness to perform', which showed some degree of stability over time. A greater understanding of horse temperament may eventually help to explain the often large differences in performance reported in learning tests (Fiske & Potter, 1979; McCall *et al.*, 1981; Marinier & Alexander, 1994). Learning performance is as likely to be influenced by factors such as fearfulness, attentiveness or motivational state, as by cognitive ability per se. Where reactivity has been directly assessed as part of the experimental procedure it is sometimes found that less reactive horses perform significantly better (Fiske & Potter, 1979; Heird *et al.*, 1986a), although care must be taken not to confound measurement of reactivity and task performance (Nicol, 2002). In a detailed study Visser *et al.* (2003) found some indication that less reactive horses were more likely to complete a series of learning trials. However, for the horses that completed learning trials, there were no significant correlations between learning performance and behavioural or physiological variables related to emotionality. The same authors did find that some horses performed better on learning tasks where they were rewarded after a correct response, while other horses performed better when they had to learn to avoid a mildly aversive stimulus. Overall, there was no evidence that performance on these unrelated tasks was correlated.

Other studies have also found low correlations between performance on different tasks, even when the tasks share common features such as the provision of a food reward. One reason may be that features perceived as common between tests by scientists are not generalized and categorized in the same way by horses. Thus, McCall *et al.* (2003) found that horses trained to use a visual stimulus as an indicator of food reward in a lever-press discrimination, made no use of the same visual stimulus when it indicated the correct route through a maze. Haag *et al.* (1980 found that maze learning in ponies was significantly associated

with their learning to avoid a mild electric shock, but most studies have reported few such links. Heird *et al.* (1986a) found little correlation between the performance of horses on two different types of discrimination task. Wolff and Hausberger (1996) found no correlation between performance on an instrumental task and learning a spatial detour. Sappington *et al.* (1997) reported a low correlation between performance of horses on a discrimination reversal learning task and their completion of two performance tasks – crossing a small bridge and jumping a cross pole. Finally, Le Scolan *et al.* (1997) found no correlations between the ability of horses to acquire or memorize an instrumental task and their performance on a spatial task. Overall, this suggests that it is unlikely that we will be able to classify horses easily on any unified scale of 'intelligence' (Nicol, 2002).

The important implication here is that we may need to consider that horses have individual learning styles. Differences in individual behavioural and physiological response to environmental challenge, producing different hormonal response can influence attention, readiness to learn and central learning mechanisms. Hormonal response to stress is associated with complex learning abilities, such as the tendency to innovate (Pfeffer *et al.*, 2002). However, this knowledge will only be useful once we know how to predict which horses are likely to show which learning style, and secondly, once we have devised, tested and validated training programmes for horses of different temperament.

Learning and training

Training a horse involves suppressing undesirable natural responses, exploiting desirable natural behaviour and instilling novel behaviour by the deliberate or accidental application of learning theory (Cooper, 1998). Some insight into learning theory can make the process of training easier and more effective for both horse and human (Mills, 1998). A potentially good example of the application of learning theory to a practical problem concerns reducing the adverse responses of ponies to clipping. Gough (1999) assessed the baseline reactions of six ponies to being clipped along the neck with a set of electric clippers. He then divided the ponies into two groups, half received a tape-recording of the noise of the clippers at feeding times, the other half received sight and sound of the clippers during feeding times, for 14 days in each case. All the ponies were less reactive to the clippers on their second exposure to clipping. The obvious problem with this study is that it is impossible to know whether the ponies may have reacted less to the second clipping even without the imposition of the 14-day treatment. But more controlled studies of this kind are desperately needed to inform decisions and improve horse husbandry and management.

An important factor that may influence the success of a training programme is the timing and organization of the lessons. There have been some attempts to examine the responses of horses to varying patterning of lessons on avoidance learning. Horses that were given one avoidance session a week took fewer trials to reach a learning criterion than horses that were given two or seven sessions a week Rubin *et al.* (1980). However, varying the number of trials within a given daily session had no significant effect on learning performance (McCall *et al.*, 1993). These results suggest that trainers using negative reinforcement should use a small or moderate number of trials within a session and should space sessions over time. Where positive reinforcement is used it may be noticed that horses appear to 'learn how to learn'. McCall *et al.* (1981), for example, found that the average number of errors made when traversing 12 different mazes declined over time in a way that could not be accounted for by maze complexity. Fiske and Potter (1979) found that there was a significant reduction in error rates in horses given successive discrimination problems, indicating that the ability to solve a certain class of problem was itself a gradually acquired ability.

The limits of equine abilities?

The mechanisms responsible for associative learning and memory can explain the behaviour of horses in a wide variety of settings. An important question for our understanding of the equine mind is whether all horse behaviour can be explained this way, or whether horses are able to deal with more abstract properties of their environment. Work on animal cognition has shown that individuals of some species are able to represent abstract categories such as time, shape, number and order, but there has been little work on equine cognition. The ability to form categories provides the basis for considerable higher cognitive processing (Nicol, 1996) so it is interesting to explore the boundaries of equine abilities in this area.

An important caveat is that although scientists can demonstrate the existence of an ability, they cannot prove its absence. Apparent limitations in animal cognitive ability may reflect the fact that researchers have designed the wrong experiments. At this early stage in the study of equine cognition, appropriate methodologies are still being developed.

Sappington and Goldman (1994) investigated the abilities of horses to categorize triangles. Although subjects were able to learn to discriminate triangles from other shapes by conventional associative learning processes, there was no evidence that they could do the same with novel (unrewarded) triangles. This suggests they had not formed an inclusive category into which all triangular stimuli could be allocated. Recent work by Hanggi (1999b) is more encouraging. Using computer generated images as discriminative stimuli, and a procedure that controlled for confounding variables such as overall areas of black and white in the stimuli, Hanggi (1999b) trained two horses to discriminate between images with open centres and solid black images. Although initial training involved instrumental conditioning of each exemplar, when novel stimuli were introduced the horses correctly chose the open-centre stimuli. In most instances the horses were correct on the first trial, before they could have formed a direct association between the new stimulus and food reward. This suggests that the horses were able to form an abstract category that was flexible and sufficiently robust to permit the instant discrimination of completely novel stimuli. The ability to form a category based on physical similarities may depend partly on the processes of stimulus generalization. It is likely to be a functionally important ability, allowing animals to develop overall categories (food, predator) permitting instant responses to novel and unpredictable situations. Categorization abilities may allow horses to make instant classifications of stimuli that they encounter for the first time, based on their physical similarity to previous exemplars. Experienced show-jumpers may well be able to adjust their stride and jump to specific types of jump (doubles, water-jumps) based on a combination of previous experience and rapid categorization.

Categorization should not be confused with concept formation (Lea, 1984). A concept is an 'idea of a class of objects' and the stimuli that are included within the same concept should be grouped on the basis that they stand for the same idea, and not that they resemble each other physically. Concept formation, especially in animals that have no language, is of great theoretical interest (Nicol, 1996) but few studies have been conducted with horses. Sappington and Goldman's (1994) work referred to concept formation but, with the definitions used here, was an example of categorization learning. Flannery (1997) attempted to determine whether horses could acquire the abstract concept of 'sameness'. She used an identity matching-to-sample procedure where subjects had to choose two similar stimuli and discriminate these from a third, different, stimulus. The horses were able to learn the required discrimination but were not given the opportunity to perform the test using novel, unrewarded stimuli. Flannery's (1997) results can therefore be explained by conventional associative mechanisms. Evidence of concept formation in the horse is currently restricted to a recent study by Hanggi (2003), who examined the ability of horses to form a concept of relational size (larger than, smaller than). An Arabian gelding was first trained with a set of six simple images always to select the larger of any pair, using four different size categories (large, medium, small and tiny). Using a new set of images and objects, he was then trained to select medium over small. The extent to which he had formed a concept of relative size was then tested by exposing him to trials where the untrained large or tiny stimuli from the new set of images or objects were presented. The horse responded as if he had developed a relative concept of size, for example by selecting the large stimulus rather than the previously rewarded medium stimulus. Two other subjects showed similar levels of performance using subsets of stimuli. The untrained stimuli in these experiments were presented on a limited number of trials to minimize direct associative learning, and selection of the larger stimulus occurred on 100% of trials for most stimuli, suggesting that transfer to the new stimuli was immediate and not dependent on further trial and error learning. However, the test phase was not run under extinction conditions, so associative learning may have contributed to the strong discrimination performance that was obtained.

The ability of horses to form abstract concepts is of more than just theoretical interest. The welfare of the stabled horse, for example, may be affected by its ability to represent time. It may or may not appreciate

when it is led into the stable that it is likely to be left there for many hours, and it would be useful to have this kind of information. If horses cannot form abstract concepts, we need to take this into account. We cannot expect horses to adapt instantly to new tasks that are 'just the same' as previously taught tasks, if we are using a concept of 'sameness or similarity' that makes no sense to the horse. We might expect a horse that has learned to enter a horse lorry to step readily into a trailer, because we classify trailers and lorries functionally as horse transporters. But horses may have no such shared functional concept, and will have to be taught separately to enter each type of transport, especially when the physical appearance of each is sufficiently different to inhibit generalization.

Conclusion

Many aspects of our interaction with horses have become more scientific in recent years, with great increases in knowledge about equine reproduction, sports physiology and disease. Training methodology, however, has not attracted the same degree of scientific interest or funding. Traditional methods may have been tried and tested for generations, but there may still be room for improvement either in the effectiveness of the training procedures, or in ways of achieving the same outcome with reduced stress for horse and trainer. Scientific scrutiny of new training 'fads', quick-fix solutions and fashionable, or money-making ventures is even more essential, to ensure that the general horse-owning public is not misled, badly advised or conned. However, rather than simply reacting to each new idea it would be good if learning theorists and horse trainers could work more closely together, and be proactive in devising training strategies and methods based on solid theoretical foundations.

Finally, regardless of the practical implications, for anyone who has ever felt a bond with a horse, work on equine cognition is an exciting new area. We have the ability and the methodology to find out much about what it is like to be a horse. Work on categorization and concept formation gives us the tools to ask horses how they perceive and classify us, the humans who control so much of their lives. It could be a salutary lesson.

References

Baer, K. L., Potter, G. D., Friend, T. H. & Beaver, B. V. (1983). Observation effects on learning in horses. *Applied Animal Ethology*, 11: 123–9.

Baker, A. E. M. & Crawford, B. H. (1986). Observational learning in horses. *Applied Animal Behaviour Science*, 15: 7–13.

Bateson, P. P. G. (1990). Is imprinting a special case? *Philosophical Transactions of the Royal Society of London, Section B*, 203: 125–31.

Bolhuis, J. J. & Bateson, P. (1990). The importance of being first: a primacy effect in filial imprinting. *Animal Behaviour*, 40: 472–83.

Bubnalittitz, H., Weinberger, H. & Windischbauer, G. (1995). Determination of the upper auditory threshold in horses by operant conditioning. *Wiener Tierarztliche Monatsschrift*, 82: 259–63.

Cairns, M. C. (2002). An investigation into the feeding behaviour and diet selection of the domestic horse. Ph.D. Thesis, De Montfort University.

Cairns, M. C., Cooper, J. J., Davidson, H. P. B. & Mills, D. S. (2002). Association in horses of orosensory characteristics of foods with their post-ingestive consequences. *Animal Science*, 75: 257–65.

Clarke, J. V., Nicol, C. J., Jones, R. & McGreevy, P. D. (1996). Effects of observational learning on food selection in horses. *Applied Animal Behaviour Science*, 50: 177–84.

Cooper, J. J. (1998). Comparative learning theory and its application in the training of horses. *Equine Veterinary Journal Supplement*, 27: 39–43.

Cooper, J. J. & Nicol, C. J. (1994). Neighbour effects on the development of locomotor stereotypies in bank voles, *Clethrionomys glareolus*. *Animal Behaviour*, 47: 214–16.

Dougherty, D. M. & Lewis, P. (1991). Stimulus generalization, discrimination learning and peak shift in horses. *Journal of the Experimental Analysis of Behaviour*, 56: 97–104.

Feh, C. & de Mazieres, J. (1993). Grooming at a preferred site reduces heart rate in horses. *Animal Behaviour*, 46: 1191–4.

Ferguson, D. L. & Rosales-Ruiz, J. (2001). Loading the problem loader: the effects of target training and shaping on trailer-loading behavior of horses. *Journal of Applied Behaviour Analysis*, 34: 409–23.

Fiske, J. C. & Potter, G. D. (1979). Discrimination reversal learning in yearling horses. *Journal of Animal Science*, 49: 583–8.

Flannery, B. (1997). Relational discrimination learning in horses. *Applied Animal Behaviour Science*, 54: 267–80.

Galef, B. G. Jr. (1996). Social enhancement of food preferences in Norway rats: A brief review. In *Social Learning in Animals: The Roots of Culture*, ed. C. M. Heyes & B. G. Galef, Jr. London: Academic Press, pp. 49–64.

Garcia, J. & Koelling, R. A. (1966). Relation of cue to consequence in avoidance learning. *Psychonomic Science*, 4: 123–4.

Garner, J. P. & Mason, G. J. (2002). Evidence for a relationship between cage stereotypies and behavioural disinhibition in laboratory rodents. *Behavioural Brain Research*, 136: 83–92.

Gough, M. R. (1999). A note on the use of behavioural modification to aid clipping ponies. *Applied Animal Behaviour Science*, 63: 171–5.

Haag, E. L., Rudman, R., and Houpt, K. A. (1980). Avoidance, maze learning and social dominance in ponies. *Journal of Animal Science*, 50: 320.

Hama, H., Yogo, M., Matsuyama, Y. (1996). Effects of stroking horses on both humans' and horses' heart rate responses. *Japanese Psychological Research*, 38: 66–73.

Hanggi, E. B. (1999a). Interocular transfer of learning in horses (*Equus caballus*). *Journal of Equine Veterinary Science*, 19: 518–24.

(1999b). Categorization learning in horses (*Equus caballus*). *Journal of Comparative Psychology*, 113: 243–52.

(2003). Discrimination learning based on relative size concepts in horses (*Equus caballus*). *Applied Animal Behaviour Science*, 83: 201–13.

Heird, J. C., Lennon, A. M. & Bell, R. W. (1981). Effects of early experience on the learning ability of yearling horses. *Journal of Animal Science*, 53: 1204–9.

Heird, J. C., Lokey, C. E. & Cogan, D. C. (1986a). Repeatability and comparison of two maze tests to measure learning ability in horses. *Applied Animal Behaviour Science*, 16: 103–19.

Heird, J. C., Whitaker, D. D., Bell, R. W., Ramsey, C. B. & Lokey, C. E. (1986b). The effects of handling at different ages on the subsequent learning ability of 2-year old horses. *Applied Animal Behaviour Science*, 15: 15–25.

Houpt, K. A., Parsons, M. S. & Hintz, H. F. (1982). Learning ability of orphan foals, of normal foals and of their mothers. *Journal of Animal Science*, 55: 1027–31.

Houpt, K. A., Zahorik, D. M. & Swartzman-Andert, J. A. (1990). Taste aversion learning in horses. *Journal of Animal Science*, 68: 2340–4.

Hyde, T. S. (1976). The effect of Pavlovian stimuli on the acquisition of a new response. *Learning and Motivation*, 7: 223–39.

Jezierski, T., Jaworski, Z. & Gorecka, A. (1999). Effects of handling on behaviour and heart rate in Konik horses: comparison of stable and forest reared youngstock. *Applied Animal Behaviour Science*, 62: 1–11.

Jones, R. & Nicol, C. J. (1998). A note on the effect of control of the thermal environment on the well-being of growing pigs. *Applied Animal Behaviour Science*, 60: 1–9.

Karrasch, S., Karrasch, V. & Newman, A. (2000). *You Can Train Your Horse to do Anything! 'On Target' Training – Clicker Training and Beyond*. Kenilworth: Kenilworth Press.

Kiley-Worthington, M. (1997). *Equine Welfare*. London: J. A. Allen.

Kratzer, D. D., Netherland, W. M., Pulse, R. E. & Baker, J. P. (1977). Maze learning in Quarter Horses. *Journal of Animal Science*, 46: 896–902.

Lea, S. E. G. (1984). In what sense do pigeons learn concepts? In *Animal Cognition*, ed. H. L. Roitblat, T. G. Bever & H. S. Terrace. New Jersey: Hillsdale, pp. 263–76.

Lehmann, J., Pryce, C. R., Bettschen, D. & Feldon, J. (1999). The maternal separation paradigm and adult emotionality and cognition in male and female Wistar rats. *Pharmacology Biochemistry and Behavior*, 64: 705–15.

Le Scolan, N., Hausberger, M. & Wolff, A. (1997). Stability over situations in temperamental traits of horses as revealed by experimental and scoring approaches. *Behavioural Processes*, 41: 257–66.

Lindberg, A. C., Kelland, A. & Nicol, C. J. (1999). Effects of observational learning on acquisition of an operant response in horses. *Applied Animal Behaviour Science*, 61: 187–99.

Lynch, J. L., Fregin, G. F., Mackie, J. B. & Monroe, R. R. (1974). Heart rate changes in the horse to human contact. *Pscyhophysiology* 11: 472–8.

Mader, D. R. & Price, E. O. (1982). Discrimination learning in horses: effects of breed, age and social dominance. *Journal of Animal Science*, 50: 962–5.

Mal, M. E. & McCall, C. A. (1996). The influence of handling during different ages on a halter training test in foals. *Applied Animal Behaviour Science*, 50: 115–20.

Mal, M. E., McCall, C. A., Cummins, K. A. & Newland, M. C. (1994). Influence of preweaning handling methods on post-weaning learning ability and manageability of foals. *Applied Animal Behaviour Science*, 40: 187–95.

Marinier, S. L. & Alexander, A. J. (1994). The use of a maze in testing learning and memory in horse. *Applied Animal Behaviour Science*, 39: 177–82.

Marinier, S. L. & Alexander, A. J. (1995). Coprophagy as an avenue for foals of the domestic horse to learn food preferences from their dams. *Journal of Theoretical Biology*, 173: 121–4.

McCall, C. A. (1989). The effect of body condition of horses on discrimination learning abilities. *Applied Animal Behaviour Science*, 22: 327–34.

  (1990). A review of learning behavior in horses and its application in horse training. *Journal of Animal Science*, 68: 75–81.

McCall, C. A. & Burgin, S. E. (2002). Equine utilization of secondary reinforcement during response extinction and acquisition. *Applied Animal Behaviour Science*, 78: 253–62.

McCall, C. A., Potter, G. D., Friend, T. H. & Ingram, R. S. (1981). Learning abilities in yearling horses using the Hebb-Williams closed field maze. *Journal of Animal Science*, 53: 928–33.

McCall, C. A., Salters, M. A. & Simpson, S. M. (1993). Relationship between number of conditioning trials per training session and avoidance learning in horses. *Applied Animal Behaviour Science*, 36: 291–99.

McCall, C. A., Salters, M. A., Johnson, K. B., Silverman, S. J., McElhenney, W. H. & Lishak, R. S. (2003). Equine utilization of a previously learned visual stimulus to solve a novel task. *Applied Animal Behaviour Science*, 82: 163–72.

McGreevy, P. D., French, N. P. & Nicol, C. J. (1995). The prevalence of abnormal behaviours in dressage, eventing and endurance horses in relation to stabling. *Veterinary Record*, 137: 36–37.

Mendl, M. (1999). Performing under pressure: stress and cognitive function. *Applied Animal Behaviour Science*, 65: 221–44.

Miller, R. (1991). *Imprint Training of the Newborn Foal*. Colorado Springs, CO: Western Horseman Inc.

Mills, D. S. (1998). Applying learning theory to the management of the horse: the difference between getting it right and getting it wrong. *Equine Veterinary Journal Supplement*, 27: 44–8.

Mirza, S. N. & Provenza, F. D. (1994). Socially induced food avoidance in lambs – direct or indirect maternal influence? *Journal of Animal Science*, 72: 899–902.

Miyashita, Y., Nakajima, S. & Imada, H. (2000). Differential outcome effect in the horse. *Journal of Experimental Analysis of Behavior*, 74: 245–54.

Nicol, C. J. (1995). The social transmission of information and behaviour. *Applied Animal Behaviour Science*, 44: 79–98.

  (1996). Farm animal cognition. *Animal Science*, 62: 375–91.

  (2002). Equine Learning: progress and suggestions for future research. *Applied Animal Behaviour Science*, 78: 193–208.

Nicol, C. J. & Pope, S. J. (1999). The effects of social status and prior foraging success on social learning in laying hens. *Animal Behaviour*, 57: 163–71.

Pfeffer, K., Fritz, J. & Kotrschal, K. (2002). Hormonal correlates of being an innovative greylag goose *Anser anser*. *Animal Behaviour*, 63: 687–95.

Pfister, J. A., Stegelmeier, B. L., Cheney, C. D., Ralphs, M. H. & Gardner, D. R. (2002). Conditioning taste aversions to locoweed (*Oxytropis sericea*) in horses. *Journal of Animal Science*, 80: 79–83.

Rubin, L., Oppegard, C. & Hintz, H. F. (1980). The effect of varying the temporal distribution of conditioning trials on equine learning behavior. *Journal of Animal Science*, 50: 1184–7.

Rushen, J., de Passille, A. M., Munksgaard, L. & Tanida, H. (2001). People as social actors in the world of farm animals. In *Social Behaviour in Farm Animals*, ed. L. J. Keeling & H. W. Gonyou. Wallingford: CABI Publishing, pp. 353–72.

Sappington, B. F. & Goldman, L. (1994). Discrimination learning and concept formation in the horse. *Journal of Animal Science*, 72: 3080–7.

Sappington, B. F., McCall, C. A., Coleman, D. A., Kuhlers, D. L. & Lishak, R. S. (1997). A preliminary study of the relationship between discrimination learning and performance tasks in yearling and 2-year old horses. *Applied Animal Behaviour Science*, 53: 157–66.

Sherwin, C. M., Nicol, C. J. & Heyes, C. M. (2002). Domestic hens use social learning to develop preferences for palatable foods but aversions for unpalatable foods. *Animal Behaviour*, 63: 933–42.

Simpson, B. S. (2002). Neonatal foal handling. *Applied Animal Behaviour Science*, 78: 303–17.

Smith, S. & Goldman, L. (1999). Color discrimination in horses. *Applied Animal Behaviour Science*, 62: 13–25.

Staddon, J. E. R. (1983). *Adaptive Behaviour and Learning*. Cambridge: Cambridge University Press.

Timney, B. & Keil, K. (1992). Visual acuity in the horse. *Vision Research*, 32: 2289–93.

Visser, E. K., van Reenan C. G., Hopster H., Schilder M. B. H., Knaap, J. H., Barneveld, A. & Blokhuis, H. J. (2001). Quantifiying aspects of young horses' temperament: consistency of behavioural variables. *Applied Animal Behaviour Science*, 74: 241–58.

Visser, E. K., van Reenan C. G., van der Werf J. T. N., Schilder M. B. H., Knaap J. H., Barneweld A. & Blokhuis, H. J. (2002). Heart rate and heart rate variability during a novel object test and a handling test in young horses. *Physiology and Behavior*, 76: 289–96.

Visser, E. K., van Reenen, C. G., Schilder, M. B. H., Barneveld, A. & Blokhuis, H. J. (2003). Learning performance in young horses using two different learning tests. *Applied Animal Behaviour Science*, 80: 311–26.

Waran, N. K. 2001. The social behaviour of horses. In *Social Behaviour in Farm Animals*, ed. L. J. Keeling & H. W. Gonyou. Wallingford: CABI Publishing, pp. 247–71.

Waters, A. J., Nicol, C. J. & French, N. P. (2002). Factors influencing the development of stereotypic and redirected behaviours in young horses: the findings of a four-year prospective study. *Equine Veterinary Journal*, 34: 455–63.

Wiepkema, P. R. & Koolhaas, J. M. (1993). Stress and animal welfare. *Animal Welfare*, 2: 195–218.

Williams, J. L. (2002). Clicker Training in Horses: Operant Conditioners and Secondary Reinforcers in Horse Training. Ph.D. Thesis, Texas A & M University.

Williams, J. L., Friend, T. H., Collins, M. N., Toscano, M. J., Sisto-Burt, A. & Nevill, C. H. (2003). The effects of an imprint training procedure conducted at birth on the reactions of foals at six months of age. *Equine Veterinary Journal*, 35: 127–32.

Williams, J. L., Friend, T. H., Toscano, M. J., Collins, M. N., Sisto-Burt, A. & Nevill, C. H. (2002). The effects of early training sessions on the reactions of foals at 1, 2 and 3 months of age. *Applied Animal Behaviour Science*, 77: 105–14.

Winskill, L. C., Waran, N. K. & Young, R. J. (1996). The effect of a foraging device (a modified Edinburgh Foodball) on the behaviour of the stabled horse. *Applied Animal Behaviour Science*, 48: 25–35.

Wolff, A. & Hausberger, M. (1996). Learning and memorisation of two different tasks in horses: the effects of age, sex and sire. *Applied Animal Behaviour Science*, 46: 137–43.

13

Horse training

Natalie K. Waran and Rachel Casey

Introduction

In the long history of the horse–human relationship, it has often been the case that horses have been attributed with almost 'magical' properties. Their highly developed senses enable them to perceive minute changes in human behaviour, and rapidly learn about their meaning, if the trainer is skilful and patient. However, for the same reasons, it is also the case that poor training can lead to problem horses that are fearful and often dangerous. 'As an artist must know what he wishes to convey by his completed work and the workman must understand how best to use his tools, so must the rider have an exact knowledge of his aim and the ways and means to obtain it', given these words of Podhajsky (1967) it is perhaps surprising such an important attribute of the horse appears to have been devoid of much scientific research for so long.

In this chapter, the various techniques that are used today in the training of competition horses are discussed. Different styles of training have evolved

over the period of the horse–human relationship and involve the mixing of different historical influences and the repeated re-evaluation and reinterpretation of earlier techniques, such as those of Xenophon documented in the 'Art of Horsemanship' over 2000 years ago. The typical manner in which horses are trained in Europe for recreational riding purposes, dressage, show-jumping, eventing and racing, and also the Western approach to training for activities such as cattle work and rodeos is described. In addition, the various new approaches to training are evaluated to determine how each is effective in terms of modifying behaviour with respect to our understanding of learning theory (see Nicol, Chapter 12).

The domestic horse has been selectively bred to meet various human requirements. The different roles of the horse, and the methods by which different types and behavioural traits have been selected for these requirements will not be discussed here since it has been covered in other chapters of this book (Hall, Chapter 2; Hausberger and Richard-Yris, Chapter 3). Similarly, the management practices for such horses will not be discussed in detail, although it is important to recognize that the intensive conditions in which most performance or sports horses are maintained, are likely to have a significant impact on the behaviour of each horse, and on the success of training. In Europe, horses maintained for recreational purposes, for example, may be housed for part of the day during the winter period, but may be kept in a field or paddock over the summer. Animals kept for sports purposes, such as dressage, show-jumping, eventing, endurance or racing, are generally housed for longer periods all year round, and experience more intensive training, controlled feeding regimes and regular transportation.

'Training' is defined as the intentional modification of the frequency and/or intensity of specific behavioural responses. Such modification can be achieved through different means using positive reinforcement, negative reinforcement and/or punishment. For a discussion of learning ability in the horse see the previous chapter.

The value of understanding what motivates the horse, and the principles of how the horse learns, should be obvious when discussing different training techniques. Effective and humane training relies on such knowledge, as does the ability to deal with any problem behaviour that may arise where training techniques go wrong (see McGreevy and McLean, Chapter 14). An understanding of learning theory should form the basis for any training method, and yet there is no mention made of the principles of learning in most horse training manuals. Indeed, although a great deal has been written in popular books for horse owners about different methods for training horses, there is relatively little scientific investigation as to whether or not such techniques are effective or humane, nor about how they actually cause behavioural changes in the horse. It is also apparent from the history of the various techniques that modern practices are in the main, based upon traditional methods that have developed since horses were domesticated (see Waran *et al.*, 2002). These systems or methods have survived and been adapted over the ages, and are often 'repackaged' and presented as 'new' methods of training.

Factors affecting training

As well as the different techniques of training, a number of other factors can influence the success or otherwise of training an individual horse. These include age, sex, breed, age at which training is started, early experiences and contact with humans, the presence or absence of other horses, current and previous management practices, physical condition, conformation and individual personality.

Age

Evidence from other species suggests that there is a greater degree of synaptic plasticity during post-natal sensitive periods for learning (Bateson, 1979). This enables young animals to have a greater ability to learn new tasks, and hence would have an impact on the effectiveness of training. Indeed, Houpt *et al.* (1982), found that foals learn faster and more consistently than their mothers, and Lindberg *et al.* (1999) found younger horses to be more active in seeking reinforcement in trial and error learning. It is also popularly believed that the ability to learn declines in older horses, although there is no evidence to support this belief. In other companion animal species, such as dogs, it is known that some aged individuals have a reduced ability to form new associative links or retain existing ones, because of pathological changes similar to those which occur in Alzheimer's disease in humans (Cummings *et al.*, 1996). Thus, it appears training is likely to be most effective when it is started when the horse is younger rather than older. Rees (1984) has advocated that most basic training should be completed before the horse reaches adulthood at five years of age.

Sex

There is evidence of differences in learning ability and memory between sexes in horses, especially with regard to spatial tests (Wolff & Hausberger, 1996). There is no direct evidence, however, of differences in learning ability between stallions and geldings, although there is often anecdotal comment that one or the other is better at performing in particular categories of competition. Observed differences in behavioural responses to learning tasks between sexes, and between entire and castrated males, are likely to be related more to differences in motivation at that stage of development rather than the actual ability to learn new tasks, although this area warrants further scientific investigation. For example, if the reinforcement used in the study is access to another horse, the differences in performance between entire and castrated males may be due to the difference in the way they perceive the other horse (i.e. competition or companion) than in their speed of learning an association or behaviour.

Breed

There is some evidence that the trainability of horses differs with breed. These differences may be related to reactivity, as those horses that are more reactive to novel stimuli score less well in learning trials (Heird *et al.*, 1986). Mader and Price (1980) reported that Thoroughbreds learn less well than Quarter Horses, although this difference may also be confounded by differences in previous experience.

Early experience of the horse

It is well established in other domestic social mammals such as dogs and cats, that the environment experienced by the young animal during the first months of life can have a great impact on adult behaviour (e.g. Appleby *et al.*, 2002). The so called 'socialization period' in these species is identified as a time when the animal's ability to learn about the components of its normal social environment and the basic aspects of social behaviour is particularly good (Scott & Marston, 1950). It is likely that there is a similar period of enhanced learning in the horse, which affects trainability later in life, although there has been little systematic study in this area (see Nicol, Chapter 12). However, whereas the dog and cat are altricial, the horse is a highly precocial species, and hence it would seem likely that the equivalent sensitive period would be at an earlier age and last for a shorter period of time. None the less it seems that the timing, type, frequency and amount of handling all influence later responses to people, as is the case in other species.

It is also highly probable that the age at which training starts affects the degree to which the trained behaviour is flexible later in life. Horses trained from a relatively early age, such as those in the racehorse industry, are considered to be more difficult to retrain to perform different patterns of behaviour later in life than those initially trained at four or five years of age.

Context in which training occurs

The context in which the training occurs is also an important factor in whether a period of training is successful. The presence of distractive elements, such as other horses, will influence the degree to which the horse attends to the trainer. In addition, the size of the area in which training occurs will influence the trainer's ability to keep the attention of the horse, and to use the animal's flight zone in training (e.g. Bayley & Maxwell, 1996). The similarity of the environment to other situations in which the horse is used will also affect the ability of the horse to generalize learned associations to other contexts (see Nicol, Chapter 12).

Current management of the horse

The environment in which the horse is managed will also affect trainability. Horses kept in a relatively barren environment with little experience of novelty may develop stereotypical behaviours (Nicol, 1999). Impoverished housing environments may also cause horses to be overly excitable when trained and ridden, such that they do not respond to the trainer. However, it is possible that when horses are maintained in such an uninteresting environment it is easier to 'channel' their motivation for forward movement, and physical energy into those activities that the trainer desires. Racehorses, for example, are kept intensively, fed concentrated high energy food and therefore probably have a strong motivation to be active. This motivation might be channelled through training, where the horse is rewarded for appropriate behaviour such as running fast, and punished if he behaves inappropriately, such as bucking, rearing and displaying other play or excitable behaviour. The possibility that a more enriched environment may reduce the motivation of a horse to display those behaviours desired by the trainer must be considered when contemplating the welfare of performance animals. The suitability of training methods and management systems should always be considered simultaneously.

'Personality'

Each horse has a unique behavioural style (Marinier & Alexander, 1994), 'personality' or 'horsonality' (Visser, 2002), which is made up from a combination of genetic factors, early experience and learned associations made throughout the life of the animal (see Hausberger and Richard-Yris, Chapter 3). This individuality must be taken into account when training each individual horse (Visser, 2002).

Physical condition and conformation

The physical condition and conformation of each individual obviously affects the horse's ability to perform each behaviour that is desired by the trainer, and hence the success of the training. Horses that have a long-backed conformation, for example, find it easier to extend than collect in dressage, and an animal that has a painful injury will be unwilling to perform exercises that involve movement of the site of injury (Casey, 2002). Further discussion of this important point is beyond the scope of this chapter but is an integral part of many classic equestrian texts, for example Decarpentry's *Academic Equitation* (translated by Bartle, 1971)

'Traditional' training techniques
Origins of training

'European' style of riding The European 'school' has remained in many ways unchanged from that developed by the ancient Greeks for training horses for transport, competition and war. Techniques were modified by the Romans and Persians, and then developed in Europe for Cavalry riders. The 'school riding' started by the Greeks is essentially the basis of training techniques for performance and recreational horses throughout Europe, and indeed many other parts of the world. The 'classical' schools of riding evolved from the military riding establishments, where accurate control of horses was highly desirable for success in battle (Richardson, 1998). Modern riding and training styles in the UK have developed from this influence, and subsequently diverged into 'hunting' styles, and those which maintained the 'school work' philosophy, such as dressage. Any advanced training of the horse beyond simple use for transport, was mainly developed by the aristocracy in the sixteenth century, which influenced the style in which competition training developed – in a formal and rather stylized fashion (Richardson, 1998). Horses were maintained at this time mainly in isolation rather than in herds, as was also the case for those used in the Americas where the 'Western' style of riding developed.

'Western' style riding Horses were reintroduced into the Americas by the Spanish in the sixteenth century, so the early influence on training style in this continent was Spanish and Moorish. As this was a large and 'frontier' country at that time, the use of the horse was generally more functional than the more formal style of use developing in Europe. Horses in America needed to be able to 'lope' over long distances, in as comfortable a gait for the rider as possible. Different breeds and types were developed for this purpose, along with methods of management and training that suited the terrain and function of the horses. Rather than being maintained in individual stables, horses were maintained in herds, and hence were largely unhandled at the initiation of training. Horsemanship skills involved catching horses from herds, staying on horses that were essentially 'wild' and controlling horses well enough for cattle work. The kinds of competitions that developed were those that 'showed off' these skills – such as rodeo. The word 'rodeo' means 'round up', and these events were probably first staged as a festival that celebrated a successful round up with horses (Richardson, 1998).

Principles of training techniques

European training method The education of the foal begins as soon as it is born, with a process of gradual habituation to handling. A head-collar (halter) or foal slip is put onto the foal in the first few days of its life to habituate it to this experience, and this is then used to teach the foal about restraint. The foal is also taught to be led by the head, using the natural response of the foal to follow the mare. The foal is either led behind the mare or, in some cases, is tied to the mare, so that it learns to move forward with the pull of the rope. The foal learns through negative reinforcement that pulling back tightens the rope and is uncomfortable, and that moving forward releases the pressure. Once accustomed to restraint and leading, the foal is usually habituated to being tied up, handled all over, grooming and having its feet lifted, and is also introduced to novel environments, such as stables (stalls) and trailers or horseboxes (floats) (Loriston-Clarke, 1995; Rose, 1977). In these cases training is brought about through the use of a combination of habituation and negative reinforcement. Once handled in the early weeks of life, foals are often turned out, initially with the mare, and then with other young stock after weaning, until specific training starts.

Specific training for riding starts at different ages, depending on the discipline for which the horse is destined. Most riding horses start training at three or four years of age, but most racehorses start training as yearlings, ready to start their racing career as a two-year-old (De Moubray, 1987). The techniques described here are those most commonly used for racing and for other contemporary purposes within the European 'school', although there are numerous variations on this approach advocated by different individuals.

Training generally starts with lungeing and/or long-lining. Both of these techniques introduce the horse to: moving forward freely, in a circle in the case of lungeing and in more varied patterns with long-reining; voice commands; pressure on the bit; moving away from the handler; the positioning of the handler (in relation to flight zone); the use of a lunge whip to encourage forward movement through the use of negative reinforcement; and the development of paces (Inderwick, 1977; Stanier 1995). Some trainers start with lungeing to introduce the principles of pressure on the bit, moving in a circle, voice control and moving forward away from the lunge whip, and then move on to long-reining to develop the horse's movement, introduce new commands and introduce the horse to new situations (e.g. MacSwiney of Mashanaglass, 1995; Rose, 1977). Long-reining has the added advantage of teaching the horse to respond to a person behind them and accustoms the horse to pressure being applied directly on the mouth. These aids then become associated with the voice commands that were used to signal the different gaits required when lungeing. The horse also learns to be driven forward from behind rather than from the side (as is the case on the lunge), and the trainer can start to develop different movements and muscle groups before the horse is even backed (mounted). In experienced hands, horses can be taught advanced dressage movements, such as piaffe and flying changes, and taught to jump in good style while on long reins or lunge (Stanier, 1995). Some trainers also use a 'touching cane' to encourage forward movement, which is applied in the place where rider's leg will normally touch, the aim being to develop an association between the leg aid, vocal aid and rein aid.

Another progression in training used by some trainers is to move from lungeing to loose schooling where the horse is set loose in a small enclosed area, and encouraged to move forward by a trainer in the centre of the school using voice commands, body position and a lunge whip (e.g. Klimke, 1985). This technique is often used in the early training of show-jumpers, where the trainer can position him- or herself to encourage the horse to move down a 'lane' in which jumps are placed. In this case loose schooling has the advantage of eliminating the risk of reins becoming caught in the jumps and/or inhibiting the free forward movement of the horse. Using several jumps at related distances ('gridwork') encourages the horse to develop skills for judging strides and points of take off as well as physical flexibility and strength.

As well as developing techniques and movements that will be used in later training, the horse is gradually introduced to a mouthing bit, a bridle and then a saddle during this stage of its training. This process is conducted one stage at a time, with the competent trainer ensuring that the horse is comfortable with one stage before progressing on to the next, and ultimately leads to the introduction of a rider. The process is started in a context with which the horse is familiar, such as the arena where it is lunged, and on each session the trainer starts with a stage with which he or she knows the horse is comfortable. It is the aim of most trainers to progress slowly with the exercise such that the horse does not show an aversive response to the addition of any of the equipment. If a horse does respond however, it is generally restrained such that it learns that attempts at avoidance or escape are not successful. The horse is habituated to a roller (a leather girth that attaches around the horse where the saddle will eventually go) first, then a saddle, and then the pressure exerted as a girth is attached. The weight of a rider is sometimes first applied with the use of a weight, or 'dumb jockey', or a lightweight rider starts by leaning over the saddle to accustom the horse to the feeling of carrying weight. The horse is then accustomed to walking while the weight is present. The rider then slowly puts his or her leg over the saddle, and slowly sits up, gradually coming into the horse's line of vision. This is carried out slowly in a progressive manner. Once the horse is accustomed to walking with the rider sitting up, the trainer starts to develop or reinforce an earlier association between the different aids – voice, leg pressure, hands and body weight.

Lungeing and long-reining are also used again in more advanced stages of training, for example, to develop greater flexibility in jumping, or to aid in developing advanced dressage movements such as passage and piaffe, which are easier for the horse to perform initially without the weight of a rider.

The European process of training, therefore, focuses on the use of gradual habituation to new equipment and experiences, combined with techniques that rely on negative reinforcement: the horse learns to perform behaviours through moving away from pressure or avoiding a (usually mildly) aversive stimulus.

Western training method As with the European style of riding, there are a range of different individual approaches to training horses in the 'Western' style. Many of the objectives of Western riding are the same as those used in the European 'school' – lightness of control, obedience, relaxation and balance. However, whereas the emphasis in the European school is on subordination, and of every movement being at the instigation of the trainer/rider, the aim of the Western school is to produce a horse that is a working animal, who is encouraged to think and act rather more independently. Well-trained cutting horses, for example, will work independently to separate calves, much as a trained sheepdog will divide sheep flocks with minimal direction. Western training also encourages free self-carriage with little 'contact' on the horse's mouth, and as a consequence the outline of the horse is more elongated than that which is desired in the European school. The style of training varies with the region, due to topographical and functional differences between regions. The source of horses to be trained probably has the largest impact on training style. Those that are bred in a human controlled environment and habituated to handling from an early age, as occurs in Europe for contemporary training, will be easier to train than those maintained in a feral or herd situation, and which are not handled prior to the commencement of training.

The traditional methods used for training unselected horses appear to involve developing a degree of learned helplessness in the animals. Feral mustangs were generally broken at four to five years of age and were unhandled until this point. They were run into a corral, then lassoed, and 'thrown' to the ground using a second lasso around the forefeet and thrown on their side. Horses were tied up, hobbled, or had their head tied to a back leg, to prevent resistance. Saddling sometimes took place when the horse was on the ground. Bucking under saddle was punished by using a flail-ended short stick (a quirt), and various other aversive techniques were used to prevent the horse from throwing the rider. Horses were also tied to posts for long periods to teach them not to pull back, and various methods were devised to discourage them from pulling away, such as attaching a rope around their back legs. Ponies were similarly broken and if the methods used were not successful, the pony was turned loose, and considered an 'outlaw' (Richardson, 1998).

The training of modern Western-style show or pleasure ('trail') horses is considerably different from that experienced during the bronc-busting days. Like European horses, these animals are usually trained through a process of gradual habituation (Strickland, 1998). The process of introduction of bridle, saddle and rider are similar, although the early training focuses much more on establishing a lightness of rein contact. The Western horse is bridled with a hackamore system of bridle that incorporates a progression of nosebands, called 'bosal'. Horses are started with a heavy bosal, which is finely balanced such that it does not touch either the nose or the chin as long as the head carriage is correct. Once the horse learns, through negative reinforcement to respond to pressure on this bosal, it is progressively changed to thinner and thinner ones, until, in the fully trained horse the trainer uses only a pencil-thin bosal that acts as a 'reminder' to the horse (Miller, 1975). Unlike their European counterparts, however, foals are not generally 'turned away' until specific training starts, but are weaned from their mother into a box, where they are kept relatively isolated and inside to facilitate training by encouraging dependence on the trainer. In some yards, the foals may be fitted with a head-collar and long rope which they trail around the stable or stall from an early age. Since stepping on this rope pulls on its head, it is thought that this will encourage the young horse to stand still when a rein is on the ground, and also to keep its head and neck low. The method could also lead to undesirable learned behaviour where young horses learn to avoid stepping on the rope by holding the head to the side and running with the head held high. In larger breeding yards weanlings may be put together until they are considered old enough to begin basic training. Most Western horses have similar basic training to that described for horses trained in the 'European style', in that they are lunged, long-reined, backed and schooled so that they are obedient to the rider and relatively supple. The specialized training for roping, cutting and ranch work is then carried out in various ways depending on the trainer's preference.

One of the differences that developed between the European and Western styles of training was due to Western trainers generally dealing with feral or unhandled animals. It is probably for this reason that the Western approach tended to use a small round corral to maintain close contact with the horse, so that the trainer could start to control the horse's movement by moving in and out of the animal's flight zone (Kevil, 2003). In Europe, where horses are traditionally handled from an early age, the emphasis was on a 'school' area or arena, as was traditional in military training establishments. Finer control of the horse in the early stages of training was established using one or two long reins (lungeing and long-reining). In the European style, there was also greater emphasis on improving the balance and collection of horse, and lightening its forehand, so that it was possible to sit on the horse's back and for the rider to experience an elegant elevated gait. The lunge lines/long-reining approach was felt to be necessary to achieve this. By contrast, Western trained horses were not required to change their style of carriage to the same degree – in fact horses were encouraged to 'lope' with much of their weight on the forehand. These Western horses were of a different type to the European cavalry-style horse, being smaller and lighter. They were probably more able to balance themselves, hence requiring less training to carry a rider in a balanced manner.

In terms of learning, however, the techniques of both approaches are similar: both rely on gradual habituation to equipment and negative reinforcement, where the horse learns to behave in the desired manner to avoid an aversive stimulus. It is interesting that the different techniques and origins of training horses should rely on similar principles. This leads us to question whether this is due to proven efficacy or whether this is merely a reflection of the inherent nature of humans and their need to 'control' the behaviour of other species.

Training to competition level Training horses so that they can jump higher, perform exciting and physically challenging movements or run faster relies on a number of ingredients. First, the horse has to have some natural talent. This is a combination of the correct build or conformation for the task, good basic training, a willingness to perform, adequate level of fitness and appropriate physical development to be able to perform the task. The second main ingredient is the skill and knowledge of the trainer in communicating the required tasks and reinforcing these appropriately. Traditionally, humans have selectively bred certain breeds for certain tasks. Interestingly the attributes of the horse bred for one task have also been found to be useful for another. For example, many of the warm-blood breeds used in competition level dressage today are derived from the European carriage horse. This is because the 'up-hill' conformation of the breed, which favours pulling weight also makes for a horse that can more easily lighten the forehand (thus shifting their centre of gravity back), something that makes collected movements like piaffe and pirouette easier to obtain.

Training to advanced levels basically relies on the refinement of the pressure–release system that is used to establish the basics, and also the chaining of associative links through classical conditioning. It is not within the remit of this chapter to discuss exactly how horses are trained to perform advanced movements in dressage or to jump higher or run faster, and there are many practical publications that provide information on this (Loriston-Clarke, 1995; Loch, 2000). However, it is clear that as horses progress in their training they become more responsive to extremely slight and specifically sited pressures applied by the rider. This is achieved through constant repetition, through use of secondary reinforcers (such as slight changes in body weight replacing the more obvious leg movements) and through well-timed releases of pressure. Eventually the fine-tuning of the aids as well as the habituation of the horse to the pressures (e.g. the new environment, the noises, smells, isolation in the ring) of competition will lead to a horse that almost seems to 'know' what is expected of him when in a given situation.

Alternative/'sympathetic' or 'natural' techniques

On examining the basis of the various 'new' or 'alternative' techniques of training, it is apparent that the origins of these techniques are based around re-inventions and combinations of techniques and ideas that have been used in the past. Ideas have come from classical practices, Moorish ideas, and practices developed by other groups, such as the Native North Americans. Skills of dealing with horses over the ages have not been lost, but, if trends and styles are traced, they can be seen to persist and undergo repeated re-evaluation and 'repackaging'. Most recently, new styles of training have integrated elements of the Western style of training into the European style and, hence, to Europeans, appear as novel techniques. A nice example of this is the re-introduction of the 'round pen' into European training techniques by trainers such as Monty Roberts and Richard Maxwell.

Types of alternative training techniques

Methods used to train horses in different cultures have been related more to the cultural attitudes towards animal welfare, ethical principles and necessity, rather than effectiveness (see Waran *et al.*, 2002 for a discussion of the history of training). More recently, various new or innovative methods have been advocated, although many of these use similar learning paradigms to the more traditional approaches. These include:

- Jeffery's 'approach and retreat' technique.
- Monty Roberts' 'join up' technique.
- Parelli's 'sympathetic horsemanship' approach.
- 'Clicker' training.
- Tellington-Jones' 'touch' method.
- Miller's 'imprint training'.

'Approach and retreat' This method is based on techniques used by the Native Americans for taming wild horses. The method involves:

- isolating the horse from all other distractions so that its attention is focused on the handler;
- using calm movements, especially when gradually approaching the horse, so that it can habituate to increasingly close contact;
- ensuring that the horse is dependent on the handler for relief from aversive stimuli that arise in the training process;
- using the instinct of the horse to move away to a certain 'flight distance' from unknown or unpleasant stimuli.

Jeffery popularized the 'approach and retreat method', using these principles in the early 1900s (Blackshaw *et al.*, 1983). The technique involves having the horse in a rectangular pen with a loose rope around its neck. As the horse moves around the enclosed space the rope is tightened when the horse moves away from the trainer and slackened when it voluntarily moves towards the trainer. The horse learns through negative reinforcement that moving closer to the trainer is effective in removing the aversive stimulus of the tightening rope. Once it is established that being close to the trainer is associated with relief, the trainer is able to build-up contact gradually, and the horse is touched, groomed and finally exposed to saddlery before being asked to carry the rider.

'Join up' The underpinning philosophy of this method of training is that the horse should regard the trainer as the leader of a two-member 'herd', rather than as a predator. Roberts (1997) emphasizes the importance of the trainer's body language in aiding the training process and this is based on using signals that horses inherently understand from their evolution. Roberts suggests that the training process should be based around disciplining the horse, as he believes would be the case in a natural herd situation, where the lead mare may attempt to exile a horse from the herd until he shows adequate signs of submission. These signs are believed to include lowering the head to the ground, chewing and licking lips. It is suggested that to send the horse away from the herd, the mare will face the offender squarely and look him in the eye. She may chase him away if necessary, but usually body language is sufficient to drive him away. By contrast when the mare is prepared to let the horse back into the herd, she will take her eye off his, and turn partially away so exposing her 'long-axis' or side. Thus, this training appears to manipulate the horse's behaviour in a controlled setting, in order to encourage it to accept the trainer as the dominant member of the dyad.

It is unclear whether the situation described above is absolutely accurate, due to the lack of scientific evidence to support it (see Boyd and Keiper, Chapter 2; Feh, Chapter 5). However, it is known that when removed from the safety of the herd, prey animals are at far greater risk of attack from predators. Therefore, as a social herbivore, the horse probably considers itself safe when it is within the herd. If the horse truly does see the relationship between itself and the trainer as similar to a herd situation, by sending the horse away, the trainer is putting the horse in an aversive situation. Initially, all training takes place in a round pen, where the horse can be contained and there are few distractions. The trainer stands in the middle of the pen with a long rope and sends the horse away. Once the horse has been sent away from the trainer it is not permitted to relax until it begins to focus its attention on the trainer. When this happens, the horse will attend to the trainer with his ear and eye 'fixed' and he may also start to chew, lick its lips and lower its head. Despite the fact that it is not clear whether this behavioural response is a sign of 'submission' or, more likely, a displacement activity, it is a reliable enough signal that the horse is likely to respond to any opportunity to display an alternative behaviour. Hence, as soon as the horse displays this behaviour the trainer ceases to signal to the horse to move away, and instead turns his body slightly and allows the horse to approach. Through repetition of this situation the horse will eventually show a 'following' response, termed 'join-up'. This is considered by Roberts (1997) to be due to the horse learning where the 'comfort zone' is. When this lesson has been learned it is argued that the horse will seek the trainer's company whenever it feels threatened. There is obviously also a process of negative reinforcement and punishment taking place in this scenario. As the round pen is relatively small, and often has a deep sandy surface, it is quite hard work for the horse to keep cantering away from the handler. Once the horse learns that attending to the trainer and then moving towards him or her prevents the 'sending away' the animal will start to approach the trainer whenever it is in that context.

Predictably, this following response makes numerous everyday horse handling tasks simple. For example, catching a horse in a large paddock becomes easy when the horse actively seeks the company of its trainer regardless of satiation.

'Sympathetic horsemanship' Similar principles to those described above for 'join-up' underlie the 'natural horsemanship' techniques popularized by Pat Parelli (1993). His training programmes involve a series of stages, termed 'games' through which the horse and handler progress. The early stages of these programmes develop a consistent relationship between horse and handler based on the use of body position to reward or 'punish' the horse for appropriate or unwanted behaviours. This system, again, combines 'stylized' intra-specific communication patterns with negative reinforcement to encourage the horse to behave in the required manner. Parelli's recipe involves seven games. Games such as the 'yo-yo'

game, where the horse learns to associate being near to the trainer with a release of pressure, and the 'squeeze' game, where the horse learns to tolerate the presence of the trainer within close proximity, enable the trainer to gradually shape the horse's behavioural response. Once the responses are established in a controlled setting, the trainer is then encouraged to generalize these to other situations to which the horse will commonly be exposed. For example, the 'squeeze' game can be used when loading a horse, since through the game, the horse has learned to move forwards within close proximity to the trainer. Again, this technique relies largely on negative reinforcement, where the horse learns to move away from a stimulus that it perceives as aversive. By controlling the environment, the trainer is able to control and direct the manner in which the horse moves away from and towards him or her. The process of games perhaps helps encourage consistency in the less experienced trainer.

'Clicker training' Originally developed for use with performing sea mammals, one commonly used form of conditioned reinforcement is clicker training. Using this approach a novel sound, produced from a plastic 'clicker' is classically conditioned to a reinforcer, generally a food reward. The use of a conditioned reinforcer like a clicker has several putative advantages in training:

- Once the classical association is made, the sound of the click is 'rewarding' for the horse. This enables the trainer to reward a behaviour rapidly after its performance, thereby increasing the chance of the horse associating the reward with the appropriate desired activity.
- The association allows the trainer to bridge the gap between the time at which an animal performs a response correctly and the arrival of a primary reinforcer (usually food). This allows training to be carried out from a distance (hence its use by dolphin trainers) and its potential value to horse riders.
- It has been suggested that conditioned reinforcement facilitates learning (McCall & Burgin, 2002).

Although, empirical evidence of these effects in the horse is currently weak (see Nicol, Chapter 12).

During the initial stages of training, the clicker sound should *always* be followed by a positive reinforcer such as a food reward. Simultaneous presentation of a reward and a novel secondary stimulus works less well, as the primary reinforcer appears to block or overshadow the new stimulus. Similarly, presentation of the secondary stimulus after the primary reinforcer is unhelpful, because although an association may grow between the two, it does not help the animal to predict the arrival of a reward. Once the association has been made, it can be maintained and strengthened via variable or intermittent reinforcement. Using a conditioned reinforcer such as a clicker, has practical advantages over using a primary reinforcer such as a food reward. The click can be used immediately after an appropriate behaviour, making it easier for the horse to associate the correct behaviour with the reward. Also, consistently using a signal to predict a food reward means that the horse will not expect food from the trainer unless it hears the click, which significantly reduces unsolicited requests (or 'mugging') for food in the early stages of training. A point to note is that although one advantage of a commercial clicker device is that the sound it makes is distinctive, any human sound could be used instead as long as it is not easily confused with words that appear in common parlance. Often termed 'clicker words', these have the advantage of being more readily available.

In contrast to the empirical investigations of Williams (2002), it has been found to be particularly useful by one author (Casey) in shaping and modifying unwanted behaviours, such as reluctance to load into a horsebox, where traditional methods have previously been misapplied or misunderstood. By deconstructing a response, the horse may come to learn that its fears have been unneccessary. Rebuilding the desired response then becomes a reasonably simple chaining (or sequencing) task.

'*Touch training*' This method is named after Linda Tellington-Jones (1985) who uses it to enable her to attain co-operation from her equine patients. The method works through habituating and desensitizing the horse to touch and possibly through the stimulation of endogenous endorphin production. Gentle circular movements of the hands are used to relax the horse, using specific sites and strokes similar to those used by horses when allo-grooming. The process may involve classical conditioning, where contextual stimuli become associated with a relaxed internal state or be an endogeonous response (Feh & de Mazieres, 1993). The relaxed state that results from the continuous touching can then be associated through classical conditioning with other stimuli, such as the voice.

Certain vocal signals, as well as touch in certain parts of the horse's body, can then be used to associate new, or previously mildly aversive stimuli, with this relaxed state. In the experience of the authors, however, the reinforcing properties of the massage alone, are insufficient to change the behaviour of a highly excited or fearful horse. The difference between this method and that advocated by Miller below is that this is usually carried out on older horses, and is not described for use on young foals.

'*Imprint training*' Taken from developmental biology, imprinting is a term used to describe the learning that takes place during a particular stage in development (sensitive period) (Bateson, 1979). The term is usually reserved for use when describing the attachment of a young bird to its parents or conspecifics after hatching. In some species, a fixed period of both species-specific and maternal identification is important for optimal survival, but there is no scientific evidence that such a rapid and irreversible sensitive period takes place among mammals, even in precocial species such as the horse. The term has, however, been used to describe a specific form of training used in horses. Good 'imprint' training is said to be advantageous because it reduces the prevalence of defensive aggression during subsequent training (Miller, 1998). As imprinting is virtually indelible, time spent working with foals in this way is thought to be efficient. Miller advocates ritualized habituation of the foal to common stimuli and then sensitization to selected performance-related stimuli. In this way he encourages passivity in the horse, while preserving a degree of responsiveness. He advocates repeated 'stimulations' of each area of the foal's body including ears, ear canals, face, upper lip, mouth, tongue and nostrils, eyes, neck, thorax, saddle area, legs, feet, rump, tail, perineum and external genitalia. These are relatively invasive and care needs to be taken while this is being carried out. Indeed, there have been concerns raised about whether the foal, unable to move away, may be extremely stressed by these stimulations. After the stimulation of the body the foal is then introduced to novel stimuli and situations. The foal may be rubbed all over using a piece of crackling plastic until no panic response is apparent. However, escape at this time of high arousal can have disastrous consequences. Desensitization of the newborn foal to gunfire, loud music, flapping flags and swinging ropes can then follow. Sensitization is used to teach the foal that resistance to pressure applied to the flanks or to the head, via a head-collar, is useless. Reward in this part of the programme is given by relieving the pressure (i.e. negative reinforcement). Apart from the welfare concerns associated with intensive handling, it is also thought that imprinting may lead some foals to become overly familiar with humans and even develop a social preference for humans over equids. Thus, the human handler should ensure that the young foal has learned the appropriate responses to cues from humans. This explains why sensitization (teaching the foal to move away on cue) is so important in so-called imprinting programmes. While it is a sensible policy, in analogy to other species, to ensure that foals are habituated to the various environments, species, situations and equipment with which they are likely to come into contact later in life, the empirical evidence in support of the benefits claimed for the technique is weak (Nicol, Chapter 12).

Problems with these approaches

Many of the so-called modern training techniques, including those described above, are popular with modern horse owners, possibly because of the style in which they are marketed. Words such as 'humane', 'natural' or 'sympathetic' are extremely seductive for horse owners who wish to have a mutually satisfactory relationship with their horse (see Mills and McNicholas, Chapter 11). Some of the new or innovative trainers also appear to promote their methods as being 'mysterious' and almost 'magical', and this is something which people may also find attractive. It does appear that many horse owners like to think that they are using something new, which may potentially provide them with a 'quick fix' for many of their equestrian problems. In addition, it is possible that people prefer to have a 'recipe' style of instructions to follow, rather than learning the underlying principles and thus being able to apply these to different situations. Most of the training techniques that fall under the banner of innovative, natural or sympathetic do appear to have the following common components:

- Isolation from other individuals, to encourage the attention of the horse towards the handler.
- Use of 'non-threatening' 'approach and retreat'.
- Quiet, calm and controlled movement.
- Placing the horse in a position of dependence on the handler for 'relief' from whatever action the

handler chooses to impose (negative reinforcement). An exception to this is clicker training where learning is through positive reinforcement of operant behaviour, as the horse is rewarded for displaying the appropriate behaviour.

One of the problems with depending on recipe-based training techniques is that when things go wrong it is almost impossible for a horse owner to determine why the problem may have arisen. Since without an understanding of learning theory and the way it is applied in training, the problem cannot be analysed and it is frequently the case that the intended meaning of the signals used by the trainer have been misunderstood by the horse, especially with the 'pressure–release' methods used in contemporary and so-called innovative training techniques as described in the previous section.

Problems can also arise as these training techniques may:

- Not generalize to other contexts. For example, a horse may perform a 'join up' within the context of a round pen, but not generalize this behaviour to different locations, such as its own field. To be able to generalize the behaviour the owner must understand the importance of continuing the training in a range of different environments and situations.
- Not be humane, for example, the use of punishers to reduce an ongoing behaviour is common in many training methods. This can place the horse in a situation where it has no control, leading to motivational conflict.
- Involve ineffective rewards; patting is unlikely to be an innately pleasurable experience for a horse and poorly timed rewards are at best meaningless, and at worst encourage inappropriate behaviours.

The management of problem behaviour is discussed in Chapter 14.

Conclusion

Current 'new' techniques do not appear to be any different from traditional methods, except perhaps where methods are borrowed from other areas of training, such as clicker training. It is however important to determine whether such new techniques are effective and humane and this can only be achieved through scientific investigation. The results of these must also be carefully and critically appraised, like the claims of proponents of 'new' techniques.

References

Appleby, D. L., Bradshaw, J. W. S. & Casey, R. A. (2002). Relationship between aggressive and avoidance behaviour by dogs and their experience in the first six months of life. *Veterinary Record*, 150: 434–8.

Bartle, N. (1971). *Academic Equitation. A Training System Based on the Methods of D'Aure, Baucher and L'Hotte – General Decarpentry*. London: J. A. Allen.

Bateson, P. (1979). How do sensitive periods arise and what are they for? *Animal Behaviour*, 27: 470–86.

Bayley, L. & Maxwell, R. (1996). *Understanding Your Horse*. Newton Abbot: David and Charles.

Blackshaw, J. K., Kirk, D. & Creiger, S. E. (1983). A different approach to horse handling, based on the Jeffery method. *International Journal Studies of Animal Problems*, 4: 2.

Casey, R. A. (2002). Clinical problems associated with the intensive management of performance horses. In *The Welfare of Horses*, ed. N. Waran. Dordrecht: Kluwer Academic Publishers, pp. 19–44.

Cummings, B. J., Head, E., Ruehl, W. W. *et al.* (1996). The canine as an animal model of human ageing and dementia. *Neurobiology and Ageing*, 17: 259–68.

De Moubray, J. (1987). *The Thoroughbred Business*. London: Hamish Hamilton Ltd.

Feh, C. & deMazieres, J. (1993). Grooming at a preferred site reduces heart rate in horses. *Animal Behaviour*, 46: 1191–4.

Heird, J. C., Lokey, C. E. & Logan, D. C. (1986). Repeatability and comparison of two maze tests to measure learning ability in horses. *Applied Animal Behaviour Science*, 16: 103–19.

Houpt, K. A., Parsons, M. S. & Hintz, H. F. (1982). Learning ability of orphan foals, of normal foals and of their mothers. *Journal of Animal Science*, 55: 1027–31.

Inderwick, S. (1977). *Lungeing the Horse and Rider*. London: David & Charles.

Kevil, M. (2003). *Starting Colts: Catching / Sacking Out / Driving / First Ride / First 30 Days / Loading*. Guilford, USA: Lyons Press.

Klimke, R. (1985). *Basic Training of the Young Horse*. London: J. A. Allen.

Lindberg, A. C., Kelland, A. & Nicol, C. J. (1999). Effects of observational learning on acquisition of an operant response in horses. *Applied Animal Behaviour Science*, 61: 187–99.

Loch, S. (2000) *Dressage in Lightness – Speaking the Horse's Language*. London: J. A Allen.

Loriston-Clarke, J. (1995). *The Young Horse: Breaking and Training*. London: David & Charles.

MacSwiney of Mashanaglass, Marquis (1987). *Training from the Ground*. London: J. A. Allen.

Mader, D. R. & Price, G. O. (1980). Discrimination learning in horses: effects of breed, age and social dominance. *Journal of Animal Science*, 50: 962–5.

Marinier, S. L. & Alexander, A. J. (1994). The use of a maze in testing learning and memory in horses. *Applied Animal Behaviour Science*, 39, 177–82.

McCall, C. A. & Burgin, S. E. (2002). Equine utilization of secondary reinforcement during response extinction and acquisition. *Applied Animal Behaviour Science*, 78, 253–62.

Miller, R. M. (1998). Imprint training the newborn foal. *Equine Veterinary Journal – Equine Clinical Behaviour* 27 (Suppl.): 63–64.

Miller, R. W. (1975). *Western Horse Behaviour and Training: Understanding, Breaking and Finishing Your Horse*. New York: Doubleday Books.

Nicol, C. J. (1999). Understanding equine stereotypies. *Equine Veterinary Journal*, 28 (Suppl.): 20–25.

Parelli, P. (1993). *Natural Horse-Man-Ship*. Western Horseman, Colorado Springs: Globe Pequot Press.

Podhajsky, A. (1967). *The Complete Training of Horse and Rider* (translated by V. D. S Williams & E. Podhajsky). London: Harrap Ltd.

Rees, L. (1984). *The Horse's Mind*. London: Stanley Paul.

Richardson, C. (1998). *The Horse Breakers*. London: J. A. Allen.

Roberts, M. (1997). *The Man Who Listens to Horses*. London: Arrow Books.

Rose, M. (1977). *Training Your Own Horse*. London: Harrap Ltd.

Scott, J. P. & Marston, M. V. (1950). Social facilitation and allelomimetic behaviour in dogs: II The effects of unfamiliarity. *Behaviour*, 5: 2.

Stanier, S. (1995). The art of long reining. London: J. A. Allen.

Strickland, C. (1998). *The Basics of Western Riding*. Pownal, Vermont: Storey Publishing.

Tellington-Jones, L. (1985). *The Tellington-Jones Equine Awareness Method*. Millwood, NY: Breakthrough Publications.

Visser, E. K. (2002). *Horsonality: a Study on the Personality of the Horse*. The Netherlands: Ponsen en Looijen.

Waran, N., McGreevy, P. & Casey, R. A. (2002). Training methods and horse welfare. In *The Welfare of Horses*, ed. N. Waran. Dordrecht: Kluwer Academic Publishers, pp. 151–80.

Williams, J. L. (2002). Clicker training in horses: operant condioners and secondary reinforcers in horse training. PhD Thesis Texas A&M University.

Wolff, A. & Hausberger, M. (1996). Learning and memorisation of two different tasks in horses: the effects of age, sex and sire. *Applied Animal Behaviour Science*, 46: 137–43.

14

Behavioural problems with the ridden horse

Paul McGreevy and Andrew McLean

Introduction

Humans influence the behaviour of horses under saddle with stimuli from their hands on the reins and their legs on the sides (ventrolateral thoracic region) of the horse as well as their seat's position and movement. Horse riding at its most humane relies on subtle interactions between horses and humans and the correct application of negative reinforcement followed by classical conditioning of stimuli (such as those coming from the seat). It is surprising therefore that so little scientific endeavour has been directed towards the effects of aversive stimuli on horses and the ways in which horses respond to stimuli of human origin (e.g. their leg pressures).

In mounting a horse, a rider relinquishes direct control of his or her own connections with the earth, and the horse subsequently loses its own autonomy to a greater or lesser extent. Most naïve horses respond to humans as they would to any predator. By moving away posturally or bodily, they behave in ways that help them avoid pressure: physical and/or

psychological. Trainers work with these basic evasive responses which can be modified successfully to produce highly responsive equine athletes or unsuccessfully to produce problem horses (see Waran and Casey, Chapter 13). Broadly, pressure may be appropriate (as part of subtle equestrian technique), while pressure, discomfort and pain may be inappropriate or unintentional (e.g. through ignorance or omission). It should be remembered that the variable pressures provided by the rider's reins on the horse's mouth and the rider's legs on the horse's body are integral to the negative reinforcement that *trains* the horse, through their removal, to behave in the various ways that are *targeted* by the trainer.

There is no evidence that horses (ridden or otherwise) learn any differently to other species (Mills, 1998a) although their capacities are related to their ecology (see Nicol, Chapter 12). Horses are highly motivated to remove physical and psychological pressure, so negative reinforcement is effective in training. Excellent equestrian technique depends on correct position, application of shaping (to reinforce successive approximations of the desired outcome) and reinforcements. The construct of a 'will-to-please' is virtually redundant and should be replaced by the 'will-to-get-relief from pressure' so that a contiguous link between pressure-release and trained responses emerges. Similarly, the concept of *desirable* responses is irrelevant to horses themselves. For horses under saddle, a perfectly round 10-m circle is as worthwhile as an ovoid as long as they get sufficient relief from the forward and turning pressures that prompted them to move in a track that ultimately formed that circular shape. It is important to appreciate that horses are mostly reinforced through the process of negative reinforcement when they learn to produce responses that are either desirable or undesirable responses to their riders. Indeed, the same is true of horses being led in-hand. During training in-hand, the handler applies pressure on the lead-rein in an anterior direction for 'go', and in a posterior direction toward the horse's neck for 'stop'. Reinforcement takes the form of pressure-release (see Waran and Casey, Chapter 13).

Negative reinforcement is used on ridden animals (namely horses, elephants and camels) for a number of reasons of which safety is perhaps the most important. Because we are on-board the animal we need the comfort of knowing that we can stop and direct it. This is why horses wear bits (a potential source of tremendous discomfort) in their mouths (a site of great sensitivity).

Past masters (Fillis, 1902; Watjen, 1958; Herbermann, 1980; de la Guérinière, 1992; Nelson, 1992) attempted to articulate their great skills but did so within an anthropomorphic framework (McLean & McGreevy, 2004). Unfortunately they showed little agreement on the details of their anthropomorphic approach. Having said that, they did acknowledge the importance of some of the guiding principles of behaviour modification but without an exact appreciation of the core processes that have emerged from the past century of empirical learning studies (these are reviewed by Nicol, Chapter 12).

We begin with a consideration of the sorts of responses that riders find undesirable. Then we examine the extent to which some of these are caused by certain humans while others are intrinsic to certain horses. Beyond this we describe the ways in which many experienced riders communicate with their mounts unintentionally and manage their horses

unhelpfully and how novice riders often confuse their horses. We also highlight ways in which undesirable reinforcement occurs as a result of particular attributes of certain horses. Since horse-related causes of undesirable behaviour can be not only innate but also acquired, we acknowledge the role of human activity in this element of our framework.

For ridden horses, pressure, discomfort and pain come in many guises. It is therefore appropriate to consider the origins of these stimuli individually and indicate the basic approaches to therapy that can be tailored for each affected horse. However, it is not our intention to over-simplify the forces that prompt and reinforce undesirable responses under saddle. Trainers should recognize that more than one mechanism may operate contemporaneously in a single animal. If schooling the horse involves the development of new responses then reinforcement is inarguably the process and in the absence of attractive stimuli (such as food, sex, water and other primary reinforcers), negative reinforcement must be at work. If, on the other hand, schooling involves the *maintenance* of behaviour then the process may incorporate classical conditioning and the application of mild pressure cues.

Human-related causes of undesirable equine behaviours can be largely attributed to the current lack of science in equitation. It is worth noting that there is evidence of confusion about terminology among horsefolk (Mills, 1998b). This presumably compounds a lack of rigour in identifying problem behaviours. The performance of the horse under-saddle and the consequent development of riding instruction tends to focus on outcomes rather than mechanisms. Additionally, riding manuals have historically bypassed the central tenets of modern learning theory. Since the ideals of equestrian technique combine art and science, students of equitation encounter a few measurable variables such as rhythm, tempo and outline alongside many more ethereal ones such as impulsion and harmony (McLean & McGreevy, 2004). This unbalanced mixture and the dearth of mechanistic substance frustrate attempts to express equestrian technique in empirical terms (Roberts, 1992) and account for some of the confusion and conflict that arise in many human–horse dyads.

Horse-related causes of 'misbehaviour' are cited more than rider error but this is probably an aberration. Horses that have been subjected to inconsistent signals or to the lack of release of pressure, often acquire the reputation of being difficult. They are sold to homes where more often than not the force used in training is escalated. This contributes to disturbing slaughter statistics. For example, in a recent French study of more than 3000 non-racing horses, some 66.4% died aged between two and seven years (Ödberg & Bouissou, 1999). Similarly Von Butler and Armbruster (1984) found that the average age at slaughter was 8.5 years. Unlike data from the racing industry (Jeffcott *et al.*, 1982; Bailey, 1998), this wastage was not attributed to orthopaedic or respiratory disease so much as to inappropriate behaviour.

Wastage would be reduced if riders and trainers were educated, while horses were properly selected for both the task intended and the rider's ability (see Hausberger and Richard-Yris, Chapter 3). Some horses are disinclined to respond to certain signals and pressures while others are incapable of responding. Therefore, it is necessary to explore the reasons for the distinction between those horses that will not and those that cannot offer desirable responses. Physical causes of undesirable behavioural responses should always be ruled out before any purely behavioural therapy is adopted. For example, it has been proposed that undiagnosed pelvic or vertebral lesions could be an important contributor to poor performance and lameness in athletic horses (Haussler *et al.*, 1999). In the absence of analgesics, it would be inhumane to persist in training these subjects. While some disorders such as 'cold-back syndrome' and the relationship between dental problems and behaviour under saddle have yet to be thoroughly explored, the treatment of many physical disorders that lead to poor performance is considered in detail in the veterinary literature.

Problems with the ridden horse

When it is acknowledged that riders constantly target locomotory responses in the horse, it is not surprising that there is considerable room for error. In the process of training, trainers hijack the natural forward and sideways movements of the horse and place them under stimulus control. For example, shaping of lateral movement in response to pressure of the rider's leg applied to one side of the horse is the first phase in developing more complex lateral movements. When training goes wrong, the horse offers changes in speed or direction that are not desired by the rider. These responses are known as resistances and, if repeated often enough, they become learned. They manifest as 'laziness', stalling to a greater or lesser extent, drifting sideways and adopting a crooked longitudinal axis or developing 'excitability'.

Because horses are adept at learning from trial and error, they may experiment with numerous hard-wired locomotory responses during times of behavioural conflict. Therefore beyond mere resistance, they may try much more dangerous behaviours such as rearing (suddenly standing on the hindlegs), bucking (lowering the head and kicking out with the hindlegs) and shying (sudden swerving of the forequarters), known as evasions. It is interesting that the terms often used to describe training dysfunctions are used in a way that implies malicious intent on the part of the horse. This is because the origins of horse training preceded any scientific evaluation of equine mentality. The same applies to the traditional term for stereotypies – vices (McGreevy & Nicol, 1998).

Evasions in the ridden horse are generally behaviours accompanied by raised flight responses and stimulation of the hypothalamic–pituitary–adrenal (HPA) axis. This explains two characteristic features of such behaviours. First, they are exhibited with added velocity, rendering them more dangerous and difficult to manage. Second, they are less subject to extinction (Le Doux, 1994) than resistances and may escalate to include other hard-wired agonistic antipredator responses such as rearing and bucking.

The development of a rearing response is generally progressive and frequently arises when horses are trapped between the rider's 'go' and 'stop' signals. Unable to resolve the pain, they may resort to shying, spinning away, leaping, or bucking, depending on their genetic make-up. Shying may evolve progressively to rearing if a horse receives sufficient practice, motivation (pressures are not resolved) and reward in the form of dislodging the rider. When horses rear, riders have difficulty maintaining their seats. If inexperienced or unprepared, they may inadvertently attempt to balance by tightening the reins and run the risk of pulling the horse backwards on to themselves. Meanwhile, bucking may escalate from simple elevation of both hindlegs, to lowering of the head between the forelegs and arching of the back while plunging, kicking out and spinning. Riders are sometimes injured by this behaviour because they can be thrown to the ground and sometimes also kicked during the dislodging process. Leaping and shying are escalated forms of resistance involving learned forward and turning responses. Both can dislodge riders.

Even though their external causal factors can involve novel or aversive stimuli, all of these behaviours are powerfully reinforced by relief from rider stimulation. In rearing, for example, riders inadvertently remove the driving signals coming from their legs, while their hands also momentarily relinquish control via the reins. It seems that such behaviours are rapidly reinforced in few trials, and through very brief moments of freedom from control. Being able to perceive and capitalize on such moments of freedom must be highly adaptive and reinforcing for prey species, such as the equids. Sometimes the resistance to go forward escalates to the point where the horse jibs (refuses to go forward) then tightens its body, leaps upwards then flips to land on its back and also sometimes the helpless rider. This is possibly the most dangerous of behaviours under saddle because riders are often unable to predict the behaviour and have little time to remove themselves from the crushing descent of the horse. It is important to differentiate flipping from rearing. Horses that have learned to flip are best removed from the ridden situation.

Horses can demonstrate various undesirable behaviours under saddle, often in combination. For the sake of simplicity, inappropriate responses are considered separately.

Inappropriate obstacle avoidance

Thorough training of the young horse involves ensuring that the forward response is as unconditional as possible so that the horse goes forward to most places without hesitation. This allows it to be directed towards places and obstacles that may have aversive qualities. However, horses vary in their perception of aversiveness. So some will move towards completely unfamiliar stimuli without hesitation, whilst others will baulk (arrest their forward movement). These variations are seen in young foals during lead training and most likely have a genetic basis that is shaped by the experience of baulking. Furthermore, horses of all ages that experience conflicts in the training of the forward response, may demonstrate their insufficient forward responses with hesitations, baulking or shying.

Horses may also inadvertently develop baulking or shying in response to surprising environmental events, such as a dog suddenly rushing out of undergrowth. During such incidents, the accidental expression of baulking and shying may be incorporated into the horse's behavioural repertoire because of the reinforcement of the loss of rider control. The reoccurrence of this behaviour is often context-specific at first but may generalize to other novel sites with continued reinforcement (see Nicol, Chapter 12).

When horses avoid places through baulking or shying, their inappropriate behaviour is reinforced by their turning away from the aversive stimuli that may be current or remembered. So, *facing away* is a common feature of such behaviours that are reinforced still further with freedom if riders allow horses to turn their forelegs away. However, the greatest reinforcement arises when horses not only turn away but also actually run away and escape from the aversive stimuli for one or more strides. Shying behaviour therefore seems to escalate through the progressive involvement of three separate body parts – the head (facing away), the forequarters (turning away) and the hindquarters (running away). Shying behaviour is most rapidly learned and incorporated into the horse's behavioural repertoire when it involves complete escape. This reinforcement is compounded by the incremental loss of control by the rider, as the horse develops worsening shying behaviour. Additionally, riders who predict the responses may inadvertently apply pressure that may increase the horses' motivation to flee.

When baulking and shying become incorporated into the behaviour repertoire of the ridden horse, they may escalate to rearing. When rearing develops, horses generally maintain the turning away component during the rear so that on landing they are not facing in the original and intended direction. Baulking, shying and rearing frequently lose context-specificity and become randomly practised. So, baulking and shying may occur anywhere and without any obvious eliciting stimulus. As horses practise this behaviour, they tend to lose some qualities of their learned 'go' and 'stop' responses. Therefore, baulking and shying can escalate to become randomly and unpredictably practised behaviours when reinforced maximally by facing away, turning away and running away.

During jump training horses learn to avoid obstacles in the jumping course through stalling at some moment in the take-off (usually during the last complete stride or the take-off stride). This stalling may be pre-empted by habitual losses of rhythm and tempo or directional line. When horses learn to stall in the jumping effort, outright refusal is generally the next step. This is why proficient jumping coaches insist on riders maintaining rhythm and line.

Hyper-reactivity responses

There is a significant variation in the reactivity of horses across breeds and within breeds (Visser *et al.*, 2001, 2002; Hausberger and Richard-Yris, Chapter 3). In general, draught horses show less reactivity than the riding breeds such as Arabians and Thoroughbreds. The less reactive breeds are known colloquially as 'cold-bloods' while the more reactive breeds are known as 'hot-bloods'. Crosses between the two, known as 'warm-bloods' currently predominate in elite dressage and jumping competitions. The less reactive breeds were probably selectively bred for lower reactivity to enable them to haul a plough or carriage without showing fear responses that would have endangered humans and equipment. On the other hand, the more reactive breeds were probably selectively bred for hair-trigger HPA responses that favour racing success. Thus we find that racing Quarter Horses are more reactive than non-racing Quarter Horses or so-called 'cutting' strains. Similarly among Thoroughbreds we find that lines of sprinters tend to be more reactive than stayers. This reflects the need for explosive activity in sprinters, and the greater requirement for trainability in stayers that must learn responses to jockeys' rein and leg signals. These characteristics account for much of the variability in reactivity among riding horses.

Hyper-reactive states are the expression of horses' futile attempts to flee stressful situations and may be precipitated by confusing signals offered by riders. As with all trial and error learning the horse may practise random responses that are not under stimulus control. Early manifestations of hyper-reactive responses in ridden horses mostly occur in two forms. First, we see extreme flexion and extension of the neck with an associated raising or lowering of the head. This represents hyper-reactivity to bit pressure that prompts the rider to lose direct contact with the horse's mouth resulting in some loss of control. Second, there may be a rapid escalation of hyper-reactivity from the contact

Figure 14.1. Typical poise of a horse showing hyper-reactivity to the bit. Note the neck and raised head.

of the rider's legs so that the rider habitually maintains no leg contact at all with the horse.

Horses may develop increasing hyper-reactivity to riders' legs to the extent that the slightest touch may precipitate hyper-reactive responses. These are characterized by the shortening and quickening of the horse's forward steps, contracting its back and increasing muscular tone. The head and neck are generally elevated dorsally while the tail is flexed ventrally but is swished regularly (Figure 14.1). These features of postural tonus may reflect a state of anxiety and increased activity of the HPA axis. Because such a flighty horse tends to lose the clarity of gaits (e.g. by jogging, trotting with a very short length of stride) and the distinction between them, it is unlikely to perform well in any test of obedience.

With repetition, this state may become a permanent feature of a horse's ridden behaviour. When the hyper-reactive states become chronic, the rider increasingly loses control as desired learned responses progressively diminish. This deterioration progresses and may result in baulking, shying, rearing, bucking, kicking-out and bolting as the horse's responses become more random with decreasing stimulus control. Riders often appear to inadvertently reinforce hyper-reactivity by:

- releasing their leg pressure when the horse gets tense, therefore rewarding the tense behaviour and thus increasing the likelihood that it will be repeated;
- failing to prevent the expression of the tension by stopping/slowing the horse.

Agonistic responses to conflict

During behavioural conflicts, for example, where fleeing the stressful situation is not possible, increasing levels of agonistic behaviour may be expressed. Horses sometimes stop, turn their heads and attempt to bite the legs of riders, or kick out with the hind legs when riders signal with their legs. Rearing behaviour, particularly in stallions, is sometimes accompanied with pawing and paddling the forelegs in the air. Bucking is often accompanied by lashing out with the hind legs.

Evidence of pain

When horses in training show signs of pain or discomfort after habituation to the girth and saddle, the causal agent is not always easily identified. Bridle lameness is a well-known phenomenon involving lameness as a result of the horse being unable to free itself of bit pressure because one or both of the reins is too tight. This constriction results in a crooked longitudinal axis of the body and irregular rhythm that can give the appearance of lameness.

In behaviour modification, pain can generally be assumed to be the prime causal agent of aberrant behaviour patterns when the resolution of such behaviour patterns by behavioural means over a period of time shows either no change or an erratic resolution pattern. It is common, however, for long-standing or strongly reinforced problems especially those associated with increased HPA activity to show slower resolution than short-lived, less reinforced patterns of behaviour.

The behavioural signs of pain and irritation may include hyper-reactivity, hypersensitivity, muscular tension around the eyes and muzzle, and undesirable responses such as baulking, shying, rearing and bucking. However, these same behaviours are frequently symptomatic of training conflicts. The use of analgesics may be helpful in unpicking the role of pain or discomfort in resistance.

Evidence of poor physical ability

Because the commercial value of a horse is often related to its ability to compete, many owners seek to push their horses to the limits of their performance. Unwillingness to accept that horses have reached these limits may prompt owners and trainers to escalate the use of force to train unattainable responses. Coercion is ineffective and inhumane when performances are perceived as unsatisfactory due to ignorance of the true limitations of the horse's ability.

Physical ability to perform particular behavioural tasks is a serious issue with performance horses. Inadequate locomotory responses may reflect pathologies, most notably fatigue, as well as a variety of innate impediments. When performance horses are continually pressed to perform movements that they simply cannot physically express, conflict behaviours typically emerge. So trainers must be clear about the factors that limit a horse's physical ability to perform and distinguish these from any disinclination to comply. The questions to be asked include: Can the horse perform the required movement? Can the horse gallop at the required speed? Can the horse jump the required height? If the horse is unable to physically perform a particular manoeuvre, it is unlikely to try it during operant training.

In dressage, many complex movements must be learned. At the higher levels (Federation Equestre Internationale test levels), these movements demand a shift in the horse's centre of gravity towards its hindquarters. The carriage of more weight by the hindquarters of the horse reflects its so-called carrying and collecting capabilities. For an individual horse, such abilities are facilitated by features of its morphological components (i.e. conformation). In particular, the horse should be taller at the withers than at the croup. The same conformation ratio favours show-jumpers.

The origin of the wither-high trait lies in draught ancestries that required horses to be able to transfer power in an ascending plane from the hindquarters to whatever was being pulled be it a plough or cart, via the collar or breastplate. A reversed wither:croup ratio is required for speed. Sprinting Thoroughbreds and Quarter Horses are typically croup high (as are zebras) because this conformation results in power

Figure 14.2. (a) The high croup of a Thoroughbred arises as a result of selection for speed. (b) The high withers of this warm-blood relate to selection for power in the shoulders. (c) The difference between the two is emphasized by the superimposition of their outlines (dotted line, Thoroughbred).

being propelled along the horizontal plane rather than in an ascending plane (Figure 14.2).

While many draught breeds are known for their elevated knee action, tripping, toe-dragging, stumbling and clumsiness are also seen as features of the heavier types that are said to be predisposed to be 'lazy'. However, this can also be seen in other breeds, such as Thoroughbreds, when they have habituated to rein pressure and are said to 'lean on their forehands'. Meanwhile, problems can emerge from a congenital propensity to disunite at the canter, as has been seen in some breeds such as Arabians where selection processes failed to evaluate the canter, for example when breeding stock are selected on performance in-hand in the show ring only at walk and trot. Sometimes, however, some components of the natural gaits can be altered through training. This is true with the trot, but less so for the walk and canter. Trainers often find that training a more elevated trot as in the passage, the suspension of the working and collected trot increases. Forging, clicking and over-reaching may result from inappropriate farriery, incorrect equitation and conformation problems. Indeed poor conformation may make many advanced responses difficult to train.

Evidence of learned helplessness

It is assumed by many trainers that horses with 'hard mouths' and 'dead sides' have accumulated scar tissue in the mouth and body respectively that prevents them perceiving the signals from the rider's reins and legs. Whilst the possible existence of scar tissue is not disputed, learning is the more plausible explanation for sluggish or inadequate responses to the rider's signals. The retraining process that trains horses to respond to subtle signals once more is not simple because the horse is unlikely to try a new response to pressure, particularly the ranges of pressures that it has experienced in the past, which it has learned to ignore. It is interesting to speculate whether learned helplessness (Lieberman, 1993) is implicated here. Established learned helplessness may compromise horse welfare since an animal in this state has undergone a critical loss of control of its environment. This means that the animal is essentially apathetic (Webster, 1994) and may appear dull and listless. Observers may assume that the horse is tolerant of pressures from the rider while actually finding them highly aversive.

If learned helplessness is occurring in the ridden horse, a number of precipitating factors should be considered. Horses with a lack of responsiveness of the sides are often the product of incorrect spur use, simply because riders have a greater potential to induce pain when wearing spurs. This can arise from leg signals of any form that are inconsistent or beyond the learning abilities of the horse. These horses may also exhibit signs of conflict with other behaviours, such as bucking and kicking out. Clinical experience suggests that, in its respective developmental stage, loss of responsiveness in the mouth results in more expressions of conflict (such as rearing, shying and hyper-reactivity) than such loss to the sides. This is most likely a consequence of the greater pain sensitivity of the horse's mouth compared to its sides.

Causes of undesirable responses in the ridden horse

Two broad sections are offered here to allow the reader to consider undesirable behavioural responses caused directly by humans, as distinct from those attributable more to specific horses than riders. Because humans are so directly involved in the first group, it is likely that interventions such as behaviour therapy are more likely to work in these cases. However, this too is an area remarkably lacking in empirical research.

Human causes of undesirable behavioural responses
Poor riding technique There is extensive literature available concerning correct riding technique (The British Horse Society, 1950; Watjen, 1958; Podhajsky, 1966). The importance of rider position cannot be over-emphasized. Unless the rider has a balanced seat, attempts to maintain or resume a centred position will have a significant effect on the horse's movement and balance. So, without a properly balanced rider, the horse is unable to perform optimally. The best designed saddles help to position all riders (novice and others) correctly.

Beyond impaired performance, we should consider the effects of a poor seat on training that can be even more significant in the long term. A rider's lack of balance allows the horse to practise behaviour patterns that are contrary to the intended signals issued by the rider's reins and legs (Figure 14.3). The more opportunities the horse has to practise behaviours that are not under stimulus control, the greater the horse perceives its freedom from the control of the rider. Thus an escalating series of random behaviours may be tried and possibly reinforced.

Figure 14.3. Riders with a poor seat are predisposed to use the reins for balance as illustrated here.

Poor application of learning theory While traditional equestrian dogma recognizes the importance of position, it is also universally guilty of failing to accept the science of equine learning and incorporate learning theory into its coaching and teaching strategies from pony club through to the highest levels of dressage. This is not to say that there are not excellent trainers available, but rather to say that these trainers mostly are unaware of the theoretical background to their practices. Thus, much of the behaviour of those that are referred to as 'problem' horses is a result of sub-optimal human–horse interactions.

Whereas mild signals or cues are easily ignored, pressure or pain provided by the rider's hands on the reins or by the legs (and spurs and whip) is not easily ignored, and the horse is unable to flee or evade the aversive stimuli. Breakdown in human–horse interactions results from stimulus–response inconsistencies during the installation of signals via negative reinforcement. Negative reinforcement is probably unavoidable in current performance horse training because the horse is ultimately controlled by approximations of various pressures provided by the rider and refined by discriminative training (see Nicol, Chapter 12). The problem, however, is not with negative reinforcement *per se*, but with misunderstand-

ings of its correct application. Trainers should learn its exact mechanism and optimal use within the context of their training endeavours.

Many horse trainers fail to comprehend adequately that the critical aspect of negative reinforcement is the removal of pressure, and that this removal reinforces any immediately preceding behaviour. It is not uncommon to see haphazard use of rein and leg pressure and its removal and, in some cases, failure to remove pressure at all. Incorrect removal of pressures commonly produces 'hard mouthed' and 'lazy' horses and horses that generally show stiffness and a lack of suppleness. In the chronic situation this may lead to learned helplessness and serious conflict behaviours such as rearing and bucking. Meanwhile the removal of the leg pressures from the hyper-reactive horse can have the effect of increasing hyper-reactivity through the same mechanism of negative reinforcement. Naturally, for safety reasons, it is frequently prudent to remove pressures during hyper-reactivity. However, it is important to remember that to do so can reinforce hyper-reactivity. Skilled equestrian 'problem solvers' recognize the importance of eliciting attempts at appropriate behaviours in such horses in response to the problematic signals. It is the removal of the rider's pressures, as well as the escape from any current

Figure 14.4. Often attempts are made to compensate for a poor understanding of learning theory with additional riding aids such as a double bridle and spurs to force compliance.

aversive environmental features, that reinforce baulking and shying.

Poor applications of learning theory manifest in horse training as: inconsistent responses resulting from specific stimuli; too many stimuli issued simultaneously and variously for a single response; overshadowing of stimuli; contradictory signals issued and maintained concurrently such as 'go' and 'stop' signals; delays in reinforcement; no shaping of responses. In performance sports, the most common cause of conflict is the constant maintenance by the rider of rein pressure. These sports must clearly define and adhere to the concepts of 'lightness' and 'self-carriage' espoused, but perhaps not always clearly or uniformly defined, in traditional training literature. This must occur to redress the welfare issues caused by such techniques (Figure 14.4).

When problems are not resolved, the welfare of the horse is jeopardized and stress responses become chronic and manifest as raised circulating corticosteroid levels that can have catabolic consequences (Wiepkema, 1987). In non-equine species chronic stress is associated with repetition and ritualization of primary conflict behaviours, development of stereotypies and injurious behaviours such as self-mutilation and increased aggression

(Stolba *et al.*, 1983; Wiepkema, 1987). Commonly in horse-training contexts, the inability of animals to obtain freedom from pain or pressure is ultimately responsible for conflict behaviours (McLean, 2003). Because links between stress and gastric ulcers have been proposed in foals (Furr *et al.*, 1992; Borrow, 1993) and established in other species (Henrotte *et al.*, 1995; Tarara *et al.*, 1995), strategies that help to reduce the emotional impact of conflict are likely to be of benefit to gastrointestinal health.

Better education of trainers will enable them to improve ridden horse welfare by identifying, avoiding and reducing stress in training. As a consequence, the industry will benefit from lower wastage rates and optimised training efficiency.

Horse-related causes of undesirable behavioural responses

Riders failing to get desired responses under saddle rarely choose to find fault in themselves. Therefore the opportunity to transfer blame has considerable appeal. Unfortunately this can sometimes obscure the core issues. For example, instead of rectifying the faults in his or her technique, a rider who habitually sits to one side of the saddle may experiment with different hardware to straighten the horse. Therefore, a series of gadgets (see later) may be used while the effect of sloppy positioning in the saddle on the horse's lumbar musculature is ignored.

When equestrian technique is excellent but riding horses perform poorly, one should consider causes that are not directly related to the rider or trainer. These include innate or acquired physical anomalies that make horses unsuitable for the work asked of them. Since many of these problems can ultimately be attributed to human intervention or omission in terms of poor stable management, the horse is rarely to blame.

Pain The application of pressure in sensitive areas is an implicit feature of equestrian technique. While humans routinely inflict pain and discomfort on ridden horses intentionally with rein and leg pressure via bits and spurs, the aim of riders should be to shrink these pressures to very light ones through contiguity of the light signal followed by the stronger signal if no response occurs. Spurs are frequently used incorrectly in horse training. They should not be used as stronger forward-driving weapons, but added

Table 14.1. *Selected examples of factors that contribute to unintentional discomfort or pain in ridden horses*

Problem area	Example
Exercise surfaces	Persistent exposure to exercise surfaces with insufficient or excessive compliance may contribute to unnecessary pain in weight-bearing joints and soft tissues of the musculoskeletal system, respectively
Hoof care	Excessive hoof trimming can contribute to solar bruising of the foot that may make horses reluctant to work, especially on resilient surfaces. Meanwhile insufficiently balanced trimming of the toe relative to the heel may predispose horses to navicular disease and chronic lameness
Malnutrition	Overfeeding in combination with under-exercise and especially confinement can increase the risk of rhabdomyolysis, a painful muscular condition that can render horses permanently unfit for ridden work. Laminitis is another painful outcome of incorrect and excessive feeding that can cause acute and chronic pain respectively and culminate in structural changes including rotation of the pedal bone
Dental anomalies	Since the upper dental arcade of horses is larger than the lower, sharp spurs and hooks can easily form on the outer surface of the molar and premolar teeth. These irregularities may make movements of the mandible uncomfortable and therefore make the horse less likely to relax its masseter muscles and 'accept' the bit. Dental disorders in the horse are dealt with comprehensively elsewhere (Pence, 2002)
Cranial pathologies	While trigeminal pain may cause typical 'head-shaking syndrome' with frenzied flexion and extension of the poll, spinose ear mites may on rare occasions cause lateral movements of the head. Either of these can render the horse unpleasant to ride, unlikely to perform well (especially in dressage competition) and in many cases simply unrideable (see Mills, Chapter 15)
Tack	Because saddles may be worn-out, poorly fitted or may have not been adjusted to accommodate changes in a horse's body condition, they may lie on the dorsal processes of the vertebral column or can pinch the dorsal lumbar musculature. This can make horses reluctant to tolerate being girthed, being mounted or making transitions into gaits that become associated with pain

later in the horse's training when it has learned to go forward from leg pressures. In traditional training, spurs should be used only to choose very specific and distinct locations on the horse's sides so as to separate the signals various movements, particularly lateral movements. Spurs as driving devices tend to make the horse retract its ribcage and tense its body. Unfortunately for the ridden horse, the shrinking of the stronger signals to very light ones is the exception rather than the rule.

Equestrian experts agree that the ideal contact (the maintained pressure of the reins and the rider's legs) is neutral in its effect and therefore should be as light as possible to avoid habituation. Unfortunately many horses endure much stronger mouth pressure during downward transitions, and many riders have never experienced the feel of downward transitions with little change in rein pressure. When riders use the reins to force the horse to go 'on the bit', they begin a process of habituation of the horse's mouth to pressure. This increases the tension in the rein that is maintained during regular contact. This is an important reason for many riders' erroneous perception that rein contact should be heavier than is necessary. Thus the horse

is forced to tolerate higher pressure to some extent or else exhibit conflict behaviours. Beyond inappropriate contact, humans also cause unintentional pain and discomfort as a result of poor management or oversight, as shown in Table 14.1.

Physical inability The demands imposed on riding horses, even at the highest level of dressage, are not beyond their capabilities because dressage movements are derived from natural movements for the species. However, physical limitations, for example due to conformation, may affect the *quality* of the training outcome. Therefore physical inabilities (as described in Table 14.2) mean that some horses will never be able to excel in some competitive equestrian domains. Rather than being misinterpreted as disobedience or the absence of a 'will-to-please', their inability should be acknowledged and, if possible and appropriate, accommodated.

Mechanical approaches

Trainers of horses that rapidly enter conflict and show resultant reluctance to comply, often experiment with increased pressure to overcome resistance. In correct

Table 14.2. *Examples of some of the ways in which horses are predisposed to poor performance*

Problem area	Example
Conformation	Inadequate height or hindquarter strength can compromise a horse's ability to jump or perform dressage manoeuvres, respectively
Gait anomalies	When asked to pace, some Standardbreds are disadvantaged because of a familial tendency to trot
Physiology	Some breeds are poorly equipped to lose body heat while others have inherently low thresholds for dehydration and fatigue. These features account for the predominance of a small number of breeds, notably Arabians, in endurance trials and the futility of conditioning many other breeds for this sort of ridden work
Age	Immature animals have rarely had the time and opportunity to learn the complex responses required for dressage and show-jumping competition. Conversely, athletic performance decreases with age, a reality that contributes to the rapid turnover of racing animals
Anomalies in perception	Partial blindness can lead to generalized wariness and has led to the adage that a totally blind horse is safer than one that is partially sighted

equestrian technique the horse should exhibit longitudinal and lateral flexion itself. The placement and relaxation of the neck is the *proof* of correct training of locomotory responses, not the other way around. Because of the perceived need to force horses into an 'outline' and bring them 'on to the bit', the tendency is to use stronger bits as the first approach to solving the problem. Mechanical restraints and stimulants may be used to magnify the pressure that riders can apply. This accounts for the multiplicity of bits on the market. As with any device, it is true to say that if there are many of them on the market, the chances are that none of them provides all the answers.

Trainers often try different bits that apply pressure to different parts of the mouth or to the same area of the mouth but with greater severity. Bit designers focus largely on ways to increase leverage. It is unfortunate that this often takes place without any attempt to remedy the rider's use of the reins. Riders who are having problems getting horses to respond to milder bits can readily acquire saw-chain bits (so-called mule bits and correction bits) even though these can and sometimes do sever the tongue (Rollin, 2000). Creeping desensitization is the likely outcome if riders continue to use excessive or persistent pressure on the reins or if they use a constant pressure with evermore severe bits.

It is appropriate for riders to ask why their horses can no longer be stopped or steered by light rein pressure. Horses are innately sensitive to foreign objects in their mouths and demonstrate this by deftly ejecting inedible particles while grazing. Significantly, horses that have recently been correctly 'broken-in' or 'foundation trained' do not have 'hard mouths'. Therefore instead of describing horses as having hard mouths – a term which implies an inherent tendency – we advocate that they should be described as having habituated to rein pressure.

Control of the head is not intended to facilitate steering and deceleration alone. It is also bound up in training the horse to flex its cervical vertebrae to adopt the desired shape or 'outline'. This is often achieved through pressuring the bit, but *disallowing* the learned response of slowing, thus creating conflict. Pressure may also be applied to the other parts of the head with training devices such as curb chains, gags, hackamores, bosals and draw reins. Unfortunately the term for such 'training devices' is often a misnomer since riders regularly develop a reliance on these extra pulleys, rather than using them solely for re-training as the name might imply. Similarly, riders may use martingales and tie-downs to apply pressure to prevent evasive raising of the head, rather than recognising raising of the head as a symptom of a training confusion. Cervical flexing should be regarded as part of the shaping process of the 'stop' response. Again, these gadgets are rarely dispensed with. Ideally, these should be considered no more than short-cut temporary solutions. Those who resort to such gadgets have usually failed to train their horses to have light accelerating or decelerating responses. So if the horse raises its neck and head then it needs training along sound equitation principles, rather than straps and pulleys (Ödberg & Bouissou, 1999; Figure 14.5).

Figure 14.5. The Pessoa training system, which connects the horse's hindlegs to its mouth in order to force the head down as the animal moves. Picture from *The Truth About Horses* by Andrew McLean (2003). (Photographed by Paul Forester © Quarto Publishing plc, used with permission.)

When horses show insufficient 'impulsion' (responsive and powerful forward movement), trainers may increase the pressure on the sides or use the long whip, where it more effectively drives the horse forward. In a similar vein, lateral movement away from the rider's leg can also be fortified by tapping the long whip. Although, for some, these stimulants are distasteful, they are not necessarily contraindicated since they can be used with subtlety and used transiently to empower the rider's leg signals.

If we agree that sound application of training principles can make horses sensitive to very subtle pressures as a result of discriminative conditioning, then we should question the use of double bridles and spurs in elite horse training.

Behaviour modification

Many traditional horse trainers assume that horses are 'understanding' their training rather than simply responding through reward. Such an anthropomorphic view of the horse is not supported by scientific investigations that have failed to demonstrate convincing evidence of many higher mental abilities (see Nicol, Chapter 12). Training horses within this traditional anthropomorphic framework has a number of potentially negative implications for horse welfare, as well as for the safety of riders and handlers. Responses should not be interpreted as reflections of human emotions and values. This is not an attempt to demean

horses but rather to recognize that there are often simpler and more plausible explanations for undesirable responses. This facilitates the development of logical and appropriate remedies that use relevant learning theory rather than the meting out of justice. Using this applied ethological approach, it becomes clearer that horses do not perform certain behaviours in a bid to 'show up' their riders and that therefore they never deserve punishment for such responses.

Punishment can work only if the aversive stimulus is contingent upon the response. It is nowadays regarded as unenlightened and it presupposes that horses know the difference between good and bad behaviour. Therefore it is increasingly inappropriate for trainers and veterinarians to advocate punishment as a response to undesirable behaviour in the horse. It also carries the risk of injury, which exposes practitioners to the risk of litigation. Furthermore it is recognized that better results tend to occur when the human–horse relationship is nurtured when the rider is both on the ground and in the saddle. This is a core principle of many modern horse training techniques that focus on consolidation of the human–horse bond (Parelli, 1995). What is sometimes overlooked is that *the best way to consolidate the horse–human bond is through the installation of clear, operant basics.*

Behaviour therapy that avoids punishment protocols can be used to interrupt and replace undesirable learned responses, with or without an innate component to their ontogeny. Examples of techniques used for behaviour modification of flight responses include counter-conditioning and habituation. These are appreciably more successful than the rather dated and less humane approach to fearful horses of so-called flooding that involved exposing them to an overwhelming amount of the fear-eliciting stimulus in the hope that their responses would become exhausted. Habituation is fundamental to overcoming fear-related responses but it can be expedited with counter-conditioning, an approach that relies heavily on the shaping of alternative responses through operant conditioning (including clicker training) (Waran *et al.*, 2002).

The cornerstone of behaviour modification is operant conditioning and this is not surprising since the same principle is at the core of basic horse training (see Waran and Casey, Chapter 13). Implicit is the assumption that conflict behaviours in horse training mostly arise from incorrect negative reinforcement. Therefore, reinstallation of the desired responses using a reasoned and tactful application of the pressure-release model of horse training is an essential component of rehabilitation and is almost always successful. However, consistent use of negative reinforcement is the key to retraining and notably achieves faster results in horses that have not habituated to pressure.

The rehabilitation of horses that have developed serious conflict behaviours is a matter of reinstalling the single responses via negative reinforcement. So it involves going back to the basics of foundation training ('breaking in') and taking time to ensure that single signals are applied to shape clear consistent responses in the training of 'go', 'stop' and 'turn' (McLean, 2003).

It is critical that trainers appreciate the importance of training just one aspect or quality of each of these responses at a time. Therefore, in the training of the 'go' response from the pressure of the rider's lower legs (calves), the horse should first be trained to elicit the 'go' response promptly, through the increase in pressure from light to stronger until this promptness occurs (followed by the instant removal of the pressure at the onset of the response). Precise timing results in the horse responding to light pressure (McLean & McLean, 2002).

The next quality to be trained in the horse is self-maintenance of its own tempo and rhythm (includ-

ing gait) through steadying or quickening signals (in difficult cases, up and down to the next gait) when the horse makes mistakes in tempo and rhythm. As before, these signals are installed with negative reinforcement that works best when sufficient attention is paid to accurate timing.

Then the adoption of and adherence to directional line (straightness) is trained. Here the horse is corrected from drifting sideways using pressures from the single reins that control the forequarters and train the 'turn' response. Again it is the removal of this pressure that trains this response. These qualities in-hand and under saddle form the basics of training. However, in dressage, other features are demanded such as alterations in the outline of the horse.

Only when the basics of rhythm and straightness are achieved should the trainer begin to train the outline of the horse (McLean, 2003). Indeed if the horse responds promptly and lightly, with a rhythm and exactly where it is pointed, it will show longitudinal flexion (neck lowering) and the early development of being 'on the bit'. Similarly, when the turns are prompt and light, conducted with a rhythm and with no deviations from the curved line, lateral flexion will develop so that the horse's atlanto-occipital joint flexes to follow the exact line of the circle. The horse is now becoming 'round'. Similarly when promptness, rhythm and line are trained in the sideways yield of the hindquarters from a single leg pressure, the rider is able further to develop 'roundness' by using what is commonly known as 'inside leg to outside rein'. The horse is now properly 'on the bit' (McLean & McLean, 2002).

The qualities of prompt responding, self-maintained rhythm and straightness are features of traditional training scales but do not appear to be sufficiently or universally attended to by horse trainers. Additionally, contemporary training and judging pay too little attention to the well-being of horses in terms of their tension levels. Trainers and judges should be acutely aware that relaxation of the body is the most accurate barometer of the clarity of the trained responses, and precise assessment of resistance responses should be incorporated into dressage judging so that marks are negatively accorded to increasing tension levels, especially at the higher levels of competition. Because competitive horses tend to develop so much conflict behaviour, especially in connection with the rein pressures, trainers should constantly test for self-carriage. This is done with the release of the reins for a couple of strides to check and demonstrate that the horse is indeed carrying itself and that its head posture is not being maintained exclusively by the rider's hands.

Once negative reinforcement has been used to re-establish the basic responses, positive reinforcement can be used to shape refined responses. For example, clicker training which involves positive secondary reinforcement can be particularly useful for shaping responses such as standing still (e.g. in the face of aversive stimuli) and refinement of dressage manoeuvres such as piaffe and passage. Clicker training can be helpful by standardizing the reinforcement but also by teaching trainers to have good timing and consistency.

Conclusion

In the absence of empirical data, understanding undesirable behaviours under saddle requires a consideration of their ethological and cognitive relevance. This sets the stage for practitioners to eliminate pain and discomfort as proximate causes and then apply learning theory to resolve learned responses. As with any training programme, behaviour modification relies on its being applied with consistency. If riders relapse into old habits, the intervention may appear to have failed. Other reasons for poor results in behavioural modification include a lack of reconciliation of the task required of the horse and its physical ability to respond.

References

Bailey, C. J. (1998). *Wastage in the Australian Thoroughbred Racing Industry*. Rural Industries Research and Development Corporation, Research Paper No. 98/52.

Borrow, H. A. (1993). Duodenal perforations and gastric ulcers in foals. *Veterinary Record*, **132**(12): 297–9.

Fillis, J. (1902). *Breaking and Riding*. London: J. A. Allen and Co., Ltd.

Furr, M. O., Murray, M. J. & Ferguson, D. C. (1992). The effects of stress on gastric ulceration, T3, T4, reverse T3 and cortisol in neonatal foals. *Equine Veterinary Journal*, **24**(1): 37–40.

de la Guérinière, F. R. (1992). *Ecole de Cavalerie*. Cleveland, OH: Xenophon Press.

Haussler, K. K., Stover, S. M. & Willits, N. H. (1999). Pathologic changes in the lumbosacral vertebrae and pelvis in Thoroughbred racehorses. *American Journal of Veterinary Research*, **60**(2): 143–53.

Henrotte, J. G., Aymard, N., Allix, M. & Boulu, R. G. (1995). Effect of pyridoxine and magnesium on stress-induced gastric ulcers in mice selected for low or high blood magnesium levels. *Annals of Nutrition and Metabolism*, **39**(5): 285–90.

Herbermann, E. (1980). *Dressage Formula*. London: J. A. Allen and Co., Ltd.

Jeffcott, L. B., Rossdale, P. D., Freestone, J., Frank, C. J. & Towers-Clark, P. F. (1982). An assessment of wastage in thoroughbred racing from conception to 4 years of age. *Equine Veterinary Journal*, **14**: 185–98.

Le Doux, J. E., (1994). Emotion, memory and the brain. *Scientific American*, June: 32–39.

Lieberman, D. A. (1993). *Learning: Behaviour and Cognition*. Pacific Grove, CA: Brooks/Cole Publishing.

McGreevy, P. D. & Nicol, C. J. (1998). Behavioural and physiological consequences associated with the short-term prevention of crib-biting in horses. *Physiology and Behaviour*, **65**(1): 15–23.

McLean, A. N. (2003). *The Truth About Horses*. Melbourne, Australia: Penguin Books.

McLean, A. N. & McGreevy P. D. (2004). Training. In *Equine Behaviour: A Guide for Veterinarians and Equine Scientists*, ed. P. D. McGreevy. London: W. B. Saunders, pp. 291–311.

McLean, A. N. & McLean, M. M. (2002). *Horse Training the McLean way – The Science Behind the Art*. Victoria, Australia: Australian Equine Behaviour Centre.

Mills, D. S. (1998a). Applying learning theory to the management of the horse: the difference between getting it right and getting it wrong. *Equine Veterinary Journal*, **27** (Suppl.), 44–8.

(1998b). Personality and individual differences in the horse, their significance, use and measurement. *Equine Veterinary Journal*, **27** (Suppl.): 10–13.

More, S. J. (1999). A longitudinal study of racing Thoroughbreds: performance during the first years of racing. *Australian Veterinary Journal*, **77**: 105–12.

Nelson, H. (1992). *Francois Baucher – The man and his method*. London: J. A. Allen and Co., Ltd.

Ödberg, F. O. & Bouissou, M. F. (1999). The development of equestrianism from the baroque period to the present day and its consequences for the welfare of horses. The role of the horse in Europe. *Equine Veterinary Journal*, **28** (Suppl.): 26–30.

Parelli, P. (1995). *Natural Horsemanship*. Colorado Springs: Western Horseman.

Pence, P. (2002). *Equine Dentistry: A Practical Guide*. Philadelphia: Lippincott Williams & Wilkins.

Podhajsky, A. (1966). *The Complete Training of Horse and Rider in the Principles of Classical Horsemanship*. London: Doubleday.

Roberts, T. (1992). *Equestrian Technique*. London: J. A. Allen and Co., Ltd.

Rollin, B. E. (2000). Equine welfare and emerging social ethics. *Animal Welfare Forum: Equine Welfare*. JAVMA, **216**(8): 1234–7.

Stolba, A., Baker, N. & Wood-Gush, D. G. M. (1983). The characterization of stereotyped behaviour in stalled sows by information redundancy. *Behaviour*, 87: 157–82.

Tarara, E. B., Tarara, R. P. & Suleman, M. A. (1995). Stress-induced gastric ulcers in vervet monkeys (*Cercopithecus aethiops*): the influence of life history factors. Part II. *Journal of Zoo and Wildlife Medicine*, 26(1). 72–5.

The British Horse Society (1950). *The Manual of Horsemanship*, ed. B. Cooper. London: The British Horse Society.

Visser, E. K., Reenen, C. G., van Hopster, H., Schilder, M. B. H., Knaap, J. H., Barneveld, A. & Blokhuis, H. J. (2001). Quantifying aspects of young horses' temperament: consistency of behavioural variables. *Applied Animal Behaviour Science*, 74(4): 241–58.

Visser, E. K., Reenen, C. G., van Werf, J. T. N., van der Schilder, M. B. H., Knaap, J. H., Barneveld, A. & Blokhuis, H. J. (2002). Heart rate and heart rate variability during a novel object test and a handling test in young horses. *Physiology and Behavior*, 76(2): 289–96.

Von Butler, I. & Armbruster, B. (1984). Struktur und abgangsursachen bei schlachtpferden. *Deutsche Tierzärtztliche Wochenschrift*, 91: 330–1.

Waran, N., McGreevy, P. & Casey. R. A. (2002). Training methods and horse welfare. In *The Welfare of Horses*, ed. N. Waran. Dordrecht: Kluwer Academic Publishers, pp. 151–80.

Watjen, R. (1958). *Dressage Riding*. London: J. A. Allen and Co., Ltd.

Webster, A. J. F. (1994). *Animal Welfare: A Cool Eye Towards Eden*. London: Blackwell Science.

Wiepkema, P. R. (1987). Behavioural aspects of stress. In *Biology of Stress in Farm Animals: An Integrative Approach*, ed. P. R. Wiepkema & P. W. M. van Adrichem. Dordrecht: Martinus Nijhoff.

15

Repetitive movement problems in the horse

Daniel S. Mills

Introduction

Many housed horses perform a variety of bizarre, repetitive behaviours such as weaving, box-walking, windsucking, crib-biting, pawing, door-kicking, head-nodding and face pulling. These behaviours are often called vices. *Chambers 20th Century English Dictionary* defines a vice as 'an immoral habit; a bad trick or habit as in a horse' and many owners appear to believe that a horse that does such behaviour is a 'bad horse', but recent evidence suggests that these behaviours are largely a response to the housed environment and associated management in predisposed individuals. The housed environment often has a predictable routine and restricts many of the activities that make up the normal time budget of the free-ranging horse (see Houpt, Chapter 6). The repetitive and apparently functionless behaviours, which are not seen in the wild state, have attracted the particular attention of many scientists. While it is tempting to ascribe the motivation for these behaviours to general emotional states such as boredom or frustration,

it is now clear that their origins are much more specific in a given case and possibly vary between individuals.

Some repetitive behaviours are tightly bound to a given stimulus that is presented repeatedly; this would include conditioned reflexes, developed through the association of rewarding events with co-incidental behaviour. The first experimental account of this relates to work by Skinner (1948), who found pigeons expressed a range of bizarre behaviours in association with the delivery of a spontaneous reward; however, closer analysis of the data suggests that much of the behaviour occurred after the delivery of the reward, suggesting that the behaviours may be related to frustration following the termination of reward (these are also called displacement or adjunctive behaviours). Thus intermittent delivery of a reward which is quickly consumed may induce repetitive behaviour both before and after its delivery. Interestingly, repetitive behaviour, especially repetitive oral activity, is also seen pre and post feeding in horses (Mills & Macleod, 2002).

The Domestic Horse: The Origins, Development, and Management of its Behaviour, ed. D. S. Mills & S. M. McDonnell.
Cambridge University Press. © Cambridge University Press 2005.

By contrast, in some individuals repetitive behaviour appears to be associated with a more general psychological predisposition for behavioural repetition, e.g. weaving (Mills *et al.*, 2002a). Various terms have been used in the scientific literature to describe these behaviours, including stereotypies (Houpt & McDonnell, 1993), stereotypes (Kiley-Worthington, 1983), obsessive–compulsive disorders (Luescher *et al.*, 1991) and compulsive disorders (Luescher *et al.*, 1998). But these terms can imply additional assumptions about the behaviour that could be misleading. For this reason and the suggestion that some of the behaviours referred to in these categories may be simple stimulus response associations, this chapter will use the purely descriptive term 'repetitive movement' unless otherwise justified.

Why repetitive behaviours are of concern

Some repetitive behaviours may be considered an unsoundness, which affects the sale value of the horse. Prince (1994) reported that cribbing may reduce a horse's value by a half and weaving by a third. However, in a recent similar internet-based repetition of this study J. Brewer and D. S. Mills, (2002, unpub. data) found a median level of reduction in value of between 10 and 20%, with 81% of horse owners (*n* − 212) reporting that they believed these behaviours would reduce the value of the horse. While median levels of reduction were the same for both leisure and professional use, the range differed between these groups with a maximum of 20% reported for leisure horses and 60% for professional-use horses. Interestingly, McBride and Long (2001) found fewer than 50% of racing and competition centre managers thought these behaviours would reduce the value of the horse, while 59% of riding school owners thought it would have an effect. There is clearly great variation between individuals in their opinion of the financial consequences of these behaviours. However, involvement in the competitive horse industry, and therefore perhaps scientific understanding of the horse, appears to be one of the important factors affecting this. It has commonly been said that these behaviours cause ill-thrift or more specific health problems; for example, that weaving makes a horse more likely to suffer from tendon problems, or horses that crib or windsuck can swallow so much air that they give themselves colic. However, the scientific evidence in support of this is either absent or leads to other conclusions. While

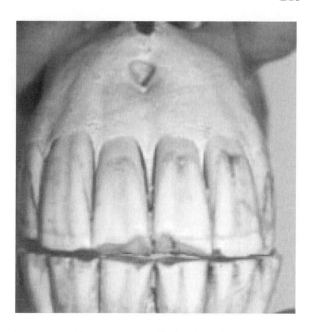

Figure 15.1. Tooth wear typically found on the incisors of a cribbing horse.

there does indeed appear to be an association between colic and crib-biting (Hillyer *et al.*, 2002), the evidence for a causal link is weak. McGreevy *et al.* (1995c) used fluoroscopy to show that very little, if any, air went down the oesophagus during cribbing. The characteristic grunt associated with the behaviour appears to be more like a burp than a gulp. Two follow-up studies have suggested a reduction in the incidence of colic following surgical treatment of the condition (Ritzberger-Matter & Kaegi, 1998; Brouckaert *et al.*, 2002) but the incidence is so low, that the reliability of these as evidence of a causal link is poor. It seems more likely that the association exists because there is a common gastrointestinal factor (such as abnormal gut pH) which is causing both problems, rather than one causing the other (McGreevy *et al.*, 2001). Cribbing can cause serious tooth wear, but this is rarely of clinical significance (Figure 15.1). There are no data on the effect of weaving on performance. None the less it has been reported that around a third of owners and managers believe that repetitive behaviours reduce performance (McBride & Long, 2001), but again the evidence for this is weak (Ritzberger-Matter & Kaegi, 1998; Brouckaert *et al.*, 2002). Of more concern, is that many of these behaviours in captive animals are typically associated with a poor environment

Figure 15.2. Spike collars, whereby prongs project onto the neck if the horse arches its neck, have been developed for the control of cribbing. Like many other types of anti-cribbing collars, it seeks to make the performance of the behaviour uncomfortable. Electric-shocks collars may also be used in an attempt to control the behaviour.

Table 15.1. *Prevalence of techniques used by owners of horses with repetitive movement problems and their reported success*

Technique	Prevalence %	Reported success %
Reduced time in stable	49.3	75.1
Stable toys	12.3	44.7
Increased exercise	1.4	100
Regular switching of stable	9.6	70.1
Increased social contact	9.6	70.1
Exercise first	5.5	100
Feed first	4.1	100
Increased hay ration	6.8	60.3
Varied view	2.7	50.0
Stable chain versus door	1.4	100
Increased size of stable	5.5	100

(Adapted from McBride & Long, 2001).

and psychological problems. This is of particular note when we consider how best to manage these behaviours, but given the common lack of awareness of our current understanding, it is perhaps not surprising that many owners seek to prevent them by whatever means they can, with varying success (see Table 15.1). These efforts at control can reach quite remarkable levels of cruelty (Figure 15.2), but this reflects the owner's misunderstanding of the nature of the behaviour rather than any ill-intent on their part, since the attitude of most owners towards their horse is very affectionate (see Mills & McNicholas, Chapter 11). Similarly, attempts at surgical intervention by the veterinary profession are often misguided, since they are not accompanied by any effort to address the motivation for the behaviour. It is now recognized by welfare scientists that either punishment or physical prevention is generally detrimental to the horse's well-being (McGreevy & Nicol, 1998a,b,c; McBride & Cuddeford, 2001).

Since 2000, there has been significant progress in our understanding of the nature of some of these behaviours through epidemiological, developmental and experimental studies, and perhaps one of the greatest advances in equine welfare would be the wider dissemination of this knowledge within the horse community.

Types of repetitive behaviour

Problematic repetitive movement behaviours are sometimes grouped together as abnormal compulsive behaviours and subdivided into oral, locomotor, sexual and grooming related activities (Luescher *et al.*, 1991). This is an inconsistent classification, using a combination of descriptive and functional criteria. As a result, motivationally distinct behaviours may be grouped together. It is also debatable whether some of the behaviours should be considered abnormal. For example, sexual arousal (often erroneously described as masturbation) is a normal and frequent repetitive act in feral animals (see McDonnell, Chapter 7) and its classification as an abnormality may reflect cultural values more than biological ones. Similarly, repetitive teeth-grinding and flank-biting are normally responses to local pain which disappear when the pain is treated. Others such as door-banging are clearly learned behaviours. It must also be born in mind that there is the potential for a given behaviour to be the 'final common pathway' of a variety of neurological inputs, thus a given behaviour may have different causes in different individuals.

Some of the factors affecting repetitive tendencies are non-specific or unquantified with regards to the behaviours they affect, while others more clearly relate to specific types of behaviour. These are both discussed below, with evidence from both epidemiological and direct experimental studies. The interpretation of both can be problematic. In the case of epidemiological studies, it must be noted that an association does not imply a causal relationship between two factors, also some specific associations may not be revealed due to the sample size used in the study. In the case of experimental studies, it must be noted that response to treatment does not necessarily imply that a causal factor has been treated.

Non-specific risk factors involving repetitive movement behaviours in the horse
Genetic influences

There have been many epidemiological studies of the more common repetitive movement behaviours of cribbing/windsucking, wood chewing, box walking and weaving and while it must be recognized that different breeds are often associated with different uses and management regimens as a result, many of these behaviours appear to be more common in

Thoroughbreds and warm-bloods compared to Standardbreds (Luescher *et al.*, 1998; Redbo *et al.*, 1998). Vecchiotti and Galanti (1986) and Marsden and Henderson (1994), through analyses of breed records, have also found an uneven distribution of the prevalence in different families, which strongly suggests a heritable component.

The effect of sex on these behaviours is less clear. Mills *et al.* (2002a) suggest that stallions are at a lower risk than mares and geldings, while Borroni and Canali (1993) found a significantly higher prevalence in stallions over two years of age. While both used only Thoroughbreds in their surveys, it may be that there are significant differences in the management of stallions in the different countries concerned, and this is a more important factor than sex per se.

It has been suggested that the heritable basis relates to the sensitivity of opiate receptors (Pell & McGreevy, 1999a). Evaluation of the genetic basis to these behaviours is complex since, it is not the behaviour itself that is inherited but the tendency to perform the behaviour, and this only becomes apparent in the presence of other immediate and developmental risk factors, such as those described below.

Diet and feeding practices

Diet and feeding regimen have been implicated as a non-specific risk factor for the expression of abnormal behaviour in the horse in several studies (McGreevy *et al.*, 1995a; Redbo *et al.*, 1998). McGreevy *et al.* (1995a) found that feeding less than 6.8 kg of forage per day, and the feeding of hay as opposed to any other forage, were both associated with an increased risk of several forms of repetitive behaviour, but especially weaving and wood-chewing. Offering forage more than three times a day was also associated with an increased risk of abnormal behaviour in this study, and Cooper and colleagues (cited in Cooper & McGreevy, 2002) found that changing the predictability of the pre-feeding routine might reduce abnormal behaviour before feeding in the short term, but not affect post-feeding levels of such behaviours. Marsden (1993) has also looked at the proximate control of several abnormal behaviours and found that time spent feeding on a diet was inversely related to the time spent performing an abnormal behaviour. Cooper *et al.* (2000), working with horses with a set routine, recorded most repetitive behaviour, especially weaving, prior to turnout, with the next peak occurring prior to feed-concentrate and a third peak prior to the feeding of forage.

It has been suggested (Marsden, 1999) that concentrates may alter both the balance of dopaminergic and serotonergic neurotransmitters which regulate the onset of these behaviours and that concentrates, through their sweet taste, enhance beta-endorphin production (Dum *et al.*, 1983). If animals at risk of repetitive behaviour possess super-sensitive opiate receptors, it is suggested that this periprandial rise may stimulate dopaminergic activity and the production of these behaviour patterns as a result. However, the neurophysiological data on the relationship between neuroendocrine function and the occurrence of repetitive behaviour is at best inconsistent (McGreevy & Nicol, 1998b; McBride, 1996; Lebelt *et al.*, 1998; Pell & McGreevy, 1999a; McBride & Cuddeford, 2001), suggesting that this must be a gross simplification of the process and cannot be the whole story.

Social influences

Although it is widely believed by many horse owners, that these behaviours can be learned, there is currently little evidence for observational learning in the horse, with all scientific studies failing to demonstrate the phenomenon. This does not mean, however, that the presence of an animal performing one of these behaviours does not affect the chance of others performing the behaviour. Both social facilitation, the tendency for individuals in a social group to synchronize their behaviour, and stimulus enhancement, might theoretically encourage an increase in repetitive behaviour. Social facilitation should not be confused with imitation, which results in the production of a new behaviour. Social facilitation is the release of already established behaviours and it might be that repetitive behaviours become incorporated into the behavioural repertoire of horses earlier in life and then become latent for a variable period of time (Mills *et al.*, 2002a). In this case, the arrival of an overtly weaving horse might theoretically release the behaviour in individuals with this predisposition. Stimulus enhancement describes the consequences of the increased attention paid to a stimulus by individuals of a population as a result of the interaction between one member of the group and the stimulus. There is some evidence of this occurring in horses (Clarke *et al.*, 1996), but it has not been shown

to specifically affect the development of repetitive behaviours. In theory, a cribbing horse, might encourage other horses to investigate the top of their own stable doors, with the effect that they then independently learn to crib, because of its own reinforcing effects. Again this would only be expected to increase the expression in a horse that is already predisposed to the behaviour. However, it should be emphasized that these are only theoretical possibilities and, to date, the only demonstrable effect of the presence of a horse with a repetitive behaviour problem on others is to reduce the behaviour (Cooper *et al.*, 2000). This effect was most apparent on weaving and nodding behaviour although a trend towards a reduction in other repetitive acts was also noted.

Other non-specific risk factors

Other factors found to influence the risk of developing several types of repetitive behaviour include social rank, with foals of dominant mares at greater risk; and weaning, with individual box weaning increasing the risk over paddock or group weaning strategies (Waters *et al.*, 2002). Time spent outside, may also be important since McGreevy *et al.* (1995a) found that time spent outside of the stable was associated with an increased risk of abnormal repetitive behaviour in dressage and eventing horses, but not endurance animals. However, the latter group tended to spend much more time outside, confounding a direct comparison of this factor across all disciplines. A similar generally higher level of turnout in Australian Thoroughbreds has also been used to explain a lack of association with time spent outside the stable in this group (Pell & McGreevy, 1999b). This would imply that confinement is only significant if it exceeds a certain threshold (around 20 hours a day) and that a simple linear relationship between these factors does not exist.

Similarly, the risk with age does not appear to be a simple one, with times of particular stress, such as weaning, being important in triggering the expression of the behaviours (Borroni & Canali, 1993). This has been confirmed by Waters *et al.* (2002) who assessed horses through their development and found a high rate of onset associated with weaning. Interestingly, another possible peak around two years of age has been suggested in Thoroughbreds, which coincides with the time when most of these animals will be prepared for riding (Mills *et al.*, 2002a), but the nature of this association requires further evaluation.

It has also been suggested (Normando *et al.*, 2002) that horses ridden Western style are at a lower risk of these behaviours than animals ridden English style. The reason for this is unclear and could be a direct effect of the style of riding or an association with other factors associated with keeping horses for these functions. It is, however, worthy of further investigation as riding imposes a range of physical and psychological stressors on the horse.

Despite some commonly held beliefs to the contrary, there is no evidence of an association between coat colour and these behaviours in the horse (Mills *et al.*, 2002a). However, these data suggest that if a horse is known to exhibit one form of repetitive movement behaviour, it is more likely to exhibit another, suggesting a general tendency to repetitiveness.

In summary, it appears that the predictability of arousing and potentially powerful reinforcing events like feeding and the potential to engage in other behaviours appear to act as general factors affecting the tendency to perform repetitive behaviour in predisposed individuals. Environmental factors, especially around the time of weaning, also appear to have major non-specific effects on the risk of an individual developing certain forms of repetitive behaviour. These effects can be explained at the mechanistic level since various forms of stress alter dopamine-receptor densities in the mesoaccumbens and nigrostriatal systems, which are involved in the expression of these behaviours (Cabib *et al.*, 1998). Genetic factors are also involved and it seems likely that this too may relate to variation in neuro-receptor sensitivities.

Repetitive locomotor behaviour
Weaving and box walking
Form and prevalence Weaving is an obvious lateral swaying movement of the head, neck, forequarters and sometimes hindquarters, (McGreevy *et al.*, 1995a), while box walking describes the repetitive tracing of a circular route within the stable. These behaviours are often considered to be related, with some justification, since it has been reported that in Thoroughbreds, if a horse either weaves or box walks, it is also much more likely to perform the converse rather than some form of oral repetitive behaviour (Mills *et al.*, 2002a). However, there are also clearly factors which differentiate the expression of one locomotor behaviour over another. For example, Luescher *et al.* (1998) in their survey of stables in south-western

Ontario recorded a higher prevalence in Arabs than any other breed group for box walking (7.32%), but this breed had one of the lowest prevalences for weaving. No weaving was recorded in Standardbreds or ponies and no box walking was recorded in Quarter Horses, warm-bloods or ponies either.

In a four-year longitudinal study, box walking had a median age of onset of 64 weeks and weaving had a similar median age of onset at 60 weeks. These values contrast with the median age of onset of cribbing which was much earlier at 20 weeks (Waters *et al.*, 2002).

Housing The housing environment is usually the focus of attention when considering these behaviours and standing stalls are associated with a lower prevalence of weaving (Luescher *et al.*, 1998). Weaving is also less common on larger (>75 horse) facilities, where they are fed a forage other than hay but where more forage is fed, and among horses that have visual and tactile contact with conspecifics and where they are bedded on straw. The potential causal relationship with the social environment is unclear from these epidemiological studies since many owners will keep affected horses away from conspecifics, for fear of others copying the behaviour (McBride & Long, 2001). None the less, increasing social contact does appear to reduce weaving (Cooper *et al.*, 2000). This effect can be replicated with mirrors (Mills & Davenport, 2002) and does not appear to be simply due to novelty, since it endures over several weeks (McAfee *et al.*, 2002). It is also worth noting that mirrors may bring about a cessation of weaving within 24 hours in subjects that have been known to weave for 15–20 years (Mills, unpub. case studies). The importance of the image of a horse in the mirror effect has been demonstrated by M. Riezebos and D. S. Mills (unpub. data), who achieved a similar response with poster images of horses, over the short term. This finding is intriguing, since horses are known to be sensitive to pictorial depth cues (Timney & Keil, 1996). They would therefore be expected to differentiate a real horse from its pictorial representation. None the less, it would seem that even abstract images of horses may evoke types of intraspecific behaviour (Grzimek, 1943). Box walking is similarly affected by the mirror and in a year-long follow-up survey of a commercial product (Mills, 2003) box walking appeared to be more consistently reduced by the mirror. It has

been suggested that in some cases stall walking may
be redirected into something more closely resembling
ambulatory grazing behaviour, through the strategic
distribution of forage around the stable (Houpt &
McDonnell, 1993).

Bedding on a substrate other than straw was found
to be associated with an increase in the risk of weaving
(McGreevy *et al.*, 1995a) and this bedding substrate
is more commonly associated with bed-eating and
more varied behaviour patterns (Mills *et al.*, 2000).
It is unclear therefore whether the provision of straw
allows other behaviours to compete with and thus
inhibit repetitive behaviour, or whether there is a critical specific factor associated with a lack of straw (e.g.
reduced forage) that predisposes animals to develop
repetitive behaviour when it is not available.

Exercise The significance of exercise to this general class of behaviour has already been mentioned.
Luescher *et al.* (1998) also found that box walking
was more common among horses turned out more
frequently in the summer and for longer in spring,
summer and autumn, with less evidence for a similar
relationship between exercise and weaving. This effect
of exercise on box walking more than weaving is further supported by the finding from another study that
endurance horses are less likely to weave than eventing or dressage horses, which both had prevalences
of around 9.5%, although box walking may be more
common in event horses (McGreevy *et al.*, 1995b).

Repetitive locomotor behaviours are reportedly
more common among horses kept in small- to
medium-sized (<1.5 ha) paddocks (Borroni & Canali,
1993) and paddocks of this size are associated with
restricted cantering (Kusunose *et al.*, 1985). Exercise
deprivation is also associated with a rebound increase
in activity when the opportunity arises again (Houpt
et al., 2001), suggesting that exercising may have a
homeostatic role in the horse. As already mentioned,
the timing of turnout may also act as a proximate
cue for a range of repetitive behaviours, but especially
weaving (Cooper *et al.*, 2000).

Evaluation of treatment-based studies Interventions
to reduce weaving such as increased social contact
and the use of mirrors, as described above, might
all act through increasing potential inhibition of
the repetitive behaviour, through the encouragement
of competing behavioural programmes. This might
also explain the efficacy of operant feeding devices

(Winskill *et al.*, 1996; Henderson & Waran 2001)
which may also remove the pre-feeding cues that
can increase arousal and trigger the behaviour. While
it must not be assumed that the efficacy of a
given method of enrichment implies that the system
stimulated is involved in the development of these
behaviours, it is logical to suggest that such methods are more welfare friendly than those such as anti-
weaving bars, which seek to physically prevent the
behaviour (McBride & Cuddeford, 2001). Even if an
intervention such as the use of a mirror in the stable does not act on the cause of the problem, it may
be an invaluable aid to the welfare of socially isolated animals by reducing the stress that has been
reported to occur during social isolation (Mal *et al.*,
1991; Jezierski, 1992) and providing a useful distrac-
tion from the cues of impossible behaviours.

Thus it seems that repetitive weaving and box walking can be explained most parsimoniously by reference to the frustration of locomotor activity in varying contexts. This results in the repetitious release of
partial or complete locomotor sequences. Both external stimuli, such as the presence of a potential social
partner nearby, and internal stimuli, such as an internally derived motivation for exercise, may regulate the
expression of the behaviour. Sensitive management
should avoid proximate cues to the behaviour, which
potentially includes an internally derived expectation
for a certain quantity or even quality of exercise, while
allowing adequate exercise and social interaction in a
controlled setting.

Headshaking and nodding

While these behaviours are superficially similar, these
behaviours are quite different in both structure and
cause.

Head nodding Head nodding is the repetitive bob-
bing of the head up and down and commonly occurs
in two contexts. In the housed horse it is often occurs
at similar times to weaving and is similarly effectively
reduced by increased social contact (Cooper *et al.*,
2000). In the exercised horse it also occurs in response
to lameness in either the fore- or hindlimb and is often
most evident at the trot. In forelimb lameness the
head is raised as the afflicted limb is placed on the
ground in order to move weight away from the forehand, while in hindlimb lameness the reverse happens
(Adams, 1974).

Headshaking

Form and prevalence The term headshaking is used to describe both a syndrome including many other behaviours and a specific behaviour within this collection. A typical headshaker is a horse which displays recurrent, intermittent, sudden and apparently involuntary bouts of head tossing which may be so extreme as to throw the horse and rider off balance. Sneezing and snorting are frequently present, accompanied by attempts by the horse to rub its nose on the ground, a foreleg or nearby objects. (Mair & Lane, 1990; Madigan, 1996). Although the condition can occur at rest, more cases are obvious during exercise, especially at trot (Cook, 1979a; Lane & Mair, 1987; Madigan *et al.*, 1995; Moore *et al.*, 1997). It can, however, be seen at any pace. The prevalence of headshaking remains unknown but many horses with the problem cannot be competed or ridden safely. Median age of onset is generally later than the other behaviours described here, being in mature animals, but it may first appear in both yearlings or senior (>20-year-old) animals (Madigan & Bell, 2001; Mills *et al.*, 2002c).

Of the repetitive behaviours seen in the horse, the aetiology of headshaking has received most attention from the veterinary profession (Cook, 1979a,b, 1980a,b, 1992) and unlike the other repetitive movement problems described here, it appears to have been largely ignored by behaviour scientists. This is perhaps because it is seen to be a response to pain or irritation although it has been suggested that, in some cases, it may be primarily a behavioural problem (Mair & Lane, 1990). The occurrence of certain specific behavioural features may suggest irritation of a specific area, for example, horses that shake their head horizontally, either in addition to or instead of the classical vertical movement, or rub their ears may be more likely to have a pain focus around the ears, such as might occur with ear mite infestation (Mair & Lane, 1990; Mayhew, 1992). 'Flipping' the nose, sneezing, snorting, 'clamping' the nostrils as if to close them and attempts by the horse to rub its face on a foreleg are all reported to be responses to naso-facial irritation or pain (Madigan & Bell, 1998). Principal component analysis has been used to analyse the collection of signs that frequently accompany headshaking further and suggests five dimensions, which might be used to describe animals with this syndrome, potentially reflecting pain in different regions of the head (Mills *et al.*, 2002c).

The condition appears to be seasonal in about 60% of cases (Madigan & Bell, 2001; Mills *et al.*, 2002c). In about a quarter of seasonal cases, the length of time the animal displays the behaviour tends to increase year on year (Mills *et al.*, 2002c). The seasonal pattern of symptoms has been used to suggest the importance of medical conditions linked to photoperiod and allergic rhinitis (Lane & Mair, 1987; Mair & Lane, 1990; Madigan *et al.*, 1995). However, many other factors also correlate with this time of year. This includes an increased tendency to ride out and higher aerial-dust burden, both of which may be implicated in irritation leading to headshaking.

Risk factors and causative agents It is frequently reported that headshaking is twice as common in geldings as in mares (Lane & Mair, 1987; Madigan & Bell, 2001; Mills *et al.*, 2002c), and Thoroughbreds appear to be at greater risk of developing the behaviour (Lane & Mair, 1987; Madigan & Bell, 2001) even though it is rarely seen in active race horses (unpublished discussions with equine practitioners). In a case controlled study, Taylor *et al.* (2001) did not find any real evidence of management factors, including bedding substrate, increasing the risk of the behaviour, nor was there any evidence of increased risk with the use of the horse or the occurrence of respiratory allergies, as has been speculated by other authors (Cook, 1979a; Mair & Lane, 1990). However, there was a suggestion that headshakers were at a greater risk of non-respiratory allergies.

Cook (1980b) suggested 58 conditions which might result in the headshaking syndrome, but a physical cause is rarely identified either ante- or post-mortem. Lane and Mair (1987), working in a referral centre, could identify a potential cause in 11% of their cases and found correction alleviated the problem in only 2%. Physical causes are therefore frequently recorded in the literature during retrospective case studies (Blythe *et al.*, 1984; Gerring & Thomsett, 1980; Gilger *et al.*, 1997; Kold *et al.*, 1982; Madigan *et al.*, 1995; Mair, 1994; McGorum & Dixon, 1990; Moore *et al.*, 1997; Walker *et al.*, 2002). In many cases without a gross physical lesion, the role of the trigeminal nerve, which provides the sensory input from a large part of the muzzle and mouth, has been of interest, since activity in this nerve is likely to result in reflex motor responses such as those seen in the headshaking syndrome.

Table 15.2. *Reported success of treatments for headshaking used by owners*

Treatment	Number of horses treated in this way	Complete success	Partial success	No success
Traditional vet. treatment	**127 (51.8)**	*8 (6.3)*	*28 (22.0)*	*93 (73.2)*
Back specialist	**50 (20.4)**	*0 (0)*	*4 (8)*	*46 (92)*
Homeopathy	**93 (38.0)**	*6 (6.4)*	*29 (31.2)*	*58 (62.4)*
Other alternative therapies	**39 (15.9)**	*2 (5.1)*	*5 (12.8)*	*32 (82.1)*
Nose veil / net	**179 (73.1)**	*48 (26.8)*	*61 (34.1)*	*70 (39.1)*
Ear net	**82 (33.5)**	*3 (3.7)*	*24 (29.3)*	*55 (67.1)*
Face net	**51 (20.8)**	*4 (7.8)*	*21 (41.2)*	*26 (51.0)*
Feed supplement	**105 (42.9)**	*5 (4.8)*	*32 (30.5)*	*68 (64.8)*
Other methods	**91 (37.1)**	*10 (11.0)*	*39 (42.9)*	*42 (46.1)*

(From Mills *et al.*, 2002b).
Figures in bold parentheses represent % of total population surveyed ($n = 245$).
Figures in italicised parentheses represent response rate for those trying this treatment.

Cook (1999, 2003) has suggested that the bit may play a pivotal role in the development of headshaking as a problem, with bit pressure damaging the mandibular branch of the nerve and the consequent pain possibly referred to other branches innervating the muzzle both externally and internally. Evidence of the reaction that the bit can cause in underlying tissues comes from the bony reaction noted in this region of the mandible in many ridden horses (Cook, 2002). This does not discount the possible role of light, which has been emphasized by some workers in the field (Madigan *et al.*, 1995) as photic sensitization of sensory aspects of the trigeminal nerve, which is a well-established phenomenon (Whitman & Packer, 1993). Nor does this hypothesis exclude the role of the posterior ethmoidal branch of the trigeminal nerve emphasized by others (e.g. Newton *et al.*, 2000) and, indeed, it helps to explain the disappointing and frequently temporary effect of surgical intervention on this nerve reported by these authors, and the similarly disappointing results following neurectomy of the infra-orbital branch of the nerve reported by others (e.g. Mair *et al.*, 1992; Mair, 1999). If the problem relates to the referral of pain to various branches of the trigeminal nerve as a result of stimulation by the bit, it is likely that the problem will persist until this stimulation is addressed, with removal of one branch of the nerve simply resulting in the problem arising from referral to another branch. It would be expected that trigger factors including light and physical irritants would probably be more significant in the

earlier stages of the condition before damage to the nerve is severe enough to result in a persistent problem. Cook (2003) further suggests that the role of the bit in this condition may explain the increased risk in males, who unlike females develop a lower canine tooth. The root of this tooth curves caudally and so this region is much more richly innervated and vascularized in the male. As a consequence, there is greater potential for bit pressure to instigate a problematic reaction within the trigeminal nerve. The parsimony and elegance of this hypothesis deserves further investigation.

Evaluation of other treatment-based studies Mills *et al.* (2002b), in a survey of 245 owners of horses that headshake, found about three-quarters of owners had consulted a veterinarian about their horse's condition, but traditional veterinary therapy was used in only 51.8% of cases. The number and proportion of owners using each treatment are given in Table 15.2. The most popular feed supplements were herbals, such as Echinacea. 'Other methods' included: inhalation of friar's balsam, vinegar and oil rubbed on the muzzle before exercise, sun-block, a leather fringe on horse's nose-band, feeding only grass and hay, keeping the horse in an open barn, riding at dawn and moving the horse to another location. While some of these may indeed help in the individual case, the range of treatments which appear to have a significant effect is notable. This could also be taken to provide further evidence of the importance of trigeminal neuralgia in the condition, since in humans unpredictable and intermittent remission is a common finding (Newton *et al.*, 2000). Coincidental remission with the use of a therapy may lead to an unfounded causal association between the two in the owner's mind. Newton *et al.* (2000) found evidence of a good response with carbamazepine, which is used diagnostically in humans to assess trigeminal neuralgia, while cyproheptadine was not helpful. However, Madigan *et al.* (1995) found cyproheptadine to be quite effective in five of our seven cases of apparent photic headshaking. It therefore seems that there may be a difference in the importance of the various trigger factors in the UK and USA, although the centrally mediated path may be similar. If cases tend to be referred for treatment at an earlier stage of development in the USA, this pattern of difference is explicable.

The most effective management aid appears to be the nose net, with more than a quarter of all users reportedly achieving complete success, suggesting that response is not coincidental. Nose nets have traditionally been thought to be effective by acting as a filter but the success of an alternative 'half-net', with a large mesh (Mills & Taylor, 2003) suggests an alternative mechanism may be involved in their success. At present the most likely explanation relates to a reduction in turbulent airflow through the nose as a result of the mesh, although other hypotheses also deserve consideration (see Mills & Taylor, 2003). Turbulent airflow might result in stimulation of sensitized branches of the trigeminal nerve within the nose. This may also explain why the problem is more common at the trot and possibly in dressage horses, since the angle of the head during these exercises may encourage a more turbulent air-flow pattern. It would also explain why the radical procedure of tracheostomy appears to have a high rate of success in relieving the signs in the small number of cases in which it has been tried (Newton *et al.*, 2000). The value of other surgical procedures has been discussed above, but it must be noted that while surgery in these cases may relieve the pain, there is a serious risk of complications which can compromise the welfare of the horse further (Mair & Lane, 1990; Newton *et al.*, 2000).

Clearly, headshaking has many causes and careful evaluation of any given case is warranted in order to determine obvious physical causes. However, it would seem that many cases reflect a hypersensitivity of one or other branch of the trigeminal nerve, which are exacerbated by environmental triggers, with the primary cause possibly relating to the use of the bit. Thus, management factors may be just as important to the understanding and humane treatment of headshaking as it is for other repetitive movement problems in the horse and this has perhaps been largely overlooked because of a false dichotomy between veterinarians and animal behaviour scientists.

Oral repetitive behaviours
Cribbing and windsucking
Form and prevalence The terms cribbing and windsucking are used to describe essentially similar behaviours. In the former the horse grasps a fixed object with its incisors, pulls back and draws air into the cranial oesophagus while emitting a characteristic pharyngeal grunt (McGreevy *et al.*, 1995c), but in the latter the behaviour is performed without grasping an object (Owen, 1982). Accordingly, in many texts the two are not differentiated functionally (Kennedy *et al.*, 1993; McGreevy & Nicol, 1998a; Mills & Macleod, 2002) and, in the absence of evidence to the contrary, the same convention will be adopted here. Surveys on different populations suggest the prevalence may range from 0% in Standardbred stables in Canada (Luescher *et al.*, 1998) to 10.5% in young UK Thoroughbreds (Waters *et al.*, 2002).

Genetic influences The potential influence of heritable factors appears stronger for cribbing than any other repetitive behaviour. Hosoda (1950) found that in certain families the prevalence of cribbing was seven times higher than the population average. A sevenfold difference in the prevalence of cribbing was also found by Redbo *et al.* (1998) between different breeds, with a higher rate in Thoroughbreds compared to Standardbreds. However, they found no difference in the prevalence of wood chewing between these breeds. Further evidence, for a heritable component to cribbing comes from the breed analysis data of Vecchiotti and Galanti (1986) and Marsden and Henderson (1994).

Dietary factors As with weaving, the environment plays a critical role in the development and expression of the behaviour. Cribbing generally has an earlier onset than the repetitive locomotor behaviours, and the feeding of concentrates after weaning has been found to increase the risk of cribbing fourfold

(Waters *et al.*, 2002). Feeding concentrates has been found to increase caecal acidity (Willard *et al.*, 1977) and crib-biting foals have a significantly lower faecal pH (Nicol *et al.*, 2002). The stomachs of crib-biting foals are also more likely to be inflamed and ulcerated than those of normal foals (Nicol *et al.*, 2002). This too may relate to the feeding of concentrates which rapidly become acidic in the stomach and reduce the amount of saliva produced each day, as a result of the reduced masticatory activity. It has been hypothesized that cribbing acts as a substitute for this chewing in an attempt to increase total saliva production and buffer against increased gastrointestinal acidity (Nicol, 1999) and it has been found that horses do indeed produce saliva during crib-biting, but less so than when chewing and that the stomach actually becomes more, not less, acidic during cribbing (K. A. Houpt pers. comm.). Murray and Eichorn (1996) conclude that confinement or intermittent feeding schedules alone may be significant in the development of gastric ulcers in horses. None the less, feeding a diet supplemented with an antacid does appear to reduce cribbing behaviour especially post feeding (Mills & Macleod, 2002). However, horses kept in paddocks measuring more than 2 ha may still crib (Borroni & Canali, 1993). They are unlikely to show repetitive locomotor behaviour though. This might suggest that cribbing, once established, is less amenable to modification, possibly due to some form of dissociation between developmental and maintenance mechanisms. It is, however, possible that there is a physical trigger for the behaviour (such as problems relating to gastrointestinal acidity or gastric ulceration), and therefore it is not necessary to evoke the concept of behavioural emancipation for the continuation of the behaviour (Cooper *et al.*, 1996).

An alternative to the 'saliva' hypothesis, is the 'sweetness' hypothesis which argues that the sweetness of feed concentrates triggers endorphin release, which leads to the cribbing behaviour. It is suggested that the antacids merely mask the sweetness of the concentrate. However, preliminary studies by Mills (2002) using a peppermint-flavoured antacid prior to the feeding of concentrate, still found a significant reduction in meal-time cribbing behaviour. The schedule of antacid delivery would not be expected to block the flavour of the subsequent concentrate and by contrast the appetizing nature of the peppermint flavour may be expected to increase endorphin

production at this time, if this hypothesis were true. By contrast Houpt and Whisher (2003) have found that using agents which mask sweetness can reduce cribbing. So it would seem sweetness may certainly exacerbate the problem.

Evaluation of other treatment-based studies It is often reported that cribbers fail to thrive, and it has been shown that these animals do tend to rest less than normal controls and may have more difficulty in maintaining body weight on comparable diets (McGreevy *et al.*, 2001). However, this is hardly a justification for the range of surgical procedures that have been developed to prevent the physical expression of the behaviour (Smith, 1924; Forssell, 1929; Karlander *et al.*, 1965; Hamm, 1977; Firth, 1980; Fraufelder, 1981; Hakansson *et al.*, 1992; de la Calle *et al.*, 2002). Not only are there risks of complications from surgery (Ritzberger-Matter & Kaegi, 1998; Schofield & Mulville, 1998), but simple prevention of the behaviour is associated with a rebound response when the behaviour is allowed and an increase in a variety of stress parameters during prevention of the behaviour (McGreevy & Nicol, 1998b). Thus it is hard to think of an ethical justification for physical prevention with collars or surgery, or the use of aversive therapy (Baker & Kear-Colwell, 1974). Given that the behaviour may have some adaptive value, it is preferable to allow the animal to continue to perform the behaviour on a suitable substrate than to simply prevent the behaviour.

There is some indirect evidence for the potential significance of non-gastrointestinal factors in the control of cribbing. Kuussaari (1983) reported on the use of acupuncture to treat the condition. Seven acupuncture points were chosen, including three specifically used to treat gastrointestinal disorders in the horse. Following treatment, 64% (14/22) of the horses with gastrointestinal symptoms improved or stopped the behaviour, while none without these signs (0/11) showed a response. The role of acupuncture in veterinary medicine is gaining wider acceptance and this apparent association is certainly worthy of note when trying to hypothesize on the nature of the behaviour.

Finally, de la Rúa-Domènech *et al.* (1997) found that cribbing and/or wood chewing was associated with an increased risk of equine motor neurone disease, and while this might reflect an association that

Table 15.3. *Developmental and releasing factors which interact to produce repetitive behaviour in the horse*

Developmental factors
Horse-centred factors
Phylogeny – evolutionary history defines certain aspects of behavioural regulation and integration, such as the tendency for certain stimuli to release goal-directed (appetitive) behaviours or certain schedules to release displacement (adjunctive) behaviours, which may become disordered if they are frequently frustrated from achieving their goal.
Ontogeny – developmental history may lead to sensitisation of relevant pathways regulating behaviour as a result of specific and general stressors, e.g. weaning and management practice.
Mechanism – inherited differences in sensitivity of neurological substrates, such as dopamine receptors in the relevant pathways, may alter the inherent risk of an individual to show repetitive behaviour in general or of a specific form.
Adaptive value – the behaviour may help the animal cope with a stressful environment and so provide internal reinforcement, e.g. cribbing may help buffer gastric acidity or reduce arousal.
Environmental factors
General risk factors – recurrent stress may result in changes in receptor sensitivity which increase the likelihood of repetitive behaviour.
Specific risk factors – specific stressors may encourage sensitisation of specific behavioural pathways.

Releasing factors
General stimulating factors – factors increasing arousal such as ongoing stable activity, reinforcement schedules etc may be associated with changes in activation threshold of receptors involved in behavioural activation.
General inhibitory factors – current physiological state may also have generalised inhibitory effects, for example a high bioavailability of tryptophan as a result of diet formulation may raise the threshold for repetitive behaviour activation.
Specific stimulating factors – the frustration of specific behaviours which are expressed in response to certain stimuli, e.g. frustration of approach behaviour towards a conspecific while contained within a stable, may result in specific repetitive acts e.g. weaving.
Specific inhibitory factors – certain environmental features may facilitate the expression of alternative behaviours at times of specific arousal, e.g. interaction with a reflected image may reduce weaving.

was also found with concentrate feeding, history should teach us not to jump to any conclusions about making assumptions of the causal nature of any relationship identified in an epidemiological study when other hypotheses remain to be tested.

Wood chewing

Form and prevalence Wood chewing, is a common repetitive behaviour, which may be exhibited by almost a third of young Thoroughbreds (Waters *et al.*, 2002) and is frequently recorded in all other captive populations. There has been debate over whether this should be considered alongside behaviours like cribbing, since, unlike the other behaviours so far described, it is quite frequently seen in the field and may be considered a normal behaviour. This perhaps reinforces the problems associated with the use of current definitions of terms like stereotypy (Mason, 1991). Wood chewing has a diurnal pattern, being more common at night (Krzak *et al.*, 1991), with this and other studies suggesting that gut-fill, or rather lack of it, may trigger the behaviour in some

circumstances (Haenlein *et al.*, 1966). Roughage in the diet appears to be the best predictor of tendency to chew wood in general (Redbo *et al.*, 1998).

Relationship to cribbing It has been suggested that woodchewing and cribbing are related, since horses that crib tend to wood chew first (Waters *et al.*, 2002) and virginiamycin appears to suppress both behaviours (Johnson *et al.*, 1998). However, this relationship is not a simple one since the provision of hard feed post-weaning is associated with a reduced risk of wood chewing, but an increased risk of crib biting. In addition, Redbo *et al.* (1998) found that flat-racing Thoroughbreds differed from Standard-bred trotters in the prevalence of crib biting (and weaving and box walking), but not in the prevalence of wood chewing. Housing foals post-weaning (as opposed to keeping them at grass) and a socially dominant mother are also clearly associated with an increased risk of wood chewing (Waters *et al.*, 2002), while the relationship between these factors and cribbing is less strong. Group housing post-weaning specifically appears to be a major risk factor in wood-chewing, but this does not appear to be the case in cribbing.

Clearly, wood chewing also has a variety of motivations, which in some cases may not be related to the stable environment. C. Feh (pers. comm.) has reported that at certain times feral horses may preferentially strip the bark of willow trees. This is particularly seen at times when the horses may be suffering gastrointestinal upset and so in this case may be a form of self medicating, since the willow is the original source of aspirin. Careful assessment of the behaviour in any context is therefore warranted rather than simple generalizations.

Thus it seems that repetitive oral-based behaviour, appears to be most accurately predicted by some form of discomfort (possibly visceral rather than musculoskeletal), which might range from hunger to acute pain. It would be useful to evaluate other potential sources of such discomfort in these animals and in those that do not respond to management of gastrointestinal pain or treatment of gastric ulceration. This contrasts markedly with the medical approach advocated in certain cases, where the use of narcotic antagonists, which block the body's natural analgesic receptors, has been proposed as a potential method of controlling some of these behaviours (Dodman *et al.*, 1987).

Conclusion Generalizations in this field of study are common but of limited use and potentially misleading. Repetitive behaviours in the horse cannot be explained simply in terms of generalizations such as learned habits or inherited behaviours nor can their motivation be usefully ascribed to similarly general terms such as boredom, frustration or coping. They are multifactorial, arising from a combination of environmental and genetic risk factors in a given individual at a given time (Table 15.3). Prevention should aim to reduce these risk factors, and humane treatment should allow the animal to adapt fully to its environment without excessive strain (Broom & Johnson, 1983). The aim of control should not be elimination of the behaviour per se, but rather the elimination of the welfare problem. The two are not the same. The emphasis should be on intelligent and sensitive management, which respects the limitations of adaptation within the species under our stewardship (Fraser *et al.*, 1997).

References

Adams, R. O. (1974) *Lameness in Horses*, 3rd edn. Philadelphia: Lea and Febiger.

Baker, G. J. & Kear-Colwell, J. (1974) Aerophagia (wind-sucking) and aversion therapy in the horse. *Proceedings of the American Association of Equine Practitioners*, 20: 127–30.

Blythe, L. L., Watrous, B. J., Schmitz, J. A. & Kaneps, A. J. (1984). Vestibular syndrome associated with temporohyoid joint fusion and temporal bone fracture in three horses. *Journal of the American Veterinary Medical Association*, 185: 775–81.

Borroni, A. & Canali, E. (1993). Behavioural problems in Thoroughbred horses reared in Italy. In *Proceedings of the 26th Congress on Applied Ethology*, ed. M. Nichelmann, H. Wierenga & S. Braun. Berlin: KTLM, pp. 43–6.

Broom, D. M. & Johnson, K. G. (1983). *Stress and Animal Welfare*. London: Chapman and Hall.

Brouckaert, K., Steenhaut, M., Martens, A., Vlaminck, L., Pille, F., Arnaerts, L. & Gasthuys F. (2002). Windsucking in the horse: results after surgical treatment – a retrospective study (1990–2000). *Vlaams Diergeneeskundig Tijdschrift*, 71: 249–55.

Cabib, S., Giardino, L., Calza, L., Zanni, M., Mele, A. & Pulisiallegra, S. (1998). Stress promotes major changes in dopamine receptor densities within the mesoaccumbens and nigrostriatal systems. *Neuroscience*, 84: 193–200.

de la Calle, J., Burba, D. J., Tetens, J. & Moore, R. M. (2002). Nd: YAG laser-assisted modified Forssell's procedure for treatment of cribbing (crib-biting) in horses. *Veterinary Surgery*, 31: 111–16.

Clarke, J. V., Nicol, C. J., Jones, R. & McGreevy, P. D. (1996). Effects of observational learning on food selection in horses. *Applied Animal Behaviour Science*, 50: 177–84.

Cook, W. R. (1979a). Headshaking in horses: part 1. *Equine Practice*, 1: 9–17.

(1979b). Headshaking in horses: part 2. History and management. *Equine Practice*, 1: 36–9.

(1980a). Headshaking in horses: part 3. Diagnostic tests. *Equine Practice*, 2: 31–40.

(1980b). Headshaking in horses: part 4. Special diagnostic procedures. *Equine Practice*, 2: 7–15.

(1992). Headshaking in horses: an afterword. *Compendium on Continuing Education for the Practising Veterinary Surgeon*, 14: 1369–72.

(1999). Pathophysiology of bit control in the horse. *Journal of Equine Veterinary Science*, 19: 196–205.

(2002). Bit-induced asphyxia in the horse. *Journal of Equine Veterinary Science*, 22: 7–14.

(2003). Bit-induced pain: a cause of fear, flight, fight and facial neuralgia in the horse. *Pferdeheilkunde*, 19: 1–8.

Cooper, J. J. & McGreevy, P. D. (2002). Stereotypic behaviour in stabled in horses. *The Welfare of Horses*, ed. N. Waran. Dordrecht: Kluwer Academic Publishers, pp. 99–124.

Cooper, J. J., McDonald, L. & Mills, D. S. (2000). The effect of increasing visual horizons on stereotypic weaving: implications for the social housing of stabled horses. *Applied Animal Behaviour Science*, 69: 67–83.

Cooper, J. J., Odberg, F. O. & Nicol, C. J. (1996). Limitations of the effect of environmental improvement in reducing stereotypic behaviour in bank voles (*Clethrionomys glareolus*). *Applied Animal Behaviour Science*, 48: 237–48.

Dodman, N. H., Shuster, L., Court, M. H. & Dixon, R. (1987). Investigation into the use of narcotic antagonists in the treatment of a stereotypic behaviour pattern (crib-biting) in the horse. *American Journal Veterinary Research*, 48: 311–19.

Dum, J., Gramsch, C. & Herz, A. (1983). Activation of hypothalamic ß-endorphin pools by reward induced by highly palatable food. *Pharmacology, Biochemistry and Behavior*, 18: 443–7.

Firth, E. C. (1980). Bilateral ventral accessory neurectomy in windsucking horses. *Veterinary Record*, 106: 30–2.

Forssell, G. (1929). Kopen, chirurgische Behandlung. *Stang-Wirth. Tierhelk. und Tierzucht*, 6: 283–7.

Fraser, D., Weary, D. M., Pajor, E. A. Milligan, B. N. (1997). A scientific conception of animal welfare that reflects ethical concerns. *Animal Welfare*, 6: 187–205.

Fraufelder, H. (1981). Treatment of crib-biting: a surgical approach in the standing horse. *Equine Veterinary Journal*, 13: 62–3.

Gerring, E. L. & Thomsett, L. R. (1980). Mites in headshaker horses. *Veterinary Record*, 106: 490.

Gilger, B. C., Davidson, M. G., Nadelstein, B. & Naisse, M. (1997). Neodymium-aluminium-garnet laser tretament of cystic granula iridica in horses: eight cases (1988–1996). *Journal of the American Veterinary Medical Association*, 211: 341–3.

Grzimek, B. (1943). Begrussung zweir pferde. Das erkennen von Phantomen und Bildern. *Zeitschrift fur Tierpsychologie*, 5: 465–80.

Hakansson, A., Franzen, P. & Petersson, H. (1992). Comparison of two surgical methods for treatment of crib-biting in horses. *Equine Veterinary Journal*, 24: 494–6.

Hamm, D. (1977). A new surgical procedure to control crib-biting. *Proceedings of the American Association of Equine Practitioners*, 23: 301–2.

Haenlein, G. F. W., Holdren, R. D. & Yoon, Y. M. (1966). Comparative response of horses and sheep to different physical forms of alfalfa hay. *Journal of Animal Science*, 25: 740–3.

Henderson, J. V. & Waran, N. K. (2001). Reducing equine stereotypies using the Equiball™. *Animal Welfare*, 10: 73–80.

Hillyer, M. H., Taylor, F. G. R., Proudman, C. J., Edwards, G. B., Smith, J. E. & French, N. P. (2002). Case control study to identify risk factors for simple colonic obstruction and distension colic in horses. *Equine Veterinary Journal*, 34: 455–63.

Hosoda, T. (1950) On the heritability of susceptibility to windsucking in horses. *Japanese Journal of Zootechnical Science*, 21: 25–8.

Houpt, K. A. & McDonnell, S. M. (1993). Equine Stereotypies. *The Compendium of Continuing Education for the Practising Veterinarian*, 15: 1265–72.

Houpt, K. A., Houpt, T. R., Johnson, J. L., Erb, H. N. & Yeon, S. C. (2001). The effect of exercise deprivation on the behaviour and physiology of straight stall confined pregnant mares. *Animal Welfare*, 10: 257–67.

Houpt, K. A. & Whisher, L. (2003). The effect of diet on cribbing. *Proceedings of the Annual Scientific Symposium of Animal Behavior of the American Veterinary Society of Animal Behaviour and the American College of Veterinary Behaviorists, Denver, Colorado, 20–21 July 2003*, p. 76.

Jezierski, T. (1992). Effects of social isolation on heart rate in horse. *Proceedings of the 26th International Congress on Applied Ethology*, ed. M. Nichelmann, H. Wierenga & S. Braun. Berlin: KTML, pp. 387–8.

Johnson, K. G., Tyrrell, J., Rowe, J. B. & Petherick, D. W. (1998). Behavioural changes in stabled horses given non-therapeutic levels of virginiamycin. *Equine Veterinary Journal*, 30: 139–42.

Karlander, S., Mansson, J. & Tufvesson, G. (1965). Buccostomy as a method of treatment for aerophagia (wind-sucking) in the horse. *Nordisk Veterinär Medicin*, 17: 455–8.

Kennedy, J. S., Schwabe, A. E. & Broom, D. M. (1993). Crib-biting and wind-sucking stereotypies in the horse. *Equine Veterinary Education*, 5: 142–7.

Kiley-Worthington, M. (1983). Stereotypies in the horse. *Equine Practice*, 5: 34–40.

Kold, S. E., Ostblom, L. C. & Philipsen, H. P. (1982). Headshaking caused by maxillary osteoma in a horse. *Equine Veterinary Journal*, 14: 167–9.

Krzak, W. E., Gonyou, H. W. & Lawrence, L. M. (1991). Wood chewing by stabled horses: diurnal pattern and effects of exercise. *Journal of Animal Science*, 69: 1053–8.

Kusunose, R., Hatakeyama, H., Kubo, K., Kiguchi, A., Asai, Y., Fujii, Y. & Ito, K. (1985). Behavioural studies on yearling horses in field environments 1. Effects of the field size on the behaviour of horses. *Bulletin of the Equine Research Institute*, 22: 1–7.

Kuussaari, J. (1983). Acupuncture treatment of aerophagia in horses. *American Journal of Acupuncture*, 11: 363–70.

Lane, J. G. & Mair, T. S. (1987). Observations on headshaking in the horse. *Equine Veterinary Journal*, 19: 331–6.

Lebelt, D., Zanella, A. J. & Unshelm, J. (1998). Physiological correlates associated with cribbing behaviour in horses; changes in thermal threshold, heart rate, plasma B-endorphin and serotonin. *Equine Veterinary Journal, Supplement*, 27: 21–7.

Luescher, U. A., McKeown, D. B. & Halip, J. (1991). Reviewing the causes of obsessive compulsive disorders in horses. *Veterinary Medicine*, 86: 527–30.

Luescher, U. A., McKeown, D. B. & Dean, H. (1998). A cross-sectional study on compulsive behaviour (stable vices) in horses. *Equine Veterinary Journal*, 27 (Suppl.): 14–18.

Madigan, J. E. (1996). Evaluation and treatment of headshaking syndrome. *Proceedings of the 1st*

International Conference on Equine Clinical Behaviour, 6–7 June, Basel, Switzerland.

Madigan, J. E. & Bell, S. A. (1998). Characterisation of headshaking syndrome – 31 cases. *Equine Veterinary Journal*, 27 (Suppl.): 28–9.

(2001). Owner survey of headshaking in horses. *Journal of the American Veterinary Medical Association*, 219: 334–7.

Madigan, J. E., Kortz, G., Murphy, C. & Rodger, L. (1995). Photic headshaking in the horse: 7 cases. *Equine Veterinary Journal*, 27: 306–11.

Mair, T. S. (1994). Headshaking associated with *Trombicula autumnalis* larval infestation in 2 horses. *Equine Veterinary Journal*, 26: 244–5.

(1999). Assessment of bilateral infra-orbital nerve blockade and bilateral infra-orbital neurectomy in the investigation and treatment of idiopathic headshaking. *Equine Veterinary Journal*, 31: 262–4.

Mair, T. S. & Lane, J. G. (1990). Headshaking in horses. *In Practice*, 9: 183–6.

Mair, T. S., Howarth, S. & Lane, J. G. (1992). Evaluation of some prophylactic therapies for the idiopathic headshaker syndrome. *Equine Veterinary Journal Supplement*, 11: 10–12.

Mal, M. E., Friend, T. H., Lay, D. C., Vogelsang, S. G. & Jenkins, O. C. (1991). Behavioural responses of mares to short-term confinement and social isolation. *Applied Animal Behaviour Science*, 31: 13–24.

Marsden, M. D. (1993). Feeding practices have greater effect than housing practices on the behaviour and welfare of the horse. In *Proceedings of the 4th International Symposium on Livestock Environment*, ed. E. Collins & C. Boon. *University of Warwick, Coventry*. Coventry: American Society of Agricultural Engineers, pp. 314–18.

(1999). Behavioral problems (stable vices). *Equine Medicine and Surgery*, ed. P. T. Colahan, A. M. Merritt, J. N. Moore & I. G. Mayhew. St Louis: Mosby, pp. 914–31.

Marsden, M. D. & Henderson J. (1994) The inheritance of susceptibility to stereotypic behaviour patterns in the horse. *Proceedings of Hereditary Diseases of Horses, Interlaken, Switzerland, September 8*.

Mason, G. J. (1991). Stereotypies: a critical review. *Animal Behaviour*, 41: 1015–37.

Mayhew, I. G. (1992) How I treat headshakers. *Proceedings of the North American Veterinary Conference, Orlando, Florida*, 6: 453–4.

McAfee, L. M, Mills, D. S. & Cooper, J. J. (2002) The use of mirrors for the control of stereotypic weaving behaviour in the stabled horse. *Applied Animal Behaviour Science*, 78: 159–73.

McBride, S. D. (1996). A comparison of physical and pharmacological treatments for stereotypic behaviour in the horse. *Proceedings of the 30th International Congress for the International Society of Applied Ethology*, ed. I. J. H. Duncan, T. M. Widowski & D. B. Haley. Guelph: Centre for the Study of Animal Welfare, p. 26.

McBride, S. D. & Cuddeford, D. (2001). The putative welfare-reducing effects of preventing equine stereotypic behaviour. *Animal Welfare*, 10: 173–89.

McBride, S. D. & Long, L. (2001). Management of horses showing stereotypic behaviour, owner perception and the implications for welfare. *Veterinary Record*, 148: 799–802.

McGorum, B. C. & Dixon, P. M. (1990). Vasomotor rhinitis with headshaking in a pony. *Equine Veterinary Journal*, 22: 220–2.

McGreevy, P. D. & Nicol, C. J. (1998a). Prevention of crib-biting: a review. *Equine Veterinary Journal*, 27 (Suppl.): 35–8.

(1998b). Physiological and behavioural consequences associated with short-term prevention of crib-biting in horses. *Physiology and Behaviour*, 65: 15–23.

(1998c). The effect of short term prevention on the subsequent rate of crib-biting in Thoroughbred horses. *Equine Veterinary Journal*, 27 (Suppl.): 30–4.

McGreevy, P. D., Cripps, P. J., French, N. P., Green, L. E. & Nicol, C. J. (1995a). Management factors associated with stereotypic and redirected behaviour in the Thoroughbred horse. *Equine Veterinary Journal*, 27: 86–91.

McGreevy, P. D., French, N. P. & Nicol, C. J. (1995b). The prevalence of abnormal behaviours in dressage, eventing and endurance horses in relation to stabling. *Veterinary Record*, 137: 36–7.

McGreevy, P. D., Richardson, J. D., Nicol, C. J. & Lane, J. G. (1995c). A radiographic and endoscopic study of horses performing an oral stereotypy. *Equine Veterinary Journal*, 27: 92–5.

McGreevy, P. D., Webster, A. J. F. & C. J. Nicol. (2001). A study of the digestive efficiency, behaviour and gut transit times of crib-biting horses. *Veterinary Record*, 148: 592–6.

Mills, D. S. (2002). Recent advances in the treatment of equine stereotypic behaviour. In *Proceedings of A Dororthy Russell Havemeyer Foundation Workshop, Horse Behavior and Welfare, 13–16 June, Holar, Iceland*, pp. 122–6. Available at http://www2.vet.upenn.edu/labs/equinebehaviour/hvnwkshp/hv02/mills.htm

(2003) Reflecting on progress: enriching equine environments. *Veterinary Times*, 33(47): 8.

Mills, D. S. & Davenport, K. (2002). The effect of a neighbouring conspecific versus the use of a mirror for the control of stereotypic weaving behaviour in the stabled horse. *Animal Science*, 74: 95–101.

Mills, D. S. & Macleod, C. A. (2002). The response of crib-biting and windsucking horses to treatment with an antacid mixture. *Ippologia*, 13: 33–41.

Mills, D. S. & Taylor, K. (2003). Field study of the efficacy of three types of nose net for the treatment of headshaking in horses. *Veterinary Record*, 152: 41–4.

Mills, D. S., Alston, R. D., Rogers, V. & Longford, N. T. (2002a). Factors associated with the prevalence of stereotypic behaviour amongst Thoroughbred horses passing through auctioneer sales. *Applied Animal Behaviour Science*, 78: 115–24.

Mills, D. S., Cook S. & Jones B., (2002b). Reported response to treatment amongst 245 cases of equine head shaking. *Veterinary Record*, 150: 311–13.

Mills, D. S., Cook, S., Taylor, K. & Jones, B. (2002c). Analysis of the variations in clinical signs shown by 254 cases of equine headshaking. *Veterinary Record*, 150: 236–40.

Mills, D. S, Eckley, S. & Cooper, J. J. (2000). Thoroughbred bedding preferences, associated behaviour differences and their implications for equine welfare. *Animal Science*, 70: 95–106.

Moore, L. A., Johnson, P. J., Messer, N. T., Kline, K. L., Crump, L. M. & Knibb, J. R. (1997). Management of headshaking in three horses by treatment for protozoal myeloencephalitis. *Veterinary Record*, 141: 264–7.

Murray, M. J. & Eichorn, E. S. (1996). Effects of intermittent feed deprivation, intermittent feed deprivation with ranitidine administration, and stall confinement with *ad libitum* access to hay on gastric ulceration in horses. *American Journal Veterinary Research*, 11: 1599–603.

Newton, S. A., Knottenbelt, D. C. & Eldridge, P. R. (2000). Headshaking in horses: possible aetiopathogenesis suggested by the results of diagnostic tests and several treatment regimes used in 20 cases. *Equine Veterinary Journal*, 32: 208–16.

Nicol, C. J. (1999). Understanding equine stereotypies. *Equine Veterinary Journal*, 28 (Suppl.): 20–5.

Nicol, C. J., Davidson, H. P. B., Harris, P. A., Waters, A. J. & Wilson, A. D. (2002). Study of crib-biting and gastric inflammation and ulceration in young horses. *Veterinary Record*, 151: 658–62.

Normando, S., Canali, E. & Verga, M. (2002). Behavioral problems in Italian saddle horses. *Journal of Equine Veterinary Science*, 22: 117–20.

Owen, R. R. (1982). Crib-biting and wind-sucking – that equine enigma. In *The Veterinary Annual 1982*, ed. C. S. G. Hill. & F. W. G. Grunsell. Bristol: Wright Scientific Publications, pp. 156–68.

Pell, S. M. & McGreevy, P. D. (1999a). The prevalence of abnormal and stereotypic behaviour in Thoroughbreds in Australia. *Australian Veterinary Journal*, 77: 678–9.

(1999b). A study of cortisol and beta-endorphin levels in stereotypic and normal Thoroughbreds. *Applied Animal Behaviour Science*, 64: 81–90.

Prince, D. (1994). Stable Vices. In *Behaviour Problems in Horses*, ed. S. McBane. Newton Abbot: David and Charles, pp. 115–22.

Redbo, I., Redbo-Torstensson, P., Odberg, F. O., Hedendahl, A. & Holm, J. (1998). Factors affecting behavioural disturbances in race-horses, *Animal Science*, 66: 475–81.

Ritzberger-Matter, G. & Kaegi, B. (1998). Retrospective analysis of the success rate of surgical treatment of aerophagia in the horse at the Veterinary Surgical Clinic, University of Zurich. *Equine Veterinary Journal*, 27 (Suppl.): 62.

de la Rúa-Domènech R., Mohammed, H. O., Cummings, J. F., Divers, T. J., de Lahunta, A. & Summers, B. A. (1997). Intrinsic, management, and nutritional factors associated with equine motor neuron disease. *Journal of the American Veterinary Medical Association*, 211: 1261–7.

Schofield, W. L. & Mulville, J. P. (1998). Assessment of the modified Forssell's procedure for the treatment of oral stereotypies in 10 horses. *Veterinary Record*, 142: 572–5.

Skinner, B. F. (1948). 'Superstition' in the pigeon. *Journal of Experimental Psychology*, 38: 168–72.

Smith, F. (1924). An attempt to deal with 'windsucking' by a surgical interference. *Veterinary Journal*, 80: 238–40.

Taylor, K. D., Cook, S. & Mills, D. S. (2001). A case controlled study investigating health, management and behavioural features of horses commonly described as headshakers. *Ippologia*, 12: 29–37.

Timney, B. & Keil, K. (1996). Horse are sensitive to pictorial depth cues. *Perception*, 25: 1121–8.

Vecchiotti, G. & Galanti, R. (1986). Evidence of heredity of cribbing, weaving and stall-walking in Thoroughbred horses. *Livestock Production Science*, 14: 91–5.

Walker, A. M., Sellon, D. C., Cornelisse, C. J., Hines, M. T., Ragle, C. A., Cohen, N. & Schott, II. C. (2002). Temporohyoid osteoarthopathy in 33 horses (1993–2000). *Journal of Veterinary Internal Medicine*, 16: 697–703.

Waters, A. J., Nicol C. J. & French N. P. (2002). The development of stereotypic and redirected behaviours in young horses: the findings of a four year prospective epidemiological study. *Equine Veterinary Journal*, 34: 572–9.

Whitman, B. W. & Packer, R. J. (1993) The photic sneeze reflex: literature review and discussion. *Neurology*, 43: 868–71.

Willard, J. G., Willard, J. C., Wolfram, S. A. & Baker, J. P. (1977). Effect of diet on cecal pH and feeding behaviour of horses. *Journal of Animal Science*, 45: 87–93.

Winskill, L. C., Young, R. J., Channing, C. E., Hurley, J. & Waran, N. K. (1996). The effect of a foraging device (a modified Edinburgh foodball) on the behaviour of the stabled horse. *Applied Animal Behaviour Science*, 48: 25–35.

16

Equine behaviour and welfare

Jonathan J. Cooper and
Melissa J. Albentosa

Introduction
Recognizing equine welfare problems

In terms of their health and welfare, horses and other equines used for sporting and recreational purposes share many of the benefits enjoyed by other companion animals. These benefits can include a close relationship with owners or keepers, relatively high financial or sentimental value and good standards of veterinary attention. However, even in these environments, situations arise where there may be a compromise between the horse's quality of life and the practical or commercial interests of the owner or keeper. For example, breeding stock may experience artificial mating and weaning practices, equine athletes may be pushed to their physical limits during training and competition, and horses may be housed in systems that are both physically and behaviourally restrictive.

Outside of the equestrian world, horses and other equines are used commercially as draught or pack animals (see Hall, Chapter 2), agricultural products (e.g.

for meat and hormone production) and as research subjects. There are also many feral or semi-feral herds that are managed to a greater or lesser degree. In general, these equines face a different range of potential welfare issues to sporting and leisure horses because of the different environments in which they live, and the priorities and financial circumstances of their owners/keepers. For example, if an equine is destined for slaughter and is viewed solely in terms of its meat value, there may be little financial incentive for the owner to provide basic good husbandry when this cost will not be recouped. In contrast, safeguarding the health of an equine used for transport may be of paramount importance to the owner's well-being, yet the knowledge, funds or services necessary to maintain the animal in good health may be absent.

To present a comprehensive review of the welfare issues of all domestic horses is a task that is beyond the scope of this chapter. Authors in preceding chapters have already highlighted welfare problems relating to a variety of areas of equine husbandry

The Domestic Horse: The Origins, Development, and Management of its Behaviour, ed. D. S. Mills & S. M. McDonnell.
Cambridge University Press. © Cambridge University Press 2005.

including their housing (see Mills, Chapter 15) and training (see McGreevy and McLean, Chapter 14). Furthermore there are several books that cover welfare issues in far more detail than can be covered here, for example that of Kiley-Worthington (1997) and a volume edited by Waran (2002). In this chapter we do not set out to reiterate these messages. Instead we set ourselves two tasks: the first is to describe a logical framework for the scientific assessment of equine welfare; and the second is to discuss, using examples, how current welfare research fits into this framework.

Understanding our concern for equine welfare

In general, we feel responsible for the welfare of animals in zoos, and domesticated species such as farm, laboratory and companion animals, but considerably less responsible for the subjective experiences of animals in their natural environment. For example, whilst we may be fascinated and intrigued by the hunting, killing and evisceration (often whilst still alive) of a zebra by a pride of lions in the wild, the feeding of live horses to lions in a zoo would be considered immoral and cruel by many people, even though the capacity to experience pain, distress or fear is likely to be similar for both domestic and wild equids. An important difference between the two situations is not the animal's capacity to suffer, but rather the moral or emotive imperative we feel towards animals in our care. In some countries, such as the UK, we are more inclined to be concerned about the welfare of equines than many other domestic animals, perhaps because this moral imperative, coupled with the high emotional and financial value we place on horses, heightens our awareness of potential welfare issues. This is illustrated by the high profile of equine welfare campaigns such as the transport of live horses from Eastern Europe prior to slaughter for food (ILPH, 2003) and the special provisions afforded to horses (in common with cats, dogs and primates) in legislation covering animal research in the UK (Abbott, 1999; Houpt & Ogilvie-Graham, 2002).

Understanding and assessing equine welfare

A commonly adopted definition of animal welfare is 'its state in its attempts to cope with environmental challenges' (Fraser & Broom, 1997). This definition is sufficiently flexible to cover both positive and negative states and can be illustrated by motivational modelling, an approach developed to explain motivational state and decision-making in animals. In motivational modelling, the animal's physiological and behavioural responses to environmental challenges are part of a homeostatic mechanism to maintain the animal's state within acceptable boundaries (McFarland, 1989; Dawkins, 1998). Deviations from these acceptable states will lead to changes in behaviour and/or physiology so as to return the animals to an acceptable state. For example, hunger will cause an animal to seek out food, eat when it finds food and accept exposure to potentially noxious stimuli, such as fear, pain or unpalatable food, in order to satisfy its nutritional requirements. From an evolutionary perspective, animals should be selected to detect and respond with priority to extreme deviations, which could lead to death, lasting harm or other threats to reproductive fitness. Consistent with this approach is the idea that the unpleasant emotions we associate with suffering, such as pain, fear or distress, would help guide animals away from potentially lethal consequences. Under many circumstances, the animal's behavioural and/or physiological responses will allow them to cope with the environmental challenge. However, if the animal is unable to respond appropriately to the challenge, or if it is maintained in an environment where the noxious challenges are inescapable, then the unresolved deviation from an acceptable welfare state may result in suffering.

Various measures have been proposed to assess a captive animal's ability to cope with its environment. These depend on knowledge of the animal's evolutionary history, the consequences of the animal's failure to adapt to the captive environment, and behavioural and physiological measures of the animal's psychological state. Whilst there is some controversy over which measures most reliably and objectively assess an animal's welfare, they can be grouped sensibly together to form three alternative approaches to animal welfare assessment. These approaches have been described by Fraser *et al.* (1997) as measures relating, first, to the naturalness of the environment, second, the biological functioning of the animal (its productivity and pathology) and, third, its psychological state or feelings. The remainder of this chapter focuses on each of these approaches in turn and, with use of relevant equine research, discusses the merits and limitations of the types of measures included within each approach.

Welfare and the natural environment

The proposal that an animal's quality of life is related to the naturalness of its environment seems fairly straightforward. Animals will be adapted to the environment in which they evolved and may not be able to adapt or cope with any significant deviations from this environment (Dawkins, 1980). This argument has been used with zoo and farm animals and was the basis of early attempts to introduce legislation to cover intensive agricultural systems (Brambell, 1965). Scientists that support the 'natural living' model of animal welfare commonly equate quality of life with performance of the full behavioural repertoire, whilst animal rights organizations may take a 'natural is right' stance more for ethical rather than purely scientific reasons. However, regardless of the basis for this approach, its value as a method of assessing animal welfare has been criticized by welfare scientists and by those directly involved in the care of animals. These criticisms are outlined below.

The first criticism of this method of assessing animal welfare is that domestic animals are not equivalent to wild animals as the former will have experienced selection through domestication, and taming through exposure to humans (Dawkins, 1980). For instance, the horse has been domesticated for thousands of years and during this time it will have been selected both intentionally and unintentionally to adapt to and perform in an artificial environment (Clutton-Brock, 1992; Levine, 1999). The second criticism is that the domestic horse will have very different environmental experiences from that of free-ranging or feral equids (Dawkins, 1980). Consequently, domestic horses may habituate to their unnatural environment or adopt alternative behavioural strategies to cope with the challenges they may face therein (Mills & Nankervis, 1999). For example, feral horses avoid humans but captive horses are exposed to humans from an early age so have plenty of opportunities to habituate to them and to learn to associate humans with rewarding experiences such as feeding.

The third reason for rejecting the simple 'natural living' approach to judging animal welfare is that allowing horses to express their full behavioural repertoire would mean exposing them to extremes of heat and cold, hunger or predatory stimuli, all of which may include negative experiences that could be completely avoided in a human-controlled environment (Dawkins, 1980; Hughes & Duncan, 1988). Quite to the contrary, domestication and modern husbandry conditions can enhance the welfare of captive horses compared with their feral counterparts. The stable may be a very unnatural habitat, but stabled horses may not need to spend time looking for food or responding to predatory stimuli for ultimate biological functioning or fitness – our care reduces exposure to potentially noxious stimuli, though perhaps at the expense of loss of control over environmental choices.

Despite these criticisms, an understanding of the adaptive behavioural responses of free-ranging horses is useful for assessing the relationship between behaviour and welfare in captivity. Studies of feral horse populations (Kownaki *et al.*, 1978; Boyd & Houpt, 1994; see also Boyd and Keiper, Chapter 4; Feh, Chapter 5) show that modern horses in free-ranging conditions adopt similar behaviour patterns to ancestral horses. For example, free-ranging horses form stable social groups, spend much of their time grazing or foraging and avoid or flee threatening or predatory stimuli. These behavioural responses to environmental challenges provide a good starting point for identifying areas where there is potential for poor welfare in captive horses. In the following section we will illustrate the value of this approach with research conducted on horses involved in the pregnant mare urine (PMU) industry.

Pregnant mare urine industry

The pregnant mare urine industry produces a supply of oestrogen, extracted from the urine of pregnant mares, for hormone replacement therapy in women. Mares are conventionally box- or tie-stalled in relatively close confinement to allow efficient collection of urine. Concerns about equine welfare have been voiced due to some practices implemented by the industry including the behavioural restriction imposed by the housing environment, the intermittent delivery of drinking water (Aubrey, 1995; Stopps, 1995) and few opportunities that horses have to exercise. Any such concerns have been refuted by the PMU industry and by some equine veterinary surgeons (King *et al.*, 1997). Research aimed at investigating each of these concerns is discussed briefly below.

Stereotypic behaviour is commonly cited as evidence of welfare problems in stabled horses (Cooper & McGreevy, 2002) yet behavioural studies of PMU horses report a relatively low incidence of stereotypic behaviour (e.g. McDonnell *et al.*, 1999; Houpt *et al.*, 2001; Flannigan & Stookey, 2002). The low incidence of stereotypy may be related to a number of factors, but the increased opportunities for social contact in stall housing and the tendency to provide PMU horses with continuous access to forage may both influence this finding (Houpt & Ogilvie-Graham, 2002).

While no limit had been placed on the total amount of water horses could drink, access to water used to be intermittent on most PMU ranches, with opportunities to drink to satiety ranging from three to twenty times per 24 hours. Water systems were turned off between periods of free access to avoid flooding and other problems. Freeman *et al.* (1999) and McDonnell *et al.* (1999) found no differences in behaviour or in the daily amount of water consumed between PMU horses on intermittent and ad libitum watering schedules, although intermittent watering had the advantage of reducing fouling of the stall thereby resulting in fewer hygiene problems. The industry now uses a custom-designed, automated system to provide near continuous access to water which also incorporates design features to minimize hygiene-related concerns.

It is usual to house PMU mares in stalls for 24 hours per day, though most farms will allow short periods of exercise at least every few days (Flannigan & Stookey, 2002). Tie stalls afford horses some opportunity to move their limbs when stepping backwards and forwards within the stall, though exercise involving more than a few steps is not possible. When Flannigan and Stookey (2002) considered the behaviour of mares housed in tie stalls they found that the behaviour was little different from free-ranging horses in terms of time spent active, feeding and foraging and other activities. Houpt *et al.* (2001) compared the behaviour of horses that were allowed turnout 30 minutes per day with that of horses allowed 30 minutes turnout per fortnight. Again, they found no difference in behaviour between horses on daily-exercise or low-exercise routines in the stalls, but horses with less frequent turnout were more active (e.g. more running was recorded) during the time at pasture, which might indicate a requirement for exercise (Houpt & Ogilvie-Graham, 2002).

According to the 'natural living' model of animal welfare, these behavioural studies suggest PMU horses have a relatively high quality of life. However, we should be wary of assuming that superficial similarity to the behaviour of free-ranging horses is a guarantee of good welfare, in much the same way as we should not assume that substantial deviation from these 'natural' patterns of behaviour is indicative of poor welfare. In order to establish if an aspect of a horse's housing or management has negative consequences for its welfare, we should also consider the impact of management on the horse's biological functioning (health, condition and performance) and its psychological state (feelings). As should become clearer through the remainder of this chapter, until we have investigated potential welfare issues using all three approaches we cannot be sure that we have eliminated all possible reasons for concern.

Biological functioning and welfare

This approach to assessing animal welfare is commonly adopted by those directly involved in the care of animals, such as farmers and veterinary surgeons, and accords significance to the animals' health and 'satisfactory' functioning of the animals' biological systems. It has also has been recommended by those welfare scientists who do not recognize the validity of measures based on feelings. There are good logical reasons for this stance (Dawkins, 1980). First, a well-designed husbandry system will protect an animal from extremes of disease, injury, malnutrition and other disturbances to biological functioning. Second, the approach relies on the use of objectively measurable parameters such as health, growth rate, fertility and mortality, which a good animal caretaker would monitor closely. For example, in the racing industry, the athletic performance, health and nutrition of the horses will be monitored closely throughout their racing lives and any remedial actions required will be taken promptly in order to maintain a horse at the peak of its performance. Paradoxically, even with this close monitoring of health and performance, athletic horses can be pushed hard and may incur many costs to their welfare as a result of over-training when young, as demonstrated by the high injury rate amongst racehorses (Williams *et al.*, 2001; Riggs, 2002). This is discussed in more detail below.

The horse-racing industry

Ethical concern about the use of horses in sports has focused on the morals of using horses for human entertainment and the risks to their welfare as a consequence of intense physical exercise. A small number of highly visible problems tend to dominate the general public's concern with the racing industry, such as 'excessive use of the whip' by jockeys or the euthanasia of horses due to catastrophic (fatal) injuries incurred during races. However, it is the costs not so apparent to the public that may be the more serious underlying welfare problems for the horses, including, for instance, catastrophic injuries and musculo-skeletal and pulmonary damage occurring as a result of intense exercise during training (Williams *et al.*, 2001; Riggs, 2002; Evans, 2002).

Musculo-skeletal damage can be divided into catastrophic injury such as broken bones (Estberg *et al.*, 1996, 1998; Williams *et al.*, 2001) and non-catastrophic injuries such as sub-clinical bone (Boston & Nunamaker, 2000), tendon (Dowling *et al.*, 2000) and ligament damage (Hill *et al.*, 2001). Catastrophic injuries during races that result in euthanasia are officially recorded in a number of countries including the UK, USA and Australia, and provide good statistical data for investigating risk factors. For example, data from all UK flat and national hunt racing between 1996 and 1998 (Williams *et al.*, 2001) show that death or euthanasia occurred in 0.29% of starts, though this incidence was higher in jump racing than in flat racing and the risk of fatal injury increased with age. Similar findings have also been reported for smaller scale surveys in Australia (Bailey *et al.*, 1997, 1998) and California, USA (Johnson *et al.*, 1994). All surveys most commonly report forelimb injury (80–90% of cases), particularly to the proximal sesamoid(s), third metacarpal and humerus, though tendon and ligament damage are also commonly diagnosed by course veterinary surgeons (Williams *et al.*, 2001).

However, incidence of catastrophic injury at racecourses underestimates the true rate of catastrophic and non-catastrophic limb injury that can lead to retirement from racing. Official records of injury during training are not maintained, but Johnson *et al.* (1994) estimated a similar rate of catastrophic injuries during training as during racing. Non-catastrophic injuries such as sub-clinical bone damage (Boston & Nunamaker, 2000), tendonitis (Dowling *et al.*, 2000) and ligament damage (Hill *et al.*, 2001) can cause lameness and under-performance. This sub-clinical damage is an important factor in wastage from the racing industry and can increase the risk of breakage

during racing or training (Boston & Nunamaker, 2000; Riggs, 2002).

A significant risk of both catastrophic and non-catastrophic limb injury is the exercise regime, particularly high-speed exercise (Estberg *et al.*, 1995, 1996; Boston & Nunamaker, 2000; Hill *et al.*, 2001). This risk can be further increased by intense exercise regimes during periods of growth and when little opportunity is available to recover from and/or adapt to exercise (Evans, 2002; Riggs, 2002). Intense exercise can also cause heart failure, leading to sudden death during racing or training (Kiryu *et al.*, 1999), and pulmonary damage (West & Mathieucostello, 1994). Exercise-induced pulmonary haemorrhage (EIPH) is a condition affecting up to 90% of horses following sprints (Evans, 2002). An overt symptom of the condition is bleeding from the nostrils, which is related to rupture of the membranes separating blood in pulmonary capillaries from alveoli and other lower airways. High pulmonary pressure during intense exercise (Meyer *et al.*, 1998) and mechanical trauma as a result of locomotory impact (Schroter *et al.*, 1998) have both been suggested as causes for this rupture. Death can occur through blood loss or by contributing to fatal falls, though this is rare in comparison to musculo-skeletal injury (Johnson *et al.*, 1994; Williams *et al.*, 2001).

If horses are to be used in racing, they will be exposed to intense periods of exercise and it is important that their training regime prepares them for this without increasing the risk of injury. Evans (2002) recommends that yearlings should be allowed access to pasture, rather than being principally stall- or stable-housed prior to training, in order to allow the natural development of bone and muscle. The use of treadmills for high-speed training rather than ridden exercise at high speed is also advised as this reduces the risk of sub-clinical injuries and has been found to increase earnings compared with horses that are only rider trained (Kobluk *et al.*, 1996). Sensitive training regimes better prepare horses for the cardio-vascular demands of race day itself, and can promote bone growth to reduce risk of injury. Generally it has been found that reducing workloads on young horses until they are mature enough to endure such mechanical pressures reduces the risk of onset of musculo-skeletal injury, whilst rest supported by non-invasive veterinary treatment such as ultrasound is most likely to return an injured horse to competitive fitness (Evans, 2002).

The preparation and competing of equine athletes illustrates the problems associated with the assessment of welfare in terms of biological functioning alone (Fraser *et al.*, 1997). The tendency is to focus on those performance criteria that are most relevant to the goals of the industry rather than the individual horse's immediate or long-term requirements. Furthermore, the emphasis on biological functioning can be selective, and greater importance can be placed on those measures that directly relate to the economic value of the animal rather than the animal's welfare. For example, surgical and pharmacological intervention can be used to mask or improve poor performance without affecting quality of life, or ignored where there is a compromise between the goals of the owner and the long-term health of the animal. Finally, and more fundamentally, whilst we may expect an association between welfare and athletic performance – as animals that were failing to cope with an adverse environment may show distress, illness or loss of condition – this relationship is not direct. Even when all an animal's primary needs appear to have been met by its artificial environment and the animal appears to be performing well, the animal itself may not realize this (Hughes & Duncan, 1988). Despite no obvious physical signs of distress, an animal may be experiencing psychological suffering (Dawkins, 1980). The final approach to assessing animal welfare, discussed below, is based on the view that the animal's welfare is entirely related to its subjective experiences or feelings.

Subjective feelings and welfare

Psychological states that relate to equine welfare will include negative states such as pain, fear, hunger, anxiety and frustration and positive states such as comfort or even pleasure. Whilst most welfare scientists accept these are important considerations when assessing an animal's quality of life, it is also recognized that there are problems with this approach in isolation (Fraser *et al.*, 1997). First, it requires acceptance that animals are sentient and capable of consciously experiencing their emotions. Second, there is a major difficulty in measuring subjective experience objectively. The biological basis for accepting that animals have the potential to experience their emotions is that feelings are part of the mechanism for

helping animals to adapt to environmental challenges (McFarland, 1989; Dawkins, 1998). However, such adaptive mechanisms operate on a continuum, and animals and even humans do not always consciously experience all environmental challenges, even if they behave as if they do. For example, when a horse is getting accustomed to the use of the bit to control its movement it is likely to experience pain (Cook, 1999), discomfort or frustration during the early stages of training. Once it has learnt the signals that predict pressure on the bit, in the hands of a skilled rider it will have learnt to behave so as to avoid any pain and may not be consciously aware of why it is behaving in response to these signals.

Faced with similar difficulties with farm animals, Fraser *et al.* (1997) suggest an integrated approach to assessing an animal's quality of life. This approach assumes that an animal's welfare is *by definition* a matter of its subjective state, but as this cannot be measured directly, we must infer this state from measures that can be taken directly. For example, physiological measures such as those used to measure stress physiology and behavioural measures such as those used to assess behavioural priorities or the consequences of behavioural deprivation can provide some information about an animal's likely psychological state. How these measures relate to animal welfare is further discussed below.

Behaviour can be related to welfare in a number of ways (Cooper & Mason, 1998). First, the behaviour can decrease welfare, for example by causing injury and associated pain. This might be seen if horses engage in self-mutilation such as biting or if inescapable antagonistic interactions lead to fighting and injury. Clearly, these are situations that should be treated, preferably by identifying and removing the root causes of the behaviour. Second, the activity may be symptomatic of some underlying welfare problem. For example, limping or inactivity may be indicative of pain or illness and would be interpreted by veterinary surgeons as symptoms of injury or disease. Third, behaviour might be a means of coping with environmental challenges. For example, performing certain behaviours may act as an outlet for an underlying behavioural motivation or may have more general de-arousing properties.

A number of physiological measures of psychological state are potentially available to welfare researchers (Broom & Johnson, 1996; Moberg &

Mench, 2000). Ultimately we may be able to track the subjective experiences of animals in terms of patterns of neural activity that are analogous to our own in terms of negative or positive emotional states (Dawkins, 2002), but such an approach is in its infancy and currently most measures of physiological state have used peripheral measures of arousal such as heart rate and blood corticosteroids. Both are involved in the adaptive syndromes that prepare animals to deal with environmental challenges. However, there can be problems with the interpretation of these measures. Handling associated with sampling can raise both heart rate and corticosteroid levels, though this problem may be overcome by using telemetric heart rate monitors or *in situ* catheters to collect samples. Alternatively, non-invasive sampling such as collection of urine, saliva or even faeces can allow estimation of circulating corticosteroids without aversive handling over longer time scales. Even if data can be sampled without disturbance, in isolation the measures cannot discriminate between situations that are arousing because they are potentially pleasurable and potentially noxious challenges. For example, both provision of highly palatable food and placing a highly palatable food out of reach are likely to elevate heart rate and circulating corticosteroids, but the former is likely to be a pleasurable experience whilst the latter may be deeply frustrating. Faced with such problems the solution is to place the physiological measures in their behavioural context.

Equine housing, stereotypic behaviour and welfare

The importance of considering behavioural and physiological measures of quality of life as a means of triangulating upon psychological state can be illustrated with work on the causes and treatment of apparently functionless, repetitive, stereotypic activities such as weaving or crib-biting in stabled horses. This evidence has been reviewed extensively elsewhere (e.g. Cooper & Mason, 1998; Nicol, 1999; Mills, Chapter 15), and has received so much attention of late by welfare scientists that it sometimes appears to be the only aspect of equine welfare that is considered. Here, we will simply summarize the evidence of an association between stereotypic behaviour and welfare in horses.

As already discussed, behaviour can be related to the welfare of the performer in a number of ways, including being the cause of further problems, as an indicator of environmental deficiencies or as a means of coping with an adverse environment. It is widely believed that equine stereotypies lead to physical injury (McBride & Long, 2001), with crib-biting associated with dental wear, wind-sucking with colic, and weaving and box-walking with lameness and loss of condition. For the majority of these beliefs, the evidence is equivocal, though cribbing and wind-sucking have been identified as risk factors associated with colic (Hillyer *et al.*, 2002). However, a causal relationship has not yet been demonstrated as wind-sucking does not lead to substantial air-ingestion (McGreevy *et al.*, 1995a), and the activities may be symptoms of underlying gut dysfunction (McGreevy *et al.*, 2001; Nicol *et al.*, 2002), rather than a causal agent.

The hypothesis that stereotypy is an indicator of environmental deficiencies has been widely discussed (e.g. Mason, 1991). There is strong evidence of a relationship between management factors, such as low forage feeding and social isolation, and the performance of stereotypic behaviour in stabled horses (e.g. Gillham *et al.*, 1994; McGreevy *et al.*, 1995b; Cooper *et al.*, 2000; Mills & Davenport, 2002). This relationship between management practices and stereotypic behaviour does not, in itself, demonstrate that repetitive activities are indicators of poor welfare. The behaviours may seem dysfunctional and visually unappealing to human observers, and perhaps suggest emotions such as boredom or frustration. We cannot, however, infer any relationship between the behaviours and the horse's quality of life, without clear evidence that they impact on the welfare of the performer, who may perceive apparently abnormal behaviour as an adequate expression of its species-typical behavioural repertoire, albeit in an artificial environment.

Stronger evidence of a relationship between behaviour and psychological state can be obtained by two routes: first, there are behavioural measures of choice, where animals demonstrate their environmental choices in terms of the value they place on resources; second, there are the behavioural and physiological consequences of preventing specific behaviours. To date, very few studies of the behavioural requirements of horses have been undertaken and consequently we do not know the degree to which they are satisfied in conventional husbandry systems. More work has been conducted on the relationship between behaviour and physiology and there is good evidence of a relationship between stereotypy and stress physiology in stabled horses. Physical prevention of stereotypic activities has been found to lead to elevation of measures associated with distress. For example, removing cribbing surfaces (McGreevy & Nicol, 1998a) or the use of cribbing straps (McBride & Cuddeford, 2001) lead to elevated corticosteroids, unless the horses are provided with an alternative means of expressing oral activity such as foraging (McGreevy & Nicol, 1998b). Similarly, preventing cribbing over the stable door by means of weaving bars elevates corticosteroids (McBride & Cuddeford, 2001), unless the animals persevere with the behaviour in an alternative form, for instance by weaving within the stable.

These findings have led to the suggestion that stereotypic behaviours in horses have a generalized coping function, protecting the animal from aversive environmental challenges. In truth, the increase in arousal may not be a special feature of stereotypy, but may be found if any part of the horse's behavioural repertoire is prevented. Nevertheless, the effects of physical prevention of stereotypic behaviour once it has developed suggest that treatment of the activities by prevention, using, for example, anti-weaving grilles or anti-cribbing collars, is at best unnecessary and at worst inhumane. Prevention can only be justified if there is clear evidence that the activities cause further welfare problems and that these are likely to be more harmful than the consequences of frustration. Alternatively, the development of more horse-friendly treatments for the activities, including enhanced social environment (Cooper *et al.*, 2000), access to forage or foraging devices (Henderson & Waran, 2001) and, more recently, dietary supplements that reduce oral stereotypic behaviour (Nicol *et al.*, 2002; Mills & MacLeod, 2002) suggests there are ways of reducing the impact of stereotypic behaviour without reducing the quality of the horse's life.

Summary: equine welfare and human responsibility

We should be concerned about the well-being of all animals in our care but the level of attention an animal receives can depend heavily on which species it is and the purpose for which the animal is kept. Owners of horses used for sport or recreation are greatly concerned about the quality of their horses' lives but welfare problems do exist within equestrianism as well as in equines reared or kept for other purposes. Decisions impacting on the welfare of sports or leisure

horses may be strongly influenced by potential benefits to the owner or rider, the economic or sentimental value of the horse and traditions or currently accepted practice, with less emphasis given to the costs to the horse. However, the method of assessing the current welfare state of any equine should be the same, regardless of the purpose for which the animal is kept. This method should be an integrated approach combining measures of the equine's biological functioning and psychological state with an understanding of the animal's ability to adapt behaviourally to the environmental challenges imposed.

Efforts to improve the welfare of sports and leisure horses are encouraging. For instance, the racing industry has taken note of performance-related welfare issues through redesign of racing courses and obstacles, monitoring of horses prior to racing for fitness or soundness, and heavy investment in research into exercise-induced disorders. However, as training programmes may condemn certain horses prior to them reaching the racetrack, it is important that training programmes are sensitive to the growth periods of the horse and the horse's adaptive biology (Evans, 2002), and improvements in horse welfare are not confined to competitive occasions.

It is also clear that horse owners are becoming more sensitive to the psychological needs of the animals in their care and many cultural misconceptions are being eroded. For example, it has become less common for stereotypic behaviour to be labelled as a 'stable vice' in equine literature – it is now more likely to be described as a sign of boredom than of malice or contrariness. Whilst the accuracy of this interpretation is questionable, it is encouraging that owners are more sensitive to their horses' potential psychological state. Further evidence for this concern is reflected in the wide range of stable 'toys' now available to occupy the stabled horse's time, though the true value of such toys after a period of habituation is not yet known.

It is, perhaps, worth noting at this point that most research on equine welfare conducted to date has concentrated on 'popular' areas of concern, such as obvious health-related issues and apparent behavioural problems. As yet we have not attempted to investigate some of the more fundamental questions relating to equine welfare, such as how being ridden affects the welfare of a horse or how different equestrian disciplines impact on the horse's psychological well-being. These areas of research should not be overlooked simply because most research is aimed at improving equine performance and management within equestrianism. Approaches adopted by researchers investigating the welfare of horses in the PMU industry have begun to follow the route of those used by farm animal welfare scientists (McDonnell *et al.*, 1999; Houpt *et al.*, 2001). It would be useful to follow this approach in other equine industries to produce more scientifically valid assessments of the horse's ability to adapt to the challenges we impose and to more convincingly refute or support the welfare concerns the general public may have for horses in these industries.

A final sign of progress in addressing equine welfare problems would be the tightening of legislation and regulation of equine industries. Welfare standards in Europe tend to be driven by national and international welfare codes imposed by government legislation as an arbitrator of public opinion and industry standards. This is particularly true for farm and laboratory animal welfare, though to a lesser extent with the equine industries. For example, whilst there are quite strict regulations concerning the use of horses as research animals and legislation covers specific practises such as docking or use of harmful aids, horses generally receive only imprecise protection through general welfare regulations covering farmed or companion animals (DEFRA, 2002). It will be interesting to see whether this remains unchanged as government legislation across Europe begins to focus on companion animals or whether all equine industries will be governed by the same legislative rules (DEFRA, 2003).

References

Abbott, E. M. (1999). The horse. In *UFAW Handbook on the Care and Management of Laboratory Animals*, 7th edn, vol. 1 *Terrestrial Vertebrates*, ed. T. Poole. Oxford: Blackwells, pp. 517–36.

Aubrey, M. E. (1995). Controversy over the use of pregnant mare urine. *Canadian Medical Association Journal*, **152**: 1745.

Bailey, C. J., Reid, S. W. J., Hodgson, D. R., Suann, C. J. & Rose, R. J. (1997). Risk factors associated with musculoskeletal injuries in Australian Thouroughbred racehorses. *Preventative Veterinary Medicine*, **32**: 47–55.

Bailey, C. J., Reid, S. W. J., Hodgson, D. R., Bourke, J. M. & Rose, R. J. (1998). Flat, hurdle and steeple racing: risk factors for musculoskeletal injury. *Equine Veterinary Journal*, **30**: 498–503.

Boston, R. C. & Nunamaker, D. M. (2000). Gait and speed as exercise components of risk factors associated with onset of fatigue injury of the third metacarpal bone in 2-year-old Thoroughbred racehorses. *American Journal of Veterinary Research*, **61**: 602–8.

Boyd, L. & Houpt, K. A. (1994). *Przewalski's Horse*. New York: State University of New York Press.

Brambell, F. W. R. (1965). *Report of the Technical Committee to Enquire into the Welfare of Animals kept under Intensive Livestock Husbandry Systems*. London: HMSO.

Broom, D. M. & Johnson, K. G. (1996). *Stress and Animal Welfare*. London: Chapman and Hall.

Clutton-Brock, J. (1992). *Horse Power: History of the Horse and Donkey in Human Societies*. Boston: Harvard University Press.

Cook, W. R. (1999). Pathophysiology of bit control in the horse. *Journal of Equine Veterinary Science*, **19**: 196–204.

Cooper, J. J. & Mason, G. J. (1998). The identification of abnormal behaviour and behavioural problems in stabled horses and their relationship to horse welfare: a comparative review. *Equine Veterinary Journal*, **27** (Suppl.): 5–9.

Cooper, J. J., McDonald, L. & Mills, D. S. (2000). The effect of increasing visual horizons on stereotypic weaving: implications for the social housing of stabled horses. *Applied Animal Behaviour Science*, **69**: 67–83.

Cooper, J. J. & McGreevy, P. D. (2002). Stereotypic and redirected behaviour in the stabled horse: causes, effects and relationship to welfare. In *The Welfare of the Horse*, ed. N. Waran. Dordrecht: Kluwer Academic Press, pp. 99–124.

Dawkins, M. S. (1980). *Animal Suffering. The Science of Animal Welfare*. London: Chapman and Hall.

  (1998). *Through Our Eyes Only: The Search for Animal Consciousness*. Oxford: Oxford University Press.

  (2002). Who needs consciousness? *Animal Welfare*, **10**: S19–S29.

DEFRA (2002). *Equine Industry Welfare Guidelines. Compendium for Horses, Ponies and Donkeys*. London: Department for Environment, Food and Rural Affairs.

  (2003). *Outline of an Animal Health and Welfare Strategy for Great Britain*. London: Department for Environment, Food and Rural Affairs.

Dowling, B. A., Dart, A. J., Hodgson, D. R. & Smith, R. K. W. (2000). Superficial digital flexor tendonitis in the horse. *Equine Veterinary Journal*, **32**: 369–78.

Estberg, L. Gardner, I. A., Stover, S. M., Johnson, B. J. Case, J. T. & Ardans, A. (1995). Cumulative racing speed exercise distance cluster as a risk factor for fatal musculoskeletal injury in thoroughbred racehorses in California. *Preventative Veterinary Medicine*, **24**: 253–63.

Estberg, L. Gardner, I. A., Stover, S. M. & Johnson, B. J. (1998). A case-crossover study of intensive racing and training schedules and risk of catastrophic musculoskeletal injury and lay-up in California Thoroughbred racehorses. *Preventative Veterinary Medicine*, **33**: 159–70.

Estberg, L., Stover, S. M., Gardner, I. A., Drake, C. M., Johnson, B. J. & Ardans, A. (1996). High-speed exercise history and catastrophic racing industry in

Thoroughbreds. *American Journal of Veterinary Research*, 57: 1549–55.

Evans, D. L. (2002). Welfare of the racehorse during exercise training and racing. In *The Welfare of the Horse*. Dordrecht: Kluwer Academic Press, pp. 181–201.

Flannigan, G. & Stookey, J. M. (2002). Day-time time budgets of pregnant mares housed in tie stalls: a comparison of draft versus light mares. *Applied Animal Behaviour Science*, 78: 125–43.

Fraser, A. F. & Broom, D. M. (1997). *Farm Animal Behaviour and Welfare*. Wallingford: CABI.

Fraser, D., Weary, D. M., Pajor, E. A. & Milligan, B. N. (1997). A scientific conception of animal welfare that reflects ethical concerns. *Animal Welfare*, 6: 187–205.

Freeman, D. A., Cymbaluk, N. F., Schott, H. C. II., Hinchcliffe, K. & McDonnell, S. M. (1999). Clinical, biochemical and hygiene assessment of stabled horses provided continuous or intermittent access to drinking water. *American Journal of Veterinary Research*, 60: 1445–50.

Gillham, S. B., Dodman, N. H., Shuster, L., Kream, R. & Rand, W. (1994). The effect of diet on cribbing behavior and plasma B-endorphin in horses. *Applied Animal Behaviour Science*, 41: 147–53.

Henderson, J. V. & Waran, N. K. (2001). Reducing equine stereotypies using the Equiball[TM]. *Animal Welfare*, 10: 73–80.

Hill, A. E., Stover, S. M., Gardner, I. A., Kane A. J., Whitcomb M. B. & Emerson A. G. (2001). Risk factors for and outcomes of non-catastrophic suspensory apparatus injury in Thoroughbred racehorses. *Journal of the American Veterinary Medical Association*, 218: 1136–44.

Hillyer, M. H., Taylor F. G. R., Proudman, C. J., Edwards, G. B., Smith, J. E. & French, N. P. (2002). Case control study to identify risk factors for simple colonic obstruction and distension colic in horses. *Equine Veterinary Journal*, 34: 455–63.

Houpt, K. A. & Ogilvie-Graham, T. S. (2002). Comfortable quarters for horses at research institutions. In *Comfortable Quarters for Laboratory Animals, 9th Edition*, ed. V. Reinhardt, A. Reinhardt, pp. 96–100. Washington: Animal Welfare Institute.

Houpt, K. A., Houpt, T. R., Johnson, J. L. Erb, H. N. & Yeon. S. C. (2001). The effect of exercise deprivation on the behaviour and physiology of straight stall confined pregnant mares. *Animal Welfare*, 10: 257–67.

Hughes, B. O. & Duncan, I. J. H (1988). The notion of ethological 'need', models of motivation and animal welfare. *Animal Behaviour*, 36: 1696–707.

ILPH (2003). International League for the Protection of Horses Report on Horse Transport. http://www.ilph.org/items.asp. (Accessed 10.11.2003.)

Johnson, B. J., Stover, S. M., Daft, B. M., Kinde, H., Read, D. H., Barr, B. C., Anderson, M., Moore, J., Woods, L., Stoltz, J. & Blanchard, P. (1994). Causes of death in racehorses over a 2-year period. *Equine Veterinary Journal*, 26: 327–330.

Kiley-Worthington, M. (1997). *Equine Welfare*. London: J. A. Allen.

King, A. B., Messer, N. T. & Roberts, C. A. (1997). *Equine Veterinarians' Consensus Report on the Care of Horses on PMU ranches*. Lexington, KY: AAEP, CVMA and ILPH Joint Publication.

Kiryu, K., Machida, N., Kashida, Y., Yoshihara, T., Amada, A. & Yamamoto, T. (1999). Pathologic and electrocardiographic findings in sudden cardiac death in racehorses. *Journal of Veterinary Medical Science*, 61: 921–8.

Kobluk, C. N., Geor, R. J., King, V. L. & Robinson, R. A. (1996). A case study of racing Thoroughbreds conditioned on a high-speed treadmill. *Journal of Equine Veterinary Science*, 16: 511–13.

Kownaki, M., Sasimowski, E., Budzynski, M., Jezieski, T., Kapron, M., Jelen, B., Jaworska, M., Dziedzic, R., Seweryn, A. & Solmka, Z. (1978). Observations of the 24 hour rhythm of natural behaviour of Polish primitive horse bred for the conservation of genetic resources in a forest reserve. *Genetica Polonica*, 19: 61–77.

Levine, M. A. (1999). Investigating the origins of horse domestication. *Equine Veterinary Journal* , 28 (Suppl.): 6–14.

Mason, G. J. (1991). Stereotypies: a critical review. *Animal Behaviour*, 41: 1015–37.

McBride, S. D. & Cuddeford, D. (2001). The putative welfare-reducing effects of preventing equine stereotypic behaviour. *Animal Welfare*, 10: 173–89.

McBride, S. D. & Long, L. (2001). Management of horses showing stereotypic behaviour, owner perception and the implications for welfare. *Veterinary Record*, 148: 799–802.

McDonnell, S. M. Freeman, D. A, Cymbaluk, N. F., Schott, H. C. II., Hinchcliffe, K. & Kyle, B. (1999). Behavior of stabled horses provided continuous or intermittent access to drinking water. *American Journal of Veterinary Research*, 60: 1451–6.

McFarland, D. (1989). *Problems of Animal Behaviour*. Harlow: Longman.

McGreevy, P. D. & Nicol, C. J. (1998a). Prevention of crib-biting: a review. *Equine Veterinary Journal*, 27 (Suppl.): 35–8.

(1998b). Physiological and behavioural consequences associated with short-term prevention of crib-biting in horses. *Physiology and Behaviour*, 65: 15–23.

McGreevy, P. D., Cripps, P. J., French, N. P., Green, L. E. & Nicol, C. J. (1995a). Management factors associated with stereotypic and redirected behaviour in the Thoroughbred horse. *Equine Veterinary Journal*, 27: 86–91.

McGreevy, P. D., Richardson, J. D., Nicol, C. J. & Lane, J. G. (1995b). A radiographic and endoscopic study of horses performing an oral stereotypy. *Equine Veterinary Journal* 27: 92–5.

McGreevy, P. D., Webster, A. J. F. & C. J. Nicol. (2001). A study of the digestive efficiency, behaviour and gut transit times of crib-biting horses. *Veterinary Record*, 148: 592–6.

Meyer, T. S., Fedde, M. R., Gaughan, E. M., Langsetmo, I. & Erickson, H. H. (1998). Quantification of exercise-induced pulmonary haemorrhage with bronchoalveolar lavage. *Equine Veterinary Journal*, **30**: 284–8.

Mills, D. S. & Davenport, K. (2002). The effect of a neighbouring con-specific versus the use of a mirror for the control of stereotypic weaving behaviour in the stabled horse *Animal Science*, **74**: 95–101.

Mills, D. S. & Macleod, C. A. (2002). The response of crib-biting and windsucking in horses to dietary supplementation with an antacid mixture. *Ippologia*, **13**: 33–41.

Mills, D. S. & Nankervis, K. (1999). *Equine Behaviour: Principles and Practice*, Oxford: Blackwell Science.

Moberg, G. P. & Mench, J. A. (2000). *The Biology of Animal Stress. Basic Principles and Implications for Animal Welfare*. Wallingford: CABI.

Nicol, C. J. (1999). Understanding equine stereotypies. *Equine Veterinary Journal*, **28** (Suppl.): 20–5.

Nicol, C. J., Davidson, H. P. B, Harris, P. A., Waters, A. J. & Wilson, A. D. (2002). Study of crib-biting and gastric inflammation and ulceration in young horses. *Veterinary Record*, **151**: 658–61.

Riggs, C. M. (2002). Fractures – a preventable hazard of racing thoroughbreds? *Veterinary Journal*, **163**: 19–29.

Schroter, R. C., Marlin, D. J. & Denny, E. (1998). Exercise-induced pulmonary haemorrhage (EIPH) in horses results from locomotory impact induced trauma – a novel, unifying concept. *Equine Veterinary Medicine*, **30**: 186–92.

Stopps, R. (1995). Pregnant mare's urine – welfare or rights for animals. *Canadian Medical Association Journal*, **153**: 520.

Waran, N. (2002). *The Welfare of the Horse*. Dordrecht: Kluwer Academic Publishers.

West, J. B. & Mathieucostello, O. (1994). Stress failure of the pulmonary capillaries as a mechanism for exercise induced pulmonary haemorrhage in the horse. *Equine Veterinary Journal*, **26**: 441–7.

Williams, R. B., Harkins, L. S., Hammond, C. J. & Wood, J. L. N. (2001). Racehorse injuries, clinical problems and fatalities recorded on British racecourses from flat racing and National Hunt racing during 1996, 1997 and 1998. *Equine Veterinary Journal*, **33**: 478–6.

Index

hyena, brown 77
hyper-reactivity 200–1, 204
hypothalamic–pituitary–adrenal (HPA)
    axis 198, 200

Iberian breeds 24, 27
Icelandic horses 117
immobilization, field 78
imprint training 145, 170–1, 193–4
imprinting 129, 136, 170
inbreeding 75
incest 75
Indian khur 84, 85
individual differences
    behaviour 33–49
    behavioural development 141–2
individual recognition 63, 85
industrialization 23, 24
infanticide 76–7
ingestive behaviour 94–100
    ontogeny 140–1
    see also drinking; feeding
injuries
    catastrophic 232
    fighting-related 65–6
    inter-male sexual behaviour 121
    non-catastrophic 232
    play restriction to avoid 150–1, 155
    play sexual behaviour 121
    racehorses 231, 232–3
    riding 162, 198
innate behaviour 140
insects 61–2, 85, 102
insemination 118
intelligence 169, 178
Iron Age 5–6
    Altai burial sites 6
    palaeopathology 15, 17
irritation 201, 219
isolation, social 123, 218
Italy 12

Japanese native horses 36
Jeffery's approach and retreat
    technique 191
Jicarilla feral horse 71, 75
join up 191–2, 194
jumping
    ability, predictors 43
    refusal 199
    training 188, 190
juvenile dispersal 71–3, 86–7, 112
juvenile horses
    differences in temperament 41, 42
    learning ability 41, 177
    trainability 185
    see also age; colts; fillies; foals;
        yearlings

Kaimanawa wild horse
    dominance hierarchies 68, 70
    habitat selection 60
    harem stability 70
    home ranges 60
    reproduction 74

Kazakhstan 5, 6, 12
khulan 84, 85
khur 84, 85
kiang 84, 85
Kladruby 27
Konik 29
kurgan burials 5, 8

labelling, maternal 129
lactation
    water intake 101
    see also nursing; suckling
lameness 218
laminitis 206
laziness 198, 203, 204
leadership 65, 87
learned behaviour 140
learned helplessness 189, 203, 204
learning
    abnormal behaviour 145, 176
    associative 172–3
        see also conditioning
    avoidance 179
    development and 170–1
    discrimination 174–5
    discrimination reversal 175
    evolutionary context 169
    food selection 171–3
    interocular transfer 175
    mechanisms 172–4
    social *see* social learning
    stimulus generalization 175–6
    stress and 177
    styles, individual 178
    taste aversion 96, 172
    theory, poor application 204–5
    training and 178–9, 185
learning ability 169–80
    behavioural tests 38, 39
    breed differences 46–7, 48, 177
    consistency over time 43
    correlation with other behavioural
        measures 43, 44
    juveniles 41, 177
    limits 179–80
    observer ratings 38
    predicting later jumping ability 43
    temperament and 177–8
leg signals 196, 198
    hyper-reactivity 200
    incorrect use 204–6
    learned helplessness response 203
    mechanical aids 208
    re-training 209
legislation 229, 236
leisure horses 29–30, 162
    housing 185
    key traits 163–4, 165
    training 189–90
    welfare issues 235
libido
    inadequate 121
    intense 121–2
life assemblage model 12–14
lifespan 74

lifestyle, healthy 165
lightness 205
Lipitsa 27
lithium chloride 172
locomotion 94, 104
    in different environments 98
locoweed 96, 172
long-reining 188, 189, 190
loose schooling 188
lope 190
lungeing 188, 189, 190
lying down *see* recumbency

maces, horse-head 7–9
machinery, driving 25
maintenance behaviour 94–106
malnutrition 206
Malopolski horse 46
management
    feral horses, effects on behaviour 78–9
    stereotypic behaviour and 234
    trainability and 184, 186
    see also exercise; feeding; housing;
        husbandry
manure 28
mare 12
    adult relationships 87
    alliances 70
    dominance hierarchies 64
    fidelity 70
    foal recognition 85
    sexual problems 123
    stallion bonding 84
    stallion-like behaviour 123
    see also fillies; maternal behaviour
mare–foal distance 131
    genetics 35–6
    maternal investment and 135–6
    ontogeny 140
    predicting adult behaviour 43
mare–foal interaction 126–37
mare–offspring bands 58
marking behaviour, elimination 67–8,
    88, 111
    domestic horses 121
    ontogeny 142
martingale 207
masturbation 120, 215
mate choice 74–5
maternal behaviour 76, 77, 126–37
    towards non-offspring 137
maternal deprivation 142–3
maternal investment 133–6
mating
    dominance and 68–9
    father–daughter 75
    half-sibling 75
    see also breeding; copulation
mating system 55
maze 175, 178
meat 11, 27
Merens 48
microsatellites 27
military use 23, 24–5
Miller, Robert 145, 193